Palgrave Series in Asia and Pacific Studies

Series Editors
May Tan-Mullins
University of Nottingham
Ningbo, Zhejiang, China

Adam Knee
Lasalle College of the Arts
Singapore

The Asia and Pacific regions, with a population of nearly three billion people, are of critical importance to global observers, academics, and citizenry due to their rising influence in the global political economy as well as traditional and nontraditional security issues. Any changes to the domestic and regional political, social, economic, and environmental systems will inevitably have great impacts on global security and governance structures. At the same time, Asia and the Pacific have also emerged as a globally influential, trend-setting force in a range of cultural arenas. The remit of this book series is broadly defined, in terms of topics and academic disciplines. We invite research monographs on a wide range of topics focused on Asia and the Pacific. In addition, the series is also interested in manuscripts pertaining to pedagogies and research methods, for both undergraduate and postgraduate levels. Published by Palgrave Macmillan, in collaboration with the Institute of Asia and Pacific Studies, UNNC.

More information about this series at
http://www.palgrave.com/gp/series/14665

Stephanie Clair Price

World Heritage Conservation in the Pacific

The Case of Solomon Islands

palgrave
macmillan

Stephanie Clair Price
University of Western Australia
Crawley, WA, Australia

Palgrave Series in Asia and Pacific Studies
ISBN 978-981-13-0601-3   ISBN 978-981-13-0602-0   (eBook)
https://doi.org/10.1007/978-981-13-0602-0

Library of Congress Control Number: 2018946655

Cover credit: Stuart Fox

Printed on acid-free paper

This Palgrave Macmillan imprint is published by the registered company Springer Nature
Singapore Pte Ltd.
The registered company address is: 152 Beach Road, #21-01/04 Gateway East, Singapore
189721, Singapore

PREFACE

East Rennell (part of the island of Rennell, in Solomon Islands) was inscribed on the World Heritage List in 1998. Its listing was a milestone in the development of the *World Heritage Convention* regime. It was the first listed World Heritage site in the independent Pacific Island States, and the first place anywhere in the world to be inscribed based on its natural heritage values and its protection under customary law. However, the threats to its outstanding universal value (OUV) have increased over time, leading the World Heritage Committee to include the site on the List of World Heritage in Danger in 2013. The Committee has repeatedly called on the Solomon Islands government (SIG) to do more to safeguard the site, including by banning logging and mining, regulating the taking of certain species, and declaring the site to be a protected area under law.

In 2013, I worked for a non-government organisation (NGO) in Solomon Islands on a project aimed at strengthening the protection of the East Rennell World Heritage site. One way that this could be achieved is through the declaration of East Rennell as a 'protected area' under the *Protected Areas Act 2010*. I had previously worked as a volunteer lawyer at Solomon Islands' Public Solicitor's Office, providing advice to customary landowners on issues related to logging, mining, and conservation.

My work in East Rennell highlighted the many dimensions of World Heritage. East Rennell is customary land, and is owned and occupied by the East Rennellese people. While working for the NGO, I participated in meetings in the East Rennell communities to discuss the process for, and implications of, establishing a protected area under the *Protected Areas Act*. Many community members supported World Heritage and were interested to hear how the Act could be used to protect their land. However, our meetings were often dominated by discussion of livelihoods, food security, infrastructure, economic development, and customary rights and governance. These issues also featured heavily in the discussions I had with people working for the SIG.

In many respects, these conversations felt far removed from the provisions of the *World Heritage Convention*, and the deliberations that the World Heritage Committee was having at the time concerning the inclusion of East Rennell on the List of World Heritage in Danger. Yet they are intrinsically linked, as the long-term conservation of the site's OUV hinges on decisions made by the East Rennellese people concerning their land and resources.

This work sparked me to research the *World Heritage Convention* in the Pacific context. I did not set out to find 'the solution' to the question of how the *Convention* can successfully be implemented in the Pacific. This would of course be impossible, given the range of economic, social, political, environmental, and cultural issues that influence whether World Heritage sites are safeguarded. It would also be inappropriate: I am not a Pacific Islander, and ultimately it is Pacific Islanders who must decide if and how they wish to implement the *Convention*. Rather, I sought to examine the opportunities and challenges associated with the use of the *Convention* to protect Pacific heritage sites. In particular, I wanted to investigate what lessons can be learned from Solomon Islands' experience to date for the protection of East Rennell and other places with similar characteristics. This book is a product of that research.

Reflecting my background as a lawyer, I undertook this research from a legal perspective. A socio-legal approach was however taken. Such an approach is warranted where there is significant variation between the form of a law and its effect in practice. This is certainly the case in Solomon Islands, where much legislation relevant to heritage protection is not routinely implemented or enforced. Using this approach, the book explores legal issues arising from the *World Heritage Convention* in their broader context.

The book is based on an analysis of primary and secondary literature, and empirical research comprising interviews conducted with people working on World Heritage matters for the SIG. All those interviewed agreed to be quoted, but only some consented to being named. The book also draws upon my work as a lawyer in Solomon Islands, through which I experienced first-hand the challenges associated with implementing and enforcing conservation legislation and laws regulating resource developments in Solomon Islands.

Part I of the book (Chap. 1) contains an overview of the *World Heritage Convention* and its implementation by the independent Pacific Island States.

Part II comprises three chapters. Many of the opportunities and challenges associated with implementation of the *Convention* in the Pacific can be linked to the nature of the region's heritage, the legal systems that govern its people, and the context within which those systems operate. Chapter 2 therefore introduces these issues and explores their relevance for World Heritage conservation. Chapter 3 examines the concept of 'World Heritage' and assesses how Pacific Island heritage 'fits' within the *Convention* regime. Chapter 4 analyses the protection regime established by the *Convention*. It covers the World Heritage Committee's changing approach to heritage conservation, and the implications of this change for Pacific Island States.

In Part III of the book, the focus narrows to Solomon Islands. Chapter 5 critically analyses the inscription of East Rennell on the World Heritage List, and explores the context for World Heritage conservation in Solomon Islands. Chapter 6 assesses the site's protection under customary law, and discusses management planning for sites subject to customary protection. Chapter 7 considers the ability and willingness of the SIG and customary landowners to utilise State legislation to protect East Rennell. The laws analysed include the *Protected Areas Act 2010*, the *Forest Resources and Timber Utilisation Act (Cap. 40)*, the *Mines and Minerals Act (Cap. 42)*, the *Environment Act 1998*, the *Fisheries Management Act 2015*, and the *Biosecurity Act 2013*.

Part IV (Chap. 8) summarises the lessons that can be learned from Solomon Islands' experience for the protection of East Rennell and other places sharing common characteristics. While recognising that heritage conservation is influenced by a range of factors, the chapter also identifies some options that could help strengthen World Heritage protection in the Pacific.

I am extremely grateful for the assistance I have received from so many people. Thank you to the people of East Rennell for allowing me to visit your incredible home, and for your wonderful hospitality. Thank you also to my colleagues in Solomon Islands, in particular Haikiu Baiabe for your guidance and assistance. I am also grateful to the people working within the SIG who agreed to be interviewed for this research, and my PhD supervisors, Professor Erika Techera and Associate Professor Catherine Kelly, for their insightful feedback and encouragement. Finally, thank you to my family, especially my parents, my brother Ivan, my partner Pete, and our children Lily and Isaac. Your unwavering support and belief in me has made this possible.

Crawley, WA, Australia   Stephanie Clair Price

CONTENTS

LIST OF ABBREVIATIONS AND ACRONYMS

Convention	*World Heritage Convention*
Convention bodies	World Heritage Committee and the Advisory Bodies
DSOCR	Desired State of Conservation for the Removal of East Rennell from the List of World Heritage in Danger
EIA	Environmental Impact Assessment
FRTU Act	*Forest Resources and Timber Utilisation Act (Cap. 40)*
Global Strategy	*Global Strategy for a Representative, Balanced and Credible World Heritage List*
ICCA	Indigenous and Community Conservation Area
ICCROM	International Centre for the Study of the Preservation and Restoration of Cultural Property
ICOMOS	International Council of Monuments and Sites
ILO 169	*International Labour Organisation's Indigenous and Tribal Peoples Convention 1989*
IUCN	International Union for the Conservation of Nature
LTWHSA	Lake Tegano World Heritage Site Association
MM Act	*Mines and Minerals Act (Cap. 42)*
MM Regulations	*Mines and Minerals Regulations 1996*
MPA	Marine Protected Area
NGO	Non Government Organisation

Operational Guidelines	*Operational Guidelines for the Implementation of the World Heritage Convention*
OUV	Outstanding universal value
PA Act	*Protected Areas Act 2010*
PA Regulations	*Protected Areas Regulations 2012*
PNG	Papua New Guinea
RAMSI	Regional Assistance Mission to the Solomon Islands
SIG	Solomon Islands government
UNCED	United Nations Conference on Environment and Development
UNCHE	United Nations Conference on the Human Environment
UNDRIP	*United Nations Declaration on the Rights of Indigenous People*
UNESCO	United Nations Educational, Scientific and Cultural Organisation
WHC	*World Heritage Convention*

LIST OF FIGURES

LIST OF TABLES

PART I

Introduction

Implementation of the *World Heritage Convention* by the Independent Pacific Island States

1.1 INTRODUCTION

The independent Pacific Island States[1] (Fig. 1.1) are home to a diverse array of heritage sites. These include impressive marine and terrestrial ecosystems, sites evidencing the development of island societies, and places of significance due to their connection with the customs of Pacific Islanders. Eight places within these States have been inscribed on the World Heritage List,[2] including East Rennell in Solomon Islands, which is the focus of this book.

East Rennell is customary land, and is owned and occupied by the East Rennellese people. It was the first place in the independent Pacific Island States to be inscribed on the World Heritage List.[3] It was also the first place anywhere in the world to be listed based on its natural heritage values and customary protection. Consequently, its listing was a landmark in

---

[1] The independent Pacific Island States are Cook Islands, Federated States of Micronesia, Fiji, Kiribati, Marshall Islands, Nauru, Niue, Palau, Papua New Guinea, Samoa, Solomon Islands, Tonga, Tuvalu, and Vanuatu (see Fig. 1.1). While this book refers to the Pacific region generally, it focuses on the independent Pacific Island States. Other States and overseas territories in the Pacific are not specifically discussed, because of their different histories, legal and governance systems, and/or territorial status.

[2] See Table 1.2.

[3] WHC Res CONF 203 VIII.A.1, WHC 22nd sess, UN Doc WHC-98/CONF/203/18 (29 January 1999) 25.

© The Author(s) 2018
S. C. Price, *World Heritage Conservation in the Pacific*,
Palgrave Series in Asia and Pacific Studies,
https://doi.org/10.1007/978-981-13-0602-0_1

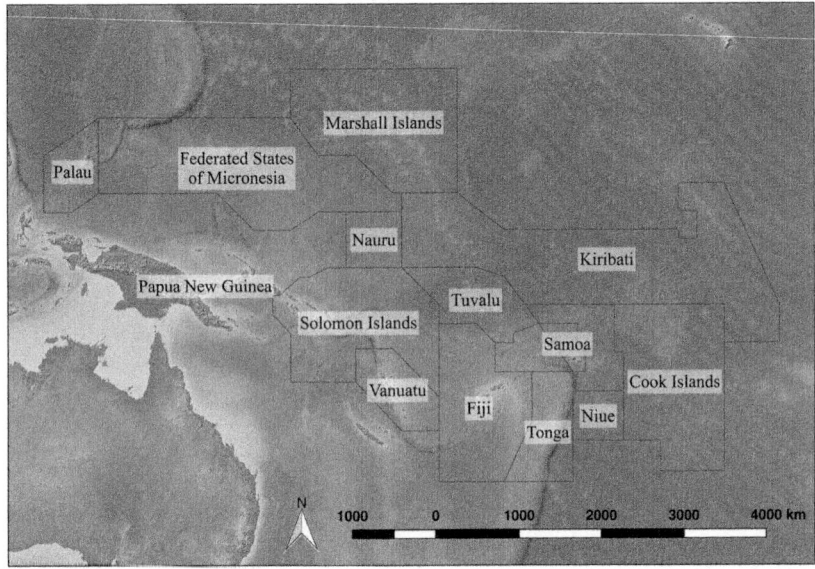

Fig. 1.1   Map of the independent Pacific Island States. Map made with data from Natural Earth. Free vector and raster map data @ naturalearthdata.com

the implementation of the *World Heritage Convention*,[4] which established an important precedent concerning the acceptance of customary law as a sufficient basis for the protection of natural sites.[5] However, East Rennell is now on the List of World Heritage in Danger,[6] threatened by the impacts of resource development, invasive species, climate change, and the over-harvesting of certain animals.[7] Addressing these threats will require a range of actions, including strengthening the site's protection under customary and State law.

[4] *Convention Concerning the Protection of the World Cultural and Natural Heritage*, opened for signature 16 November 1972, 1037 UNTS 151 (entered into force 17 December 1975) ('*World Heritage Convention*').

[5] T Badman et al, *Outstanding Universal Value: Standards for Natural World Heritage* (IUCN, 2008) 24.

[6] WHC Res 37 COM 7B.14, WHC 37th sess, UN Doc WHC-13/37.COM/20 (5 July 2013) 68.

[7] See, for example, Paul Dingwall, *Report on the Reactive Monitoring Mission to East Rennell, Solomon Islands, 21–29 October 2012* (IUCN, 2013). The threats to East Rennell are discussed in Sect. 5.3.1.

The *World Heritage Convention* requires State parties to implement the legal measures needed to protect the World Heritage within their borders,[8] but does not mandate what form that legislation must take. It therefore allows State parties to tailor their World Heritage protection laws to suit their context. This creates an opportunity for the *Convention* to be utilised by Pacific Island States in a manner that is consistent with the nature of their heritage, land tenure, and legal systems. Despite this, developing and implementing effective legislation remains challenging for many such States, including Solomon Islands. If East Rennell is to retain its World Heritage listing, its legal protection must be improved. In addition, if the representation of Pacific heritage on the World Heritage List is to increase, and if the *Convention* is to be successfully used to conserve significant heritage sites, greater understanding of its application in the Pacific is required.

This book therefore explores the *World Heritage Convention* regime in the Pacific context, to identify the opportunities and challenges it presents for the protection of the region's heritage. Solomon Islands' implementation of the *Convention* is critically analysed, revealing lessons that could improve World Heritage protection in that country and elsewhere. The book begins here with an introduction to the *Convention* and its implementation in the Pacific.

1.2 THE *WORLD HERITAGE CONVENTION* REGIME

The *World Heritage Convention* was adopted by the General Conference of the United Nations Educational, Scientific and Cultural Organisation (UNESCO) in November 1972.[9] Among other things, it was a response to growing international concern about the impacts of human activities on cultural sites and wilderness areas. It also reflected increasing appreciation of the interrelationship between culture and nature, and the need to preserve heritage for future generations (discussed further in Sect. 3.2.1).

The drafters of the *Convention* wanted the treaty to apply to sites of significance to humankind, rather than places possessing only local or national value.[10] Thus, sites only fall within the ambit of the *Convention* if

---

[8] *World Heritage Convention* arts 4–5.

[9] UNESCO, *Records of the General Conference – volume 1*, 17th sess (17 October–21 November 1972) 135.

[10] Sarah M Titchen, 'On the Construction of "Outstanding Universal Value": Some Comments on the Implementation of the 1972 UNESCO World Heritage Convention' (1996) 1 *Conservation and Management of Archaeological Sites* 235, 236.

they have 'outstanding universal value' (OUV).[11] State parties have the primary responsibility to safeguard such places, and must take 'effective and active' measures to achieve that end.[12] However, as the deterioration of World Heritage constitutes a 'harmful impoverishment of the heritage of all the nations of the world',[13] the *Convention* also establishes a system of international assistance to help State parties comply with their duties.[14]

The *World Heritage Convention* has never been amended, and this would be a 'long and risky' task[15] as there are now 193 State parties.[16] Despite this, the *Convention* regime has evolved, because the *Convention* document itself only establishes a framework. It creates the key structural elements of the regime, namely:

- the World Heritage Committee (an executive decision-making body comprising 21 State parties);
- the World Heritage List (a list of sites that the World Heritage Committee considers have OUV, and has decided to include in the List on that basis); and
- the World Heritage Fund (a fund administered by the World Heritage Committee, used to assist State parties and others to identify and protect World Heritage) (Table 1.1).

It also gives an advisory role to three international non-government organisations (NGOs): the International Union for the Conservation of Nature (IUCN), the International Council on Monuments and Sites (ICOMOS), and the International Centre for the Study of the Preservation and Restoration of Cultural Property (ICCROM). However, the *Convention* gives the World Heritage Committee, the Advisory Bodies, and State parties substantial discretion to determine how it should be implemented.

To facilitate the implementation of the *Convention*, the World Heritage Committee has adopted the *Operational Guidelines for the Implementation*

---

[11] *World Heritage Convention* arts 1–2.

[12] Ibid., arts 4–5.

[13] Ibid., preamble para 2.

[14] Ibid., arts 6–7.

[15] Ian Strasser, 'Putting Reform into Action: Thirty Years of the World Heritage Convention: How to Reform a Convention without Changing its Regulations' (2002) 11(2) *International Journal of Cultural Property* 215, 233.

[16] UNESCO, *State Parties Ratification Status* http://whc.unesco.org/en/statesparties/.

Table 1.1   Key features of the *World Heritage Convention* regime

Feature of the regime	Explanation	Key provisions of the World Heritage Convention	Key provisions of the Operational Guidelines 2016
World Heritage	Sites (including monuments, groups of buildings and natural features) that meet the definitions of 'cultural heritage' and/or 'natural heritage' in Articles 1 and 2 of the Convention. Essentially sites meet these definitions if they have outstanding universal value.	Articles 1–2	Part II.A
The World Heritage Committee	An executive body established under the Convention, comprising 21 State parties elected for 6 year terms. The Committee's decision-making powers include determining whether sites should be inscribed on the World Heritage List or the List of World Heritage in Danger, whether States should receive international assistance, and administering the World Heritage Fund. The Committee also examines the state of conservation of listed World Heritage Sites through a monitoring and reporting system.	Articles 8–10	Part I.E
The World Heritage List	A list of sites that the World Heritage Committee considers meet the definitions of cultural heritage and natural heritage in Articles 1 and 2 of the Convention, and has decided to include in the List. Before a site can be listed, it must be nominated by the State party within which it is located. It must also have been included in the State party's Tentative List.	Articles 11(2)–(3), (5)	Parts II.D–II.F, III
Tentative List	A national inventory prepared by a State party and submitted to the World Heritage Committee, of the World Heritage within the State.	Article 11(1)	Part II.C
The List of World Heritage in Danger	A list of sites on the World Heritage List compiled by the World Heritage Committee, which are threatened by serious and specific danger and which require major operations in order to be conserved.	Article 11(4)–(5)	Part IV.B

(*continued*)

Table 1.1   (continued)

Feature of the regime	Explanation	Key provisions of the World Heritage Convention	Key provisions of the Operational Guidelines 2016
The World Heritage Fund	A fund established under the Convention, comprising (among other things) compulsory and voluntary contributions from the State parties.	Articles 15–18	Part VII
The Advisory Bodies	The International Centre for the Study of the Preservation and Restoration of Cultural Property (ICCROM), the International Council of Monuments and Sites (ICOMOS) and the International Union for Conservation of Nature and Natural Resources (IUCN). They monitor the state of conservation of sites on the World Heritage List, and (in the case of ICOMOS and IUCN) make recommendations to the World Heritage Committee concerning properties nominated for inclusion on that list.	Articles 8(3), 13(7)	Part I.G

of the World Heritage Convention.[17] These address matters such as the preparation of nominations for the World Heritage List,[18] monitoring and reporting,[19] and the provision of international assistance to State parties.[20]

[17] UNESCO, *Operational Guidelines for the Implementation of the World Heritage Convention*, UN Doc WHC.16/01 (26 October 2016) ('*Operational Guidelines 2016*'). For an explanation of the history of the *Operational Guidelines*, see Sarah M Titchen, *On the Construction of Outstanding Universal Value: UNESCO's World Heritage Convention (Convention Concerning the Protection of the World Cultural and Natural Heritage, 1972) and the Identification and Assessment of Cultural Places for Inclusion in the World Heritage List* (PhD Thesis, Australian National University, 1995) 104–108. The World Heritage Committee has decided that the *Operational Guidelines* should be restricted to operational guidance, and a new policy document should be prepared to capture the policies that the Committee and the General Assembly have adopted. Work is underway to prepare this 'Policy Compendium'. It will likely lead to substantial changes to the *Operational Guidelines*. For discussion, see *Progress Report on the Draft Policy Compendium*, WHC 42nd sess, UN Doc WHC/18/42.COM/11 (28 May 2018).

[18] *Operational Guidelines 2016*, UN Doc WHC.16/01, part III.A.

[19] Ibid., parts IV–V.

[20] Ibid., part VII.

Importantly, the *Operational Guidelines* also prescribe the requirements that a site must meet before the Committee will consider it eligible for World Heritage listing.[21] These involve consideration of the site's value and significance, as well as its protection and management.[22]

Although the *Operational Guidelines* are not legally binding, they are critically important because they underlie much of the Committee's decision-making.[23] By amending the *Operational Guidelines*, the Committee has influenced how the *Convention* is implemented in response to changes in the international community's views towards heritage and its protection.[24] As will be explored in Chaps. 3 and 4, through this process the *Convention* regime has evolved to better facilitate the recognition and conservation of Pacific heritage.

1.3   WORLD HERITAGE IN THE INDEPENDENT PACIFIC ISLAND STATES

Soon after sites began to be inscribed on the World Heritage List, the composition and balance of the List became a topic of discussion among members of the World Heritage Committee and the Advisory Bodies. In response to growing concern about the under-representation of certain regions and types of heritage sites, in 1994 the Committee adopted the *Global Strategy for a Representative, Balanced and Credible World Heritage List*.[25] The *Global Strategy* is a framework and operational methodology for the implementation of the *Convention*. Among other things, it involves encouraging States in under-represented regions to sign the *Convention*, and to prepare Tentative Lists and nominations.[26] It also led to the adoption of a priority system for the assessment of nominations, which favours

---

[21] Ibid., part II.

[22] Ibid., paras 77–78. The requirements for World Heritage listing are analysed in Sects. 3.3 and 4.3.3 of this book.

[23] Strasser, above n 15, 245–246.

[24] See, for example, Titchen, above n 10, 240; Sophia Labadi, *UNESCO, Cultural Heritage and Outstanding Universal Value* (AltaMira Press, 2013) 31; Lynn Meskell, 'UNESCO's World Heritage Convention at 40: Challenging the Economic and Political Order of International Heritage Conservation' (2013) 54(4) *Current Anthropology* 483, 486.

[25] WHC Res CONF 003 X.10, WHC 18th sess, UN Doc WHC-94/CONF.003/16 (31 January 1995) 41–44. See also *Operational Guidelines 2016*, UN Doc WHC.16/01, paras 55–58.

[26] *Operational Guidelines 2016*, UN Doc WHC.16/01, para 60.

those that will help improve the balance of the List.[27] As the Pacific has always been an under-represented region, it is a focus of the *Global Strategy*. The *Global Strategy* has had some positive outcomes in the Pacific. As will be explored in Chaps. 3 and 4, it contributed to the Committee broadening its interpretation of the concept of 'World Heritage' in recognition of the diverse range of heritage sites that exist around the world. These include 'cultural landscapes' (sites that reflect the interaction between humans and their environment), which are common in the Pacific, and are now recognised as a category of World Heritage site.[28] The *Global Strategy* also encouraged the acceptance of different forms of heritage protection, such as that offered by customary law.[29] This is highly significant for the Pacific, where a high proportion of land is under customary tenure, and customary legal systems remain relevant to many people.

More generally, workshops and studies conducted as part of the *Global Strategy* increased awareness of and interest in the *Convention* regime in the Pacific. Twelve of the 14 independent Pacific Island States are now signatories, and eight sites within these countries have been listed (see Table 1.2). In addition, the *Global Strategy* created impetus for the development of the *Pacific 2009 World Heritage Programme*, which was adopted by the Committee in 2003.[30] This was a significant development, as it was the first initiative specifically focused on World Heritage in the region. It provided a framework for efforts to improve implementation, including through awareness raising and capacity building.[31] It has been superseded by the *Pacific World Heritage Action Plan 2016–2020*.[32] This Action Plan (discussed in Chap. 8) identifies regional- and national-level

[27] Ibid., para 61.

[28] Ibid., para 47.

[29] Badman et al, above n 5, 27.

[30] WHC Res 27 COM 6A, WHC 27th sess, UN Doc WHC-03/27.COM/24 (10 December 2003) 7, 8. For a discussion of the history of the Pacific Programme, see Anita Smith, 'The World Heritage Pacific 2009 Programme' in Anita Smith (ed), *World Heritage in a Sea of Islands: Pacific 2009 Programme*, World Heritage Papers 34 (UNESCO, 2012) 2.

[31] UNESCO, *World Heritage – Pacific 2009 Programme* http://whc.unesco.org/en/pacific2009.

[32] This plan superseded the *Pacific World Heritage Action Plan 2010–2015*. The 2016–2020 plan was adopted by delegates at a regional meeting in Suva, Fiji in December 2015. It was updated at a regional workshop in Palau in August/September 2017. See UNESCO Office for the Pacific States, *Final Report: Pacific Heritage Workshop, Koror, Palau, 30 August–1 September 2017* (UNESCO, 2018).

Table 1.2 Overview of the implementation of the *World Heritage Convention* by the independent Pacific Island States

State party	Date of signature of the Convention	Name of World Heritage Site	Year of inscription on the World Heritage List	Cultural, natural or mixed site	Relevant criteria (para 77 of the 2016 Operational Guidelines)	Brief summary of the site's World Heritage values	Land tenure and human occupation of the World Heritage Site
Cook Islands	2009	—	—	—	—	—	—
Federated States of Micronesia	2002	Nan Madol: Ceremonial Centre of Eastern Micronesia	2016	Cultural	(i), (iii), (iv) and (vi)	Comprises more than 100 islets containing the remains of stone palaces, temples, tombs and residential domains built between 1200 and 1500. Evidence of the complex social and religious practices of island societies during that time.	Customary tenure. Uninhabited.[a]
Fiji	1990	Levuka Historical Port Town	2013	Cultural	(ii) and (iv)	First colonial capital of Fiji. Example of the architecture of European colonisation. Demonstrates interactions between Pacific islanders and colonisers that occurred as part of colonisation process.	5.95 ha privately owned freehold land, 0.5 ha owned by the State.[b]

(continued)

Table 1.2 (continued)

State party	Date of signature of the Convention	Name of World Heritage Site	Year of inscription on the World Heritage List	Cultural, natural or mixed site	Relevant criteria (para 77 of the 2016 Operational Guidelines)	Brief summary of the site's World Heritage values	Land tenure and human occupation of the World Heritage Site
Kiribati	2000	Phoenix Islands Protected Area	2010	Natural	(vii) and (ix)	Comprises over 400,000 sq km of marine and terrestrial habitats. A relatively pristine mid-ocean environment, hosting high marine biodiversity. Contains numerous sea mounts, and is a breeding site for many species.	State owned. No permanent inhabitants. About 50 people live there, associated with park management.[c]
Marshall Islands	2002	Bikini Atoll Nuclear Site	2010	Cultural	(iv) and (vi)	Provides tangible evidence of the birth of the Cold War and the nuclear arms race. Demonstrates the effect of nuclear testing on island populations.	Customary ownership. State owns area below high water mark. Approximately 25 inhabitants.[d]
Nauru	–	–	–			–	–
Niue	2001	–	–			–	–

Country		Site		Category	Criteria	Description	Ownership
Palau	2002	Rock Islands Southern Lagoon	2012	Mixed	(iii), (v), (vii), (ix) and (x)	A large lagoon including 445 islands. Remains of burial sites and rock art bear testimony to the island communities that existed there over some three millennia. Has high conservation value because of its spectacular marine and terrestrial biodiversity.	Under customary tenure. Ownership of the site has been the subject of several court cases. No permanent inhabitants.[c]
Papua New Guinea	1997	Kuk Early Agricultural Site	2008	Cultural	(iii) and (iv)	Comprises archaeological evidence of transformation of agricultural practices around 6500 years ago.	Customary tenure. Subject to government lease. Approximately 150 inhabitants.[f]
Samoa	2001	–	–	–	–	–	–
Solomon Islands	1992	East Rennell	1998	Natural	(ix)	Encompasses Lake Tegano, dense forest and a marine area. An important site for the study of island biogeography because of the speciation processes that have occurred there. Hosts several endemic species.	Customary tenure. Approximately 750 inhabitants.[g]
Tonga	2004	–	–	–	–	–	–
Tuvalu	–	–	–	–	–	–	–

(continued)

Table 1.2 (continued)

State party	Date of signature of the Convention	Name of World Heritage Site	Year of inscription on the World Heritage List	Cultural, natural or mixed site	Relevant criteria (para 77 of the 2016 Operational Guidelines)	Brief summary of the site's World Heritage values	Land tenure and human occupation of the World Heritage Site
Vanuatu	2002	Chief Roi Mata's Domain	2008	Cultural	(iii), (v) and (vi)	Cultural landscape comprising three sites associated with the life and death of the last Paramount Chief of Roi Mata. Reflects the continuing chiefly system, and the connection between people and their environment.	Customary land. No residents within the site. Around 670 residents in the buffer zone around the site.[h]

Except as set out below, information in this table is drawn from UNESCO, *World Heritage List* http://whc.unesco.org/en/list; UNESCO, *State Parties Ratification Status* http://whc.unesco.org/en/statesparties/

[a]ICOMOS, *Evaluations of Nominations of Cultural and Mixed Properties to the World Heritage List*, WHC 40th sess, UN Doc WHC/16/40.COM/INF.8B1 (July 2016) 103 (Nan Madol, Federated States of Micronesia, Advisory Body Evaluation 1503) 107

[b]Republic of Fiji, *Nomination of Levuka Historical Port Town to the World Heritage List* (2013) 213

[c]Government of Kiribati, *Phoenix Islands Protected Area, Kiribati, Nomination for a World Heritage Site* (2009) 13, 115

[d]Republic of the Marshall Islands, *Bikini Atoll Nomination by the Republic of the Marshall Islands for Inscription on the World Heritage List* (2010) 64–65. This nomination dossier states that most Bikinians were relocated from the site before it was used to conduct nuclear tests from 1946: at 58

[e]Republic of Palau, *The Rock Islands Southern Lagoon Nomination for Inscription on the World Heritage List* (2012) 91, 105–107

[f]Government of Papua New Guinea, *Kuk Early Agricultural Site Cultural Landscape – A Nomination for Consideration as World Heritage Site* (2007) 29, 60–62

[g]Elspeth J Wingham, *Nomination of East Rennell, Solomon Islands by the Government of Solomon Islands for Inclusion in the World Heritage List Natural Sites* (1997) 5

[h]Republic of Vanuatu, *Chief Roi Mata's Domain – Nomination by the Republic of Vanuatu for Inscription on the World Heritage List* (2007) 93, 96

actions designed to address the challenges associated with identifying and protecting Pacific World Heritage.

Despite the successes of the *Global Strategy*, imbalances in the World Heritage List have increased since it was adopted.[33] Today less than 1% of all listed World Heritage sites are located in the independent Pacific Island States (see Fig. 1.2). While a perfect regional balance is neither desirable nor achievable,[34] the magnitude of the imbalance suggests that impediments to the listing of Pacific sites remain.

Several factors influence the composition of the World Heritage List, including the politicisation of the listing process[35] and the composition of the Committee.[36] Fundamentally however, the Pacific is under-represented because sites can only be listed if they are first nominated by the relevant State party,[37] and to date the rate of nomination by Pacific nations has been low. There are many reasons for this. Most Pacific countries only signed the *Convention* within the last 15 years (see Table 1.2), giving them less time than others to prepare nominations. They have also (at least historically) had less interest and involvement in the *Convention*, in part because they were not involved with its drafting (see Sect. 3.2.1). The lack of expert resources, including comprehensive inventories of Pacific heritage

---

[33] See, for example, Lasse Steiner and Bruno S Frey, 'Correcting the Imbalance of the World Heritage List: Did the UNESCO Strategy Work?' (2012) 3 *Journal of International Organisation Studies* 25, 38; Lynn Meskell, Claudia Liuzza and Nicholas Brown, 'World Heritage Regionalism: UNESCO from Europe to Asia' (2015) 22 *International Journal of Cultural Property* 437, 438.

[34] ICOMOS, *The World Heritage List: Filling the Gaps – An Action Plan for the Future* (ICOMOS, 2004) 19; *Joint ICOMOS-IUCN Paper and Papers by ICOMOS and IUCN on the Application of the Concept of Outstanding Universal Value*, WHC 30th sess, UN Doc WHC-06/30.COM/INF (29 June 2006) 12, 38.

[35] See, for example, Craig Forrest, *International Law and the Protection of Cultural Heritage* (Routledge, 2011) 247; Lynn Meskell, 'The Rush to Inscribe: Reflections on the 35th Session of the World Heritage Committee, UNESCO Paris, 2011' (2012) 37(2) *Journal of Field Archaeology* 145; Bruno S Frey, Paolo Pamini and Lasse Steiner, 'Explaining the World Heritage List: An Empirical Study' (2013) 60 *International Review of Economics* 1; Lynn Meskell, 'States of Conservation: Protection, Politics and Pacting within UNESCO's World Heritage Committee' (2014) 87(1) *Anthropological Quarterly* 217; Enrico E Bertacchini and Donatella Saccone, 'Toward a Political Economy of World Heritage' (2012) 36 *Journal of Cultural Economics* 327.

[36] See, for example, Meskell, above n 24, 489; Bruno S Frey and Lasse Steiner, 'World Heritage List: Does it Make Sense?' (2011) 17(5) *International Journal of Cultural Policy* 555, 560.

[37] *World Heritage Convention* art 11(3).

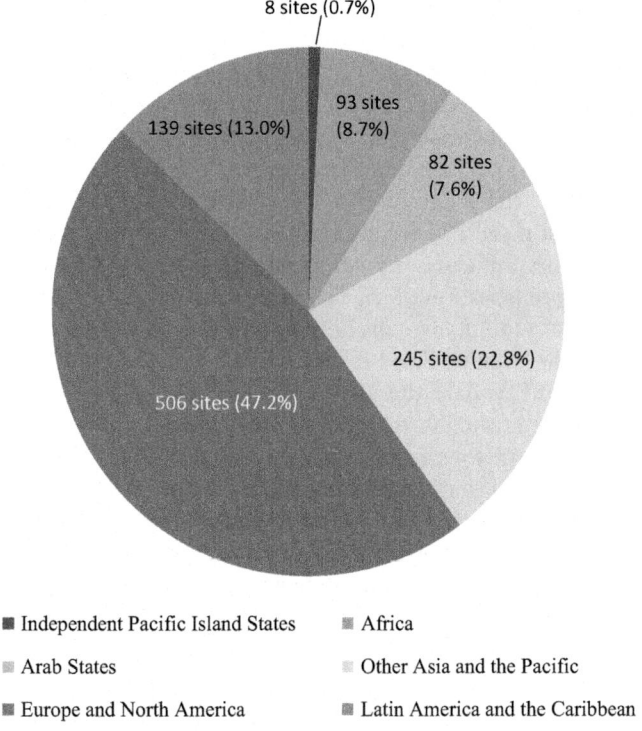

Fig. 1.2 Regional distribution of World Heritage sites. Data sourced from UNESCO, *World Heritage List Statistics* http://whc.unesco.org/en/list/stat#s1

places, also impedes the development of nominations.[38] While two thematic studies conducted as part of the *Global Strategy* have alleviated this problem,[39] many Pacific Island governments still lack the resources to prepare a nomination dossier with the requisite level of detail.[40]

---

[38] Anita Smith, 'Context for the Thematic Study' in Anita Smith and Kevin L Jones (eds), *Cultural Landscapes of the Pacific Islands* (ICOMOS, 2007) 5, 5.

[39] The two thematic studies are: Anita Smith and Kevin L Jones (eds), *Cultural Landscapes of the Pacific Islands* (ICOMOS, 2007); Ian Lilley (ed), *Early Human Expansion and Innovation in the Pacific: Thematic Study* (ICOMOS, 2010).

[40] The requirements for a nomination dossier are prescribed in the *Operational Guidelines 2016*, UN Doc WHC.16/01, part III.B, annex 5. As noted by Bertacchini and Saccone, preparing nomination dossiers is very costly: see Bertacchini and Saccone, above n 35, 331.

Furthermore, economic and social development is often a higher priority than heritage conservation for Pacific Island governments, particularly in Least Developed Countries such as Solomon Islands (see Sect. 2.5.1).

Another likely contributor to the low rate of nominations is that many Pacific Island States lack strong legal frameworks for heritage protection.[41] To be eligible for World Heritage listing, the World Heritage Committee considers that a site must be adequately managed and protected[42] (see Sect. 4.3.3.1). While a site may meet this requirement because of its customary protection,[43] custom is seldom able to deal with all contemporary threats to a site.[44] Consequently, additional legislative measures will often be required. The lack of effective heritage legislation in many Pacific Island States thus contributes to the region's under-representation on the World Heritage List, as well as directly hampering protection at a local level.

The *Pacific 2009 World Heritage Programme* aimed to build the capacity of Pacific Island States to implement the *Convention*. However, the programme did not substantially improve 'the institutional capacity of Pacific Island governments to protect and manage their heritage or to support customary owners to do so'.[45] As such, there remains a critical need to strengthen World Heritage protection in the Pacific. In recognition of this, increasing the effectiveness and coordination of policy and legislation is one of the key aims of the *Pacific World Heritage Action Plan 2016–2020*.[46]

---

[41] Smith, above n 38, 5; Anita Smith, 'Building Capacity in Pacific Island Heritage Management: Lessons from Those Who Know Best' (2007) 3(3) *Archaeologies* 335, 347.

[42] *Operational Guidelines 2016*, UN Doc WHC.16/01, paras 78, 97.

[43] Ibid., para 97. The Committee's decision to recognise customary protection of World Heritage sites is analysed in Sect. 4.3.3.

[44] See, for example, Smith, above n 30, 5; Chris Ballard and Meredith Wilson, 'Unseen Monuments: Managing Melanesian Cultural Landscapes' in Ken Taylor and Jane L Lennon (eds), *Managing Cultural Landscapes* (Routledge, 2012) 130, 132; Pepe Clarke and Charles Taylor Gillespie, *Legal Mechanisms for the Establishment and Management of Terrestrial Protected Areas in Fiji* (IUCN, 2009) 2.

[45] Anita Smith, 'East Rennell World Heritage Site: Misunderstandings, Inconsistencies and Opportunities in the Implementation of the World Heritage Convention in the Pacific Islands' (2011) 17(6) *International Journal of Heritage Studies* 592, 604.

[46] *Pacific World Heritage Action Plan 2016–2020* (2016) 7.

1.4   World Heritage in Solomon Islands

1.4.1   *The East Rennell World Heritage Site*

Solomon Islands is an independent Pacific Island nation, comprising around 1000 islands stretching across 1450 km between Bougainville and the northern islands of Vanuatu (see Fig. 1.3). It became a signatory to the *World Heritage Convention* in 1992. East Rennell (its only listed World Heritage site) was inscribed on the World Heritage List in 1998.[47]

The East Rennell World Heritage site encompasses the south-eastern part of the island of Rennell, 236 km south of Honiara (the nation's capital) in the province of Rennell and Bellona. It includes the marine area extending three nautical miles into the sea (see Fig. 1.4).

The World Heritage site is dominated by the expansive Lake Tegano,[48] which covers 18% of Rennell, making it the largest lake in the Pacific Islands[49] (Fig. 1.5). The remainder of the terrestrial part of the site is predominantly dense, low-stature forest that supports many unique species[50] (Fig. 1.6). The marine area includes extensive fringing coral reefs, hosting diverse invertebrate, fish, and benthic marine life[51] (Fig. 1.7).

East Rennell is customary land, and is owned and occupied by the East Rennellese people pursuant to their customary tenure system. Their ancestors arrived on the island from the Wallis and Futuna group[52] and thus the East Rennellese are of Polynesian descent. Today, approximately 750 people live within the World Heritage site,[53] mainly in four villages

---

[47] WHC Res CONF 203 VIII.A.1, WHC 22nd sess, UN Doc WHC-98/CONF/203/18 (29 January 1999) 25.

[48] Tegano is sometimes spelled Teganno or Te Nggano.

[49] See, for example, Elspeth J Wingham, *Nomination of East Rennell, Solomon Islands by the Government of Solomon Islands for Inclusion in the World Heritage List Natural Sites* (1997) 10.

[50] See, for example, *Adoption of Retrospective Statements of Outstanding Universal Value*, WHC 36th sess, UN Doc WHC-12/36.COM/8E (15 June 2012) 55 (East Rennell, Solomon Islands). East Rennell's Statement of Outstanding Universal Value was adopted by the World Heritage Committee pursuant to WHC Res 36 COM 8E, WHC 36th sess, UN Doc WHC-12/36.COM/19 (June–July 2012) 225.

[51] See, for example, Simon Albert et al, *Survey of the Condition of the Marine Ecosystem within the East Rennell World Heritage Area, Solomon Islands* (University of Queensland, Solomon Islands Marine Ecology Laboratory, Griffith University and WWF-Solomon Islands, 2013).

[52] Wingham, above n 49, 23.

[53] Solomon Islands Government, *Volume I Report on 2009 Population and Housing Census: Basic Tables and Census Description*, Statistical Bulletin 6/2012 (Solomon Islands Government,

Fig. 1.3 Map of Solomon Islands. Map made with data from Natural Earth. Free vector and raster map data @ naturalearthdata.com

located on the southern boundary of the lake (Tebaitahe, Nuipani, Tegano, and Hutuna) (Figs. 6.1 and 6.2). They live predominantly subsistence lifestyles, relying on fish from the lake and sea, and resources from

2012) 24. Population estimates for the site do however vary, in part reflecting permanent and/or temporary migration away from the site. The site's World Heritage nomination dossier stated that in 1997 the population was approximately 1500 but declining: see Wingham, above n 49, 26. IUCN's estimate in its review of the nomination dossier was 800: see IUCN, *Evaluations of Nominations of Natural and Mixed Properties to the World Heritage List*, WHC 22nd sess (1998) 79, 80. Wein estimated the population at 700: see Laurie Wein, *East Rennell World Heritage Site Management Plan* (Solomon Islands National Commission for UNESCO, 2007) 12. Anita Smith has estimated the population at around 700 people: see Smith, above n 45, 594. Gabrys and Heywood stated that the population was approximately 600 people: see Kasia Gabrys and Mike Heywood, 'Community and Governance in the World Heritage Property of East Rennell' in Anita Smith (ed), *World Heritage in a Sea of Islands: Pacific 2009 Programme*, World Heritage Papers 34 (UNESCO, 2012) 60, 60. The Statement of OUV for the site adopted by the World Heritage Committee in 2012 says the population is approximately 1200: see *Adoption of Retrospective Statements of Outstanding Universal Value*, WHC 36th sess, UN Doc WHC-12/36.COM/8E (15 June 2012) 55; WHC Res 36 COM 8E, WHC 36th sess, UN Doc WHC-12/36.COM/19 (June–July 2012) 225.

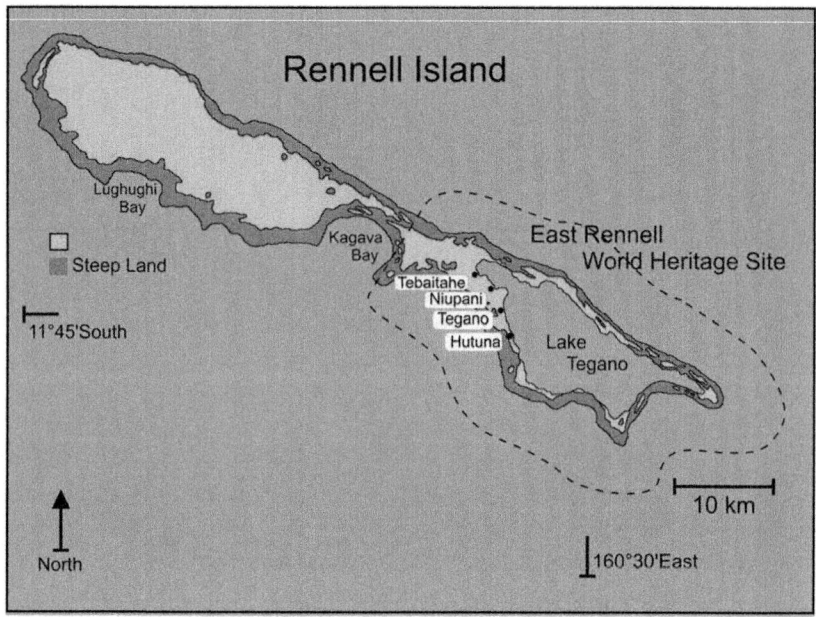

Fig. 1.4   Map of the East Rennell World Heritage site. Source: Laurie Wein, *East Rennell World Heritage Site Management Plan* (Solomon Islands National Commission for UNESCO, 2007)

the forests and their gardens.[54] As will be explained throughout this book, the conservation of the World Heritage site is intrinsically linked with their customs and livelihoods.

East Rennell was inscribed on the World Heritage List as a 'natural' site. The World Heritage Committee considered that it warranted listing due to the island's role as a 'stepping stone in the migration and evolution of species in the region', and because of the speciation processes that have occurred there.[55] The site's cultural significance was not recognised in the

---

[54] See, for example, Wingham, above n 49, 27.

[55] *Adoption of Retrospective Statements of Outstanding Universal Value*, WHC 36th sess, UN Doc WHC-12/36.COM/8E (15 June 2012) 55 (East Rennell, Solomon Islands); WHC Res 36 COM 8E, WHC 36th sess, UN Doc WHC-12/36.COM/19 (June–July 2012) 225.

Fig. 1.5 View from Lake Tegano (Stephanie Price, 2012)

listing, which has ongoing implications for the site's protection (see Sects. 5.2.1 and 8.3.1).

The Committee considered that the 'protection and management' requirements for World Heritage listing were met because East Rennell enjoyed protection under the customary legal system of the East Rennellese people.[56] To supplement this customary protection, the Committee called upon the Solomon Islands government (SIG) to implement a management plan and legislation to ensure the long-term conservation of the area.[57] While a management plan was prepared in 2007, it has not been effective, and today the site is only weakly protected under State law (discussed in Chaps. 6 and 7).

[56] WHC Res 36 COM 8E, WHC 36th sess, UN Doc WHC-12/36.COM/19 (June–July 2012) 225. See Sects. 1.6.2 and 1.6.3 for discussion of the meaning of the terms 'customary legal system' and 'customary protection'.

[57] Ibid.

Fig. 1.6 View from limestone cliffs along the south/east coast of East Rennell, looking eastwards over dense forest towards Lake Tegano (Michael Woodward, 2011)

East Rennell is now threatened by logging and mining, which is being carried out in West Rennell[58] and which may commence within the World Heritage site in the near future.[59] Invasive species, climate change, and the over-harvesting of coconut crabs and marine species could also damage the site's OUV[60] (see Sect. 5.3.1). As a result of these threats, the Committee has put the site on the List of World Heritage in Danger.[61] As explored in Part III of this book, safeguarding East Rennell's OUV in the long term will require a range of actions, including strengthening the protection of the site under customary and State law.

[58] The term 'West Rennell' is used here to describe all parts of the island of Rennell other than East Rennell.
[59] See, for example, Dingwall, above n 7, 4.
[60] Ibid., 13–24.
[61] The Committee placed the site on the List of World Heritage in Danger in 2013: see WHC Res 37 COM 7B.14, WHC 37th sess, UN Doc WHC-13/37.COM.20 (5 July 2013) 68. The site has been retained on that list at all subsequent meetings.

Fig. 1.7 View from limestone cliffs along the south/east coast of East Rennell, looking down on the marine area within the southern side of the World Heritage site (Michael Woodward, 2011)

1.4.2   Solomon Islands' Tentative List

Solomon Islands' Tentative List was submitted to the World Heritage Committee in 2008 and refers to two sites. The first is the 'Marovo-Tetepare Complex', which encompasses large marine areas and several islands in the west of the country.[62] This site includes Marovo Lagoon (one of the world's largest coral reef lagoons), which was identified as a possible candidate for World Heritage listing in the early 1990s[63] (see

---

[62] UNESCO, *Marovo – Tetepare Complex* http://whc.unesco.org/en/tentativelists/5414/.

[63] John McKinnon, *Solomon Islands World Heritage Site Proposal: Report on a Fact Finding Mission (4–22 February 1990)* (Victoria University of Wellington, 1990); Charles d'E Darby, *Rennell Island and Marovo Lagoon: A Proposal by Solomon Islands for World Heritage Site*

Fig. 1.3). Marovo was not nominated at that time, in part because the size of the resident population (approximately 8500 people) made conducting community consultations logistically difficult.[64] The SIG contends that the 'Marovo-Tetepare Complex' has OUV as a mixed site because of its outstanding marine and terrestrial environments, which are connected to the cultural identity and spiritual lives of the local peoples.[65]

The second site is referred to as 'Tropical Rainforest Heritage of Solomon Islands', and comprises rainforest areas in Makira-Ulawa, Choiseul, and Western and Central provinces. It has been included in the Tentative List based on its outstanding natural environment, in particular because of the many unique bird species found there.[66]

The difficulties SIG is experiencing in relation to East Rennell may dissuade it from nominating these sites, at least in the short term. However, if they are nominated, lessons learned from the East Rennell experience should be heeded. Like East Rennell, the proposed protection regimes for these sites involve customary systems supplemented by management plans and legally recognised protected areas.[67] Thus, many of the issues identified in this book will apply to these sites.

1.5   STRENGTHENING WORLD HERITAGE PROTECTION IN THE PACIFIC

The Pacific Island States have a history of regional cooperation, as evidenced by numerous regional organisations[68] and treaties.[69] Pacific regionalism presents a significant opportunity for strengthening World Heritage

Listing as the Basis of a Sustainable Rural Development Programme (Conservation Development Services, 1989).

[64] Elspeth J Wingham, *World Heritage/Ecotourism Programme: Draft Project Implementation Document, August 1998*, attached as attachment 3 to Elspeth J Wingham, *Nomination of East Rennell, Solomon Islands by the Government of Solomon Islands for Inclusion in the World Heritage List Natural Sites* (1997) 7.

[65] UNESCO, above n 62.

[66] UNESCO, *Tropical Rainforest Heritage of Solomon Islands* http://whc.unesco.org/en/tentativelists/5416/.

[67] Ibid.; UNESCO, above n 62.

[68] There are now more than 300 regional organisations in the Pacific focused on a range of issues, including economic, religious, commercial, educational, technical, professional, cultural, sporting, and environmental issues: Ron Crocombe, *The South Pacific* (University of the South Pacific, 2001) 591.

[69] See, for example, *Convention for the Protection of Natural Resources and Environment of the South Pacific Region (Noumea Convention)*, opened for signature 25 November 1986, 26

protection. It has been fostered by meetings and workshops held in the Pacific as part of the implementation of the *Global Strategy*, which have provided Pacific islanders with opportunities to meet and discuss common issues.[70] Importantly, regional cooperation has helped the Pacific Island States to clearly articulate their views to the World Heritage Committee. The most significant example of this was the *Pacific Appeal*, which was presented to the Committee by representatives of the Pacific Island States in 2007.[71] That document brought the vision of Pacific islanders concerning their heritage and the *Convention* to the world stage. It explained that the Pacific 'contains a series of spectacular and highly powerful spiritually-valued natural features and cultural places', unlike other regions which comprise extensive monumental heritage.[72] Furthermore, Pacific islander heritage is 'holistic, embracing all life, both tangible and intangible' and is understood through cultural traditions.[73] The implementation of the *Convention* in the region must be considered in the context of these types of heritage places. Importantly, the *Pacific Appeal* also highlighted that the protection of this heritage 'must be based on respect for and understanding and maintenance of the traditional cultural practices, indigenous knowledge and systems of land and sea tenure' in the region.[74] This includes recognition of customary legal systems, which continue to govern many aspects of the lives of Pacific islanders. These systems not only form part of the heritage of the Pacific, they have been utilised to manage natural resources and culturally significant places for millennia. Therefore, they can contribute to the preservation of World Heritage.

The Pacific Island States exhibit 'legal pluralism', in part because both State and customary legal systems operate there[75] (see Sect. 2.4).

ILM 38 (entered into force 22 August 1990); *Convention on Conservation of Nature in the South Pacific (Apia Convention)*, opened for signature 12 June 1976, [1990] ATS 41 (entered into force 28 June 1990).

[70] For example, the regional World Heritage workshop held in Suva, Fiji in December 2015. For details of other meetings and workshops, see, for example, Smith, above n 30.

[71] *Presentation of the World Heritage Programme for the Pacific*, WHC 31st sess, UN Doc WHC-07/31.COM/11C (10 May 2007) annex I (Appeal to the World Heritage Committee from the Pacific Island State Parties).

[72] Ibid., annex I para 11.

[73] Ibid., annex I para 9.

[74] Ibid., annex I para 13.

[75] Legal pluralism is commonly referred to as the existence of two or more legal orders in the same social field: Sally Engle Merry, 'Legal Pluralism' (1988) 22 *Law and Society Review* 869, 870; John Griffiths, 'What is Legal Pluralism?' (1986) 24 *Journal of Legal Pluralism* 1, 12.

Developing and implementing heritage protection legislation in a legally plural context can be challenging. As Smith has noted:

> In many Pacific countries a tension remains between national legislation for protection of World Heritage properties (in compliance with the State party's obligations under the World Heritage Convention) and the rights of customary land owners. Developing legal protection for Pacific Island heritage that recognizes the rights of customary owners and satisfies international standards established in very different social, cultural and political systems, remains a great challenge and will require flexibility and cultural sensitivity in the World Heritage system.[76]

This challenge is exacerbated by the fact that in the Pacific region there is 'limited financial and human resources, skills and capacities within communities, and institutions to adequately manage the region's cultural and natural heritage'.[77] Consequently, most Pacific Island States do not have well-established frameworks for the protection of culturally significance places. In addition, while many have legislation for the protection of natural areas, such laws are rarely consistently implemented and enforced.[78] To improve this situation, greater understanding of the role of, and the relationship between, State and customary laws in the context of World Heritage protection is needed. This book provides new insights into these issues.

1.6   KEY TERMINOLOGY USED IN THIS BOOK

1.6.1   *World Heritage*

The term 'World Heritage' is not defined in the *Convention*, and in fact only appears in the treaty's preamble.[79] The *Convention* instead applies to 'cultural heritage' and 'natural heritage', terms which are defined in Articles 1 and 2, respectively (see Sect. 3.1). Essentially, to meet these definitions a heritage site must possess OUV.

---

[76] Smith, above n 30, 9.
[77] *Pacific World Heritage Action Plan 2016–2020* (2016) 3.
[78] Smith, above n 30, 9–10.
[79] Para 6.

In common parlance, the terms 'World Heritage' and 'World Heritage site' are often used to refer to a place inscribed on the World Heritage List. However, despite the visibility of that List, the *Convention* does not just apply to listed sites.[80] Pursuant to the *Convention*, State parties have obligations with respect to the protection of *all* properties that fall within the definitions in Articles 1 and 2, irrespective of whether those sites have been nominated for or inscribed on the World Heritage List.[81]

In recognition of this, the terms 'World Heritage' and 'World Heritage site' are used in this book to refer to all heritage sites falling within the definitions of 'cultural heritage' and 'natural heritage' in Articles 1 and 2 of the *Convention*, not just those on the World Heritage List. Where necessary for clarity, places that meet the definitions in Articles 1 and 2 and that have been inscribed on the List are referred to as 'listed sites' or 'listed World Heritage sites'.

1.6.2 Customary Legal Systems, Customary Laws, Customs, and Kastoms

This book uses the term 'customary legal system'. Adopting Forsyth's description of a '*kastom*[82] system', a 'customary legal system' encompasses 'traditional norms of behaviour that are backed up by a sanction of some description (either positive or negative) administered by a member or members of the local community, or a chief at some level of the chiefly hierarchy' as well as the processes by which disputes are dealt with.[83] The system therefore involves customary norms, governance bodies, and dispute resolution processes.

Customary norms are variously described in different contexts as 'customs' (or *kastoms*) or 'customary laws'. These terms are used interchange-

---

[80] Guido Carducci, 'Articles 4–7 National and International Protection of the Cultural and Natural Heritage' in Francesco Francioni (ed), *The 1972 World Heritage Convention: A Commentary* (Oxford University Press, 2008) 103, 113.

[81] *World Heritage Convention* arts 4–5, 12. For analysis of these provisions, see generally Carducci, above n 80; Federico Lenzerini, 'Article 12 Protection of Properties Not Inscribed on the World Heritage List' in Francesco Francioni (ed), *The 1972 World Heritage Convention: A Commentary* (Oxford University Press, 2008) 201.

[82] '*Kastom*' is the pijin term for 'custom'.

[83] Miranda Forsyth, 'Beyond Case Law: *Kastom* and Courts in Vanuatu' (2004) 35 *Victoria University of Wellington Law Review* 427, 431.

ably in common parlance[84] and some academic literature.[85] They are broad terms, subject to numerous definitions. One definition of *kastom* is that it encompasses 'indigenous ideologies, relationship to and management of land, moral frameworks, dispute management, gender relations and social organisation'.[86]

There is some debate about where the boundary between 'custom' (or *kastom*) and 'customary law' lies. It is commonly argued that a custom becomes law through uniform practice and the peoples' subjective belief that the norm must be complied with.[87] However, in practice determining whether a custom has reached that threshold is difficult[88] (discussed further in Sect. 2.4.1). No attempt is made here to further this debate. In this book, the terms are used interchangeably to describe the norms that form part of a customary legal system.

1.6.3   Customary Protection and Traditional Protection

The term 'customary protection' is used in this book to describe the protection provided to a heritage place through the operation of a customary legal system. The *Operational Guidelines* and other literature use the term

---

[84] Allen et al, *Justice Delivered Locally: Systems, Challenges and Innovations in Solomon Islands* (World Bank, 2013) 34; Sue Farran, 'Is Legal Pluralism an Obstacle to Human Rights? Considerations from the South Pacific' (2006) 52 *Journal of Legal Pluralism and Unofficial Law* 77, 100.

[85] For discussion of this issue, see Jennifer Corrin Care and Jean G Zorn, 'Legislating pluralism: Statutory "Developments" in Melanesian Customary Law' (2001) 46 *Journal of Legal Pluralism* 49, 52–3.

[86] Geoffrey M White, 'Three Discourses of Custom' (1993) 6(4) *Anthropological Forum* 475, 492. See also Ton Otto, 'Transformations of Cultural Heritage in Melanesia: From Kastom to Kalsa' (2015) 21(2) *International Journal of Heritage Studies* 117. Writing about Manus in Papua New Guinea, Otto states that 'kastom refers to a wide range of things and practices, including traditional leadership and conflict mediation, ceremonial exchange and transition rituals, traditional rights to land and sea, and beliefs about illness and spirits': at 122. See also David Akin, 'Ancestral Vigilance and the Corrective Conscience: Kastom as Culture in a Melanesian Society' (2004) 4(3) *Anthropological Theory* 299. Akin says that kastom denotes 'ideologies and activities formulated in terms of empowering indigenous traditions and practices': at 299.

[87] T W Bennett and T Vermeulen, 'Codification of Customary Law' (1980) 24(2) *Journal of African Law* 206, 215; Francesco Parisi, 'The Formation of Customary Law' (Paper presented at the 96th Annual Conference of the American Political Science Association, Washington DC, August 31–September 3, 2000) 4.

[88] Farran, above n 84, 93; Jennifer Corrin Care, 'Wisdom and Worthy Customs: Customary Law in the South Pacific' (2002) 80 *Reform* 31, 32.

'traditional protection' to mean the same thing.[89] In this book, the 'fortress'-style approach to protected area management (which was prevalent in Western countries in the early years of the implementation of the *Convention*) is referred to as the 'traditional' approach (see Chap. 4). Thus, to avoid confusion, the term 'customary protection' is used here rather than 'traditional protection'.

1.6.4 Customary Land, Customary Ownership, and Customary Owners

'Customary land' is land held pursuant to customary law. Rights over customary land depend on the applicable customary laws, which vary throughout the Pacific. Like most other relevant literature, this book uses the terms 'customary ownership' and 'customary owners'. However, it is acknowledged that customary tenure is better thought of as a complex and flexible system of rights and obligations, rather than a system of ownership.[90] Thus, people who have the right to occupy and/or use customary land do not 'own' that land in the Western sense of that word. While it is acknowledged that references to 'customary ownership' and 'customary owners' misrepresent the true nature of Pacific land tenure, those terms are used for convenience purposes (see Sect. 2.4.5 for further discussion).

REFERENCES

ARTICLES, BOOKS AND REPORTS

Akin, David, 'Ancestral Vigilance and the Corrective Conscience: Kastom as Culture in a Melanesian Society' (2004) 4(3) *Anthropological Theory* 299

Albert, Simon, Peter Ramohia, Andrew Olds, Tingo Leve and Charlotte Kvennefors, *Survey of the Condition of the Marine Ecosystem within the East Rennell World Heritage Area, Solomon Islands* (University of Queensland, Solomon Islands Marine Ecology Laboratory, Griffith University and WWF-Solomon Islands, 2013)

Allen, Matthew, Sinclair Dinnen, Daniel Evans and Rebecca Monson, *Justice Delivered Locally: Systems, Challenges and Innovations in Solomon Islands* (World Bank, 2013)

---

[89] See, for example, *Operational Guidelines 2016*, UN Doc WHC.16/01, para 97.

[90] Jim Fingleton (ed), *Privatising Land in the Pacific: A Defence of Customary Tenures*, Discussion Paper 80 (The Australia Institute, 2005) ix.

Badman, T, B Bomhard, A Fincke, J Langley, P Rosabal and D Sheppard, *Outstanding Universal Value: Standards for Natural World Heritage* (IUCN, 2008)

Ballard, Chris and Meredith Wilson, ' Unseen Monuments: Managing Melanesian Cultural Landscapes' in Ken Taylor and Jane L Lennon (eds), *Managing Cultural Landscapes* (Routledge, 2012) 130

Bennett, T W and T Vermeulen, 'Codification of Customary Law' (1980) 24(2) *Journal of African Law* 206

Bertacchini, Enrico E and Donatella Saccone, 'Toward a Political Economy of World Heritage' (2012) 36 *Journal of Cultural Economics* 327

Carducci, Guido, 'Articles 4–7 National and International Protection of the Cultural and Natural Heritage' in Francesco Francioni (ed), *The 1972 World Heritage Convention: A Commentary* (Oxford University Press, 2008) 103

Clarke, Pepe and Charles Taylor Gillespie, *Legal Mechanisms for the Establishment and Management of Terrestrial Protected Areas in Fiji* (IUCN, 2009)

Corrin Care, Jennifer, 'Wisdom and Worthy Customs: Customary Law in the South Pacific' (2002) 80 *Reform* 31

Corrin Care, Jennifer and Jean G Zorn, 'Legislating pluralism: Statutory 'Developments' in Melanesian Customary Law' (2001) 46 *Journal of Legal Pluralism* 49

Crocombe, Ron, *The South Pacific* (University of the South Pacific, 2001)

d'E Darby, Charles, *Rennell Island and Marovo Lagoon: A Proposal by Solomon Islands for World Heritage Site Listing as the Basis of a Sustainable Rural Development Programme* (Conservation Development Services, 1989)

Dingwall, Paul, *Report on the Reactive Monitoring Mission to East Rennell, Solomon Islands, 21–29 October 2012* (IUCN, 2013)

Farran, Sue, 'Is Legal Pluralism an Obstacle to Human Rights? Considerations from the South Pacific' (2006) 52 *Journal of Legal Pluralism and Unofficial Law* 77

Fingleton, Jim (ed), *Privatising Land in the Pacific: A Defence of Customary Tenures*, Discussion Paper 80 (The Australia Institute, 2005)

Forrest, Craig, *International Law and the Protection of Cultural Heritage* (Routledge, 2011)

Forsyth, Miranda, 'Beyond Case Law: *Kastom* and Courts in Vanuatu' (2004) 35 *Victoria University of Wellington Law Review* 427

Frey, Bruno S and Lasse Steiner, 'World Heritage List: Does it Make Sense?' (2011) 17(5) *International Journal of Cultural Policy* 555

Frey, Bruno S, Paolo Pamini and Lasse Steiner, 'Explaining the World Heritage List: An Empirical Study' (2013) 60 *International Review of Economics* 1

Gabrys, Kasia and Mike Heywood, 'Community and Governance in the World Heritage Property of East Rennell' in Anita Smith (ed), *World Heritage in a Sea of Islands: Pacific 2009 Programme*, World Heritage Papers 34 (UNESCO, 2012) 60

Government of Kiribati, *Phoenix Islands Protected Area, Kiribati, Nomination for a World Heritage Site* (2009)

Government of Papua New Guinea, *Kuk Early Agricultural Site Cultural Landscape – A Nomination for Consideration as World Heritage Site* (2007)

Griffiths, John, 'What is Legal Pluralism?' (1986) 24 *Journal of Legal Pluralism* 1

ICOMOS, *The World Heritage List: Filling the Gaps – An Action Plan for the Future* (ICOMOS, 2004)

Labadi, Sophia, UNESCO, *Cultural Heritage and Outstanding Universal Value* (AltaMira Press, 2013)

Lenzerini, Federico, 'Article 12 Protection of Properties Not Inscribed on the World Heritage List' in Francesco Francioni (ed), *The 1972 World Heritage Convention: A Commentary* (Oxford University Press, 2008) 201

Lilley, Ian (ed), *Early Human Expansion and Innovation in the Pacific: Thematic Study* (ICOMOS, 2010)

McKinnon, John, *Solomon Islands World Heritage Site Proposal: Report on a Fact Finding Mission (4–22 February 1990)* (Victoria University of Wellington, 1990)

Merry, Sally Engle, 'Legal Pluralism' (1988) 22 *Law and Society Review* 869

Meskell, Lynn, 'The Rush to Inscribe: Reflections on the 35th Session of the World Heritage Committee, UNESCO Paris, 2011' (2012) 37(2) *Journal of Field Archaeology* 145

Meskell, Lynn, 'UNESCO's World Heritage Convention at 40: Challenging the Economic and Political Order of International Heritage Conservation' (2013) 54(4) *Current Anthropology* 483

Meskell, Lynn, 'States of Conservation: Protection, Politics and Pacting within UNESCO's World Heritage Committee' (2014) 87(1) *Anthropological Quarterly* 217

Meskell, Lynn, Claudia Liuzza and Nicholas Brown, 'World Heritage Regionalism: UNESCO from Europe to Asia' (2015) 22 *International Journal of Cultural Property* 437

Otto, Ton, 'Transformations of Cultural Heritage in Melanesia: From Kastom to Kalsa' (2015) 21(2) *International Journal of Heritage Studies* 117

Parisi, Francesco, 'The Formation of Customary Law' (Paper presented at the 96th Annual Conference of the American Political Science Association, Washington DC, August 31–September 3, 2000)

Republic of Fiji, *Nomination of Levuka Historical Port Town to the World Heritage List* (2013)

Republic of Palau, *The Rock Islands Southern Lagoon Nomination for Inscription on the World Heritage List* (2012)

Republic of the Marshall Islands, *Bikini Atoll Nomination by the Republic of the Marshall Islands for Inscription on the World Heritage List* (2010)

Republic of Vanuatu, *Chief Roi Mata's Domain – Nomination by the Republic of Vanuatu for Inscription on the World Heritage List* (2007)

Smith, Anita, 'Building Capacity in Pacific Island Heritage Management: Lessons from Those Who Know Best' (2007) 3(3) *Archaeologies* 335

Smith, Anita, 'Context for the Thematic Study' in Anita Smith and Kevin L Jones (eds), *Cultural Landscapes of the Pacific Islands* (ICOMOS, 2007) 5

Smith, Anita, 'East Rennell World Heritage Site: Misunderstandings, Inconsistencies and Opportunities in the Implementation of the World Heritage Convention in the Pacific Islands' (2011) 17(6) *International Journal of Heritage Studies* 592

Smith, Anita, 'The World Heritage Pacific 2009 Programme' in Anita Smith (ed), *World Heritage in a Sea of Islands: Pacific 2009 Programme*, World Heritage Papers 34 (UNESCO, 2012) 2

Smith, Anita and Kevin L Jones (eds), *Cultural Landscapes of the Pacific Islands* (ICOMOS, 2007)

Solomon Islands Government, *Volume I Report on 2009 Population and Housing Census: Basic Tables and Census Description*, Statistical Bulletin 6/2012 (Solomon Islands Government, 2012)

Steiner, Lasse and Bruno S Frey, 'Correcting the Imbalance of the World Heritage List: Did the UNESCO Strategy Work?' (2012) 3 *Journal of International Organisation Studies* 25

Strasser, Ian, 'Putting Reform into Action: Thirty Years of the World Heritage Convention: How to Reform a Convention without Changing its Regulations' (2002) 11(2) *International Journal of Cultural Property* 215

Titchen, Sarah M, 'On the Construction of 'Outstanding Universal Value': Some Comments on the Implementation of the 1972 UNESCO World Heritage Convention' (1996) 1 *Conservation and Management of Archaeological Sites* 235

UNESCO Office for the Pacific States, *Final Report: Pacific Heritage Workshop, Koror, Palau, 30 August–1 September 2017* (UNESCO, 2018)

Wein, Laurie, *East Rennell World Heritage Site Management Plan* (Solomon Islands National Commission for UNESCO, 2007)

White, Geoffrey M, 'Three Discourses of Custom' (1993) 6(4) *Anthropological Forum* 475

Wingham, Elspeth J, *Nomination of East Rennell, Solomon Islands by the Government of Solomon Islands for Inclusion in the World Heritage List Natural Sites* (1997)

Wingham, Elspeth J, *World Heritage/Ecotourism Programme: Draft Project Implementation Document, August 1998*, attached as attachment 3 to Elspeth J Wingham, *Nomination of East Rennell, Solomon Islands by the Government of Solomon Islands for Inclusion in the World Heritage List Natural Sites* (1997)

CONVENTIONS

Convention Concerning the Protection of the World Cultural and Natural Heritage, opened for signature 16 November 1972, 1037 UNTS 151 (entered into force 17 December 1975)
Convention for the Protection of Natural Resources and Environment of the South Pacific Region (Noumea Convention), opened for signature 25 November 1986, 26 ILM 38 (entered into force 22 August 1990)
Convention on Conservation of Nature in the South Pacific (Apia Convention), opened for signature 12 June 1976, [1990] ATS 41 (entered into force 28 June 1990)

UNITED NATIONS DOCUMENTS

Adoption of Retrospective Statements of Outstanding Universal Value, WHC 36th sess, UN Doc WHC-12/36.COM/8E (15 June 2012) 55 (East Rennell, Solomon Islands)
ICOMOS, *Evaluations of Nominations of Cultural and Mixed Properties to the World Heritage List*, WHC 40th sess, UN Doc WHC/16/40.COM/INF.8B1 (July 2016) 103 (Nan Madol, Federated States of Micronesia, Advisory Body Evaluation 1503)
IUCN, *Evaluations of Nominations of Natural and Mixed Properties to the World Heritage List*, WHC 22nd sess (1998) 79 (East Rennell, Solomon Islands)
Joint ICOMOS-IUCN Paper and Papers by ICOMOS and IUCN on the Application of the Concept of Outstanding Universal Value, WHC 30th sess, UN Doc WHC-06/30.COM/INF (29 June 2006)
Presentation of the World Heritage Programme for the Pacific, WHC 31st sess, UN Doc WHC-07/31.COM/11C (10 May 2007) annex I (Appeal to the World Heritage Committee from the Pacific Island State Parties)
Progress Report on the Draft Policy Compendium, WHC 42nd sess, UN Doc WHC/18/42.COM/11 (28 May 2018)
UNESCO, *Records of the General Conference – volume 1*, 17th sess (17 October–21 November 1972) 135
UNESCO, *Operational Guidelines for the Implementation of the World Heritage Convention*, UN Doc WHC.16/01 (26 October 2016)

WHC Res CONF 003 X.10, WHC 18th sess, UN Doc WHC-94/CONF.003/16 (31 January 1995) 41

WHC Res CONF 203 VIII.A.1, WHC 22nd sess, UN Doc WHC-98/CONF/203/18 (29 January 1999) 25

WHC Res 27 COM 6A, WHC 27th sess, UN Doc WHC-03/27.COM/24 (10 December 2003) 7

WHC Res 36 COM 8E, WHC 36th sess, UN Doc WHC-12/36.COM/19 (June–July 2012) 225

WHC Res 37 COM 7B.14, WHC 37th sess, UN Doc WHC-13/37.COM/20 (5 July 2013) 68

INTERNET MATERIALS

UNESCO, *Marovo – Tetepare Complex* http://whc.unesco.org/en/tentativelists/5414/

UNESCO, *State Parties Ratification Status* http://whc.unesco.org/en/statesparties/

UNESCO, *Tropical Rainforest Heritage of Solomon Islands* http://whc.unesco.org/en/tentativelists/5416/

UNESCO, *World Heritage List* http://whc.unesco.org/en/list

UNESCO, *World Heritage – Pacific 2009 Programme* http://whc.unesco.org/en/pacific2009

THESES

Titchen, Sarah M, *On the Construction of Outstanding Universal Value: UNESCO's World Heritage Convention (Convention Concerning the Protection of the World Cultural and Natural Heritage, 1972) and the Identification and Assessment of Cultural Places for Inclusion in the World Heritage List* (PhD Thesis, Australian National University, 1995)

OTHER

Pacific World Heritage Action Plan 2016–2020 (2016)

PART II

World Heritage Conservation in the Pacific

CHAPTER 2

The Pacific Context

2.1 INTRODUCTION

As will be explored throughout this book, many of the opportunities and challenges associated with World Heritage conservation in the Pacific can be linked to the nature of the region's heritage, the legal systems that govern its people, and the context within which those systems operate. This chapter therefore explores those issues and examines their relevance to the protection of World Heritage. It does not aim to provide a comprehensive analysis of all characteristics of Pacific Island States that impact on heritage protection, and indeed, it would not be possible to do so within one chapter. Rather, the chapter identifies key issues that help explain the context within which Pacific Island States are attempting to implement the *World Heritage Convention*.[1]

The chapter begins by examining the types of heritage sites prevalent in the Pacific, including natural environments, landscapes reflecting the settlement and development of island societies, and places associated with European and American contact with the region (Sect. 2.2). The key threats to such places are also noted (Sect. 2.3).

---

[1] *Convention Concerning the Protection of the World Cultural and Natural Heritage*, opened for signature 16 November 1972, 1037 UNTS 151 (entered into force 17 December 1975) ('*World Heritage Convention*').

© The Author(s) 2018
S. C. Price, *World Heritage Conservation in the Pacific*,
Palgrave Series in Asia and Pacific Studies,
https://doi.org/10.1007/978-981-13-0602-0_2

The chapter continues by explaining how a legacy of colonialism in the Pacific is the creation of legally plural States, in which both customary and State laws apply (Sect. 2.4). After briefly outlining the development of Pacific legal systems, the chapter demonstrates how customary legal systems have been shaped by outside influences, but nevertheless remain integral to the lives of most Pacific Islanders. Laws concerning customary land tenure are discussed in some detail, because many heritage places in the region are under customary ownership.

The potential for customary and State legal norms to regulate matters relevant to heritage conservation is then assessed, including exploring the economic, social, and political context within which those legal systems operate (Sect. 2.5). The chapter argues that greater understanding of how customary and State legal systems operate and interact is needed to strengthen the protection of the region's spectacular natural and cultural sites.

2.2   HERITAGE SITES OF THE PACIFIC ISLANDS

Few inventories of heritage sites in the Pacific have been prepared, and those that exist are limited in scope and/or reflect the interests of foreign researchers rather than Pacific Islanders.[2] Smith and Jones' 2007 study of cultural landscapes[3] and Lilley's 2010 study of early human expansion in the region[4] significantly enhanced the body of knowledge concerning Pacific heritage. However, the character and diversity of culturally significant sites have not yet been comprehensively documented.[5] Similarly, few ecosystems in the Pacific have been thoroughly researched.[6] Despite these

---

[2] Ian Lilley and Christophe Sand, 'Thematic Frameworks for the Cultural Values of the Pacific' in Anita Smith (ed), *World Heritage in a Sea of Islands: Pacific 2009 Programme*, World Heritage Papers 34 (UNESCO, 2012) 22, 24, 26.

[3] Anita Smith and Kevin L Jones (eds), *Cultural Landscapes of the Pacific Islands* (ICOMOS, 2007).

[4] Ian Lilley (ed), *Early Human Expansion and Innovation in the Pacific: Thematic Study* (ICOMOS, 2010).

[5] Lilley and Sand, above n 3, 24.

[6] See, for example, Hugh Govan et al, *Status and Potential of Locally-Managed Marine Areas in the South Pacific: Meeting Nature Conservation and Sustainable Livelihood Targets Through Wide-Spread Implementation of LMMAs* (SPREP/WWF/WorldFish-Reefbase/CRISP, 2009), 16; Gunnar Keppel et al, 'Isolated and Vulnerable: The History and Future of Pacific Island Terrestrial Biodiversity' (2014) 20(2) *Pacific Conservation Biology* 136, 141; Matt McIntyre, *Pacific Environment Outlook* (United Nations Environment Programme and the Secretariat of the Pacific Regional Environment Programme, 2005) 1, ch 2.

gaps, the available literature demonstrates that the region's heritage places are diverse and face a range of threats, so no one form of heritage protection legislation will be appropriate and effective at all sites.[7]

2.2.1   The 'Natural' Environment of the Pacific Islands

The Pacific region comprises diverse marine and terrestrial ecosystems. Its marine areas range from deep ocean trenches to coral reefs and large enclosed lagoons,[8] and support more marine biodiversity than any other region.[9] Within the expansive Pacific Ocean lie thousands of islands, with varied geologies, topographies, ecologies, and climates.[10] They include 'continent'-like landmasses, high volcanoes, atolls, and raised coral limestone islands.[11] Many are home to a variety of terrestrial species, some of which are endemic (i.e. unique to that place). Biodiversity and endemism are particularly high in the west of the region (including in Solomon Islands),[12] but much lower in areas where islands are smaller and more remote.[13] Three places in the Pacific Island States have been inscribed on the World Heritage List based on their natural heritage values: East Rennell in Solomon Islands, the Phoenix Islands Protected Area in Kiribati, and the Rock Islands Southern Lagoon in Palau (discussed in Sect. 3.5.2).

While the terrestrial and marine environments of the Pacific comprise the natural heritage of the region, few are pristine. Direct and indirect

---

[7] Intangible heritage is not covered here, despite its importance to Pacific Islanders, because purely intangible heritage does not fall within the scope of the *World Heritage Convention*. See Sect. 3.2.2 for discussion of the scope of the *Convention*.

[8] See, for example, Richard Herr, 'Environmental Protection in the South Pacific: The Effectiveness of SPREP and its Conventions' in Olav Schram Stokke and Øystein B Thommessen (eds), *Yearbook of International Co-operation on Environment and Development 2002/2003* (Earthscan Publications, 2002) 41, 43.

[9] See, for example, Govan et al, above n 7, 16.

[10] See, for example, Anita Smith, 'The Cultural Landscapes of the Pacific Islands' in Anita Smith and Kevin L Jones (eds), *Cultural Landscapes of the Pacific Islands* (ICOMOS, 2007) 17, 18.

[11] See, for example, Paul Dingwall, 'Pacific Islands World Heritage Tentative Lists' in Anita Smith (ed), *World Heritage in a Sea of Islands: Pacific 2009 Programme*, World Heritage Papers 34 (UNESCO, 2012) 28, 30; Stuart Chape, 'Natural World Heritage in Oceania: Challenges and Opportunities' in Anita Smith (ed), *World Heritage in a Sea of Islands: Pacific 2009 Programme*, World Heritage Papers 34 (UNESCO, 2012) 40, 40.

[12] See, for example, Barry Cox and Peter Moore, *Biogeography: An Ecological and Evolutionary Approach* (Oxford, 1980) 109–11.

[13] See, for example, Smith, above n 10, 18.

human influences on island environments began when the region was first settled.[14] Some settlers caused environmental change by introducing new plants (such as coconut, banana, taro, yam, cassava, paw paw, and bread-fruit) and animals (including pigs, dogs, and chickens)[15] to make their new island homes more 'familiar and manageable'.[16] Further changes were caused by settlers clearing and burning forest,[17] cultivating land,[18] constructing permanent features, altering fresh water resources,[19] and hunting native fauna species.[20]

On some islands, settlers caused considerable environmental degradation. For example, the clearing and torching of land to allow for shifting cultivation and garden crops altered island vegetation, and increased erosion and soil degradation.[21] Island animals were vulnerable to the introduction of fauna species and other human activities because they evolved in areas with few terrestrial predators.[22] Consequently, settlers caused the extinction of some fauna species, particularly ground-dwelling birds.[23] Marine creatures were also often depleted due to over-harvesting.[24]

Pacific Island settlers not only modified their environment to suit their livelihoods, but also developed customary laws regulating the use and management of their land and natural resources. Today, many Pacific Islanders still possess 'deep traditional knowledge about their sea and

[14] See, for example, Patrick D Nunn, 'Nature-society interactions in the Pacific Islands' (2013) 85(4) *Geografiska Annaler, Series B, Human Geography* 219, 222; Frank R Thomas, 'The Precontact Period' in Moshe Rapaport (ed), *The Pacific Islands: Environment and Society* (University of Hawai'i Press, 2013) 125, 133–134.

[15] See, for example, Nunn, above n 14, 219; Smith, above n 10, 28.

[16] John R McNeill, 'Of Rats and Men: A Synoptic Environmental History of the Island Pacific' (1994) 5(2) *Journal of World History* 299, 304.

[17] See, for example, Patrick V Kirch, 'Late Holocene Human-Induced Modifications to a Central Polynesian Island Ecosystem' (1996) 93 *Proceedings of the National Academy of Sciences* 5296, 5296.

[18] See, for example, Chape, above n 11, 40.

[19] See, for example, Smith, above n 10, 28.

[20] See, for example, Nunn, above n 14, 219.

[21] See, for example, McNeill, above n 16, 306–307; Keppel et al, above n 7, 138.

[22] See, for example, McNeill, above n 16, 302; Keppel et al, above n 7, 136.

[23] See, for example, David W Steadman, 'Prehistoric Extinctions of Pacific Island Birds: Biodiversity Meets Zooarchaeology' (1995) 267 *Science* 1123; Stacy Jupiter, Sangeeta Manguhai and Richard T Kingsford, 'Conservation of Biodiversity in the Pacific Islands of Oceania: Challenges and Opportunities' (2014) 20(2) *Pacific Conservation Biology* 206, 206; McNeill, above n 16, 305–307.

[24] See, for example, McNeill, above n 16, 305.

forests and elaborate traditional practices expressed through dances and customary rites of their environment', which evidence their close connection with their environment.[25]

In regions such as the Pacific, where Indigenous people continue to possess cultural and spiritual connections with their environment, the concepts of 'nature' and 'culture' may overlap.[26] A key characteristic of Pacific Island heritage is therefore that the distinction between 'cultural heritage' and 'natural heritage' is often blurred.[27] This presents a challenge for the implementation of the *World Heritage Convention*, which deals separately with cultural and natural sites (see Sect. 3.3.1). It also raises questions about the appropriateness of Pacific sites being recognised as natural World Heritage sites (see Sect. 3.5.2).

2.2.2   Sites Reflecting the Settlement and Development of Pacific Island Societies

Large-scale monuments are relatively rare in the Pacific region.[28] More commonly, Pacific heritage places exemplify the settlement of the islands and the development of islander societies.

---

[25] Eric L Kwa, 'Climate Change and Indigenous People in the South Pacific' (Paper presented at IUCN Academy of Environmental Law Conference on 'Climate Law in Developing Countries Post-2012: North and South Perspectives', Ottawa, Canada, 26–28 September 2008) 3.

[26] Darrell Addison Posey, 'Introduction: Culture and Nature – The Inextricable Link' in Darrell Addison Posey (ed), *Cultural and Spiritual Values of Biodiversity* (UNEP, 1999) 1, 7.

[27] See, for example, Paige West and Dan Brockington, 'An Anthropological Perspective on Some Unexpected Consequences of Protected Areas' (2006) 20(3) *Conservation Biology* 609, 611; Giovanni Boccardi, 'The World Heritage Pacific 2009 Programme: Addressing the Aims of the Global Strategy in the Pacific Regions' in Anita Smith (ed), *World Heritage in a Sea of Islands: Pacific 2009 Programme*, World Heritage Papers 34 (UNESCO, 2012) 12, 12; Chris Ballard and Meredith Wilson, 'Unseen Monuments: Managing Melanesian Cultural Landscapes' in Ken Taylor and Jane L Lennon (eds), *Managing Cultural Landscapes* (Routledge, 2012) 130, 134; Anita Smith and Cate Turk, 'Customary Systems of Management and World Heritage in the Pacific Islands' in Sue O'Connor, Denis Byrne and Sally Brockwell (eds), *Transcending the Culture-Nature Divide in Cultural Heritage: Views from the Asia-Pacific Region* (ANU E Press, 2012) 22, 29; *Identification of World Heritage Properties in the Pacific: Second World Heritage Global Strategy Meeting for the Pacific Islands Region* (Port Vila, Vanuatu, 24–27 August 1999) preamble para 6. For discussion of the link between cultural and natural heritage generally, see Ben Boer and Stefan Gruber, 'Heritage Discourses' in Brad Jessup and Kim Rubenstein (eds), *Environmental Discourses in Public and International Law* (Cambridge University Press, 2012) 375, 376–377.

[28] Ballard and Wilson, above n 27, 130.

Some commonalities and differences that exist across the Pacific can be explained with reference to the three geo-cultural regions: Melanesia, Micronesia, and Polynesia[29] (see Fig. 2.1). It is acknowledged that such an analysis risks masking significant variation within the regions, and characteristics attributed to one region may be found elsewhere in the Pacific. The geo-cultural divisions do however help explain some important characteristics. For example, as discussed below, the regions were settled at different times and from different sources, contributing to the cultural and ethnic diversity of Pacific Islanders.[30]

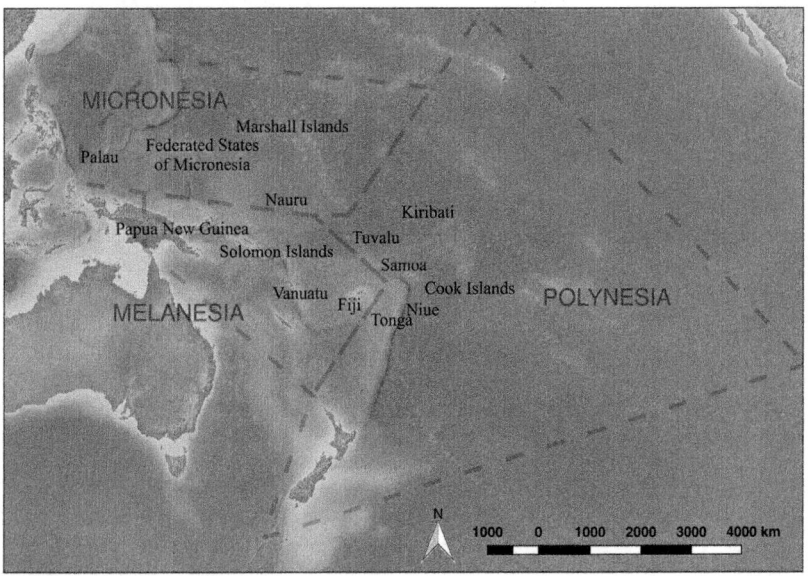

Fig. 2.1   Map of the Pacific showing geo-cultural regions (Melanesia, Polynesia, and Micronesia). Map made with data from Natural Earth. Free vector and raster map data @ naturalearthdata.com

---

[29] Of the independent Pacific Island States, Fiji, Papua New Guinea, Solomon Islands, and Vanuatu are within Melanesia; Cook Islands, Niue, Samoa, Tonga, and Tuvalu are within Polynesia; and Federated States of Micronesia, Kiribati, Marshall Islands, Nauru, and Palau are within Micronesia.

[30] For more comprehensive discussion of the settlement of the region, see Donald Denoon, 'Human Settlement' in Donald Denoon, Malama Meleisea, Stewart Firth, Jocelyn Linnekin and Karen Nero (eds), *The Cambridge History of Pacific Islanders* (Cambridge University Press, 2008) 37.

The first phase of settlement of Melanesia occurred between 30,000 and 50,000 years ago, and involved the settlement of 'Near Oceania' (New Guinea, the Bismarck Archipelago, and Solomon Islands).[31] These migrants, often referred to as Papuans, lacked the technology to migrate any further than the Solomon Islands, so settlement stalled there for thousands of years.[32] Around 4000 years ago, Austronesians (a Southern Mongoloid population from southern China) arrived in the region.[33] Their technologies enabled sailing crews to survive longer at sea, allowing them to settle the eastern parts of Papua New Guinea (PNG).[34] From there, settlement expanded multi-directionally,[35] with the Austronesians reaching outer Solomon Islands, Vanuatu, New Caledonia, Fiji, Tonga, and Samoa by around 3000 years ago.[36] Settlement paused there for over 1000 years.[37]

During the next phase of settlement, which started between AD 700 and AD 1000, settlers continued their expansion beyond Samoa, to settle eastern Polynesia.[38] In the same era, some Polynesians ventured westward, establishing settlements on the outlying islands in Melanesia and Micronesia. These islands are now referred to as 'Polynesian Outliers',[39] and include Rennell in Solomon Islands, which was settled by people from the Wallis and Futuna group.[40] Hence, while most Solomon Islanders are of Melanesian decent, the Rennellese are Polynesian. One consequence of this is that some customary laws of the Rennellese people (including their land tenure system) differ significantly from those in other parts of Solomon Islands. Some implications of this are discussed in Sect. 7.2.2 and Sect. 7.3.1(B).

---

[31] See, for example, Geoffrey Irwin, 'Navigation and Seafaring' in Ian Lilley (ed), *Early Human Expansion and Innovation in the Pacific: Thematic Study* (ICOMOS, 2010) 47, 51; Smith, above n 10, 22.

[32] See, for example, Ron Crocombe, *The South Pacific* (University of the South Pacific, 2001) 44.

[33] See, for example, ibid., 45.

[34] See, for example, Irwin, above n 31, 51.

[35] See, for example, Steven Roger Fischer, *A History of the Pacific Islands* (Palgrave Macmillan, 2nd ed, 2013), 16.

[36] See, for example, Nunn, above n 14, 220.

[37] See, for example, Thomas, above n 14, 127.

[38] See, for example, Irwin, above n 31, 52.

[39] Smith, above n 10, 24.

[40] See, for example, Elspeth J Wingham, *Nomination of East Rennell, Solomon Islands by the Government of Solomon Islands for Inclusion in the World Heritage List Natural Sites* (1997) 23.

Settlement of Micronesia began around 3500 years ago, but some islands were only settled during the last millennium.[41] Current evidence suggests settlers arrived from several sources, including early movements from South East Asia, and later movements from Melanesia and Polynesia,[42] contributing to the considerable cultural diversity within that region.[43]

Pacific heritage places can help us understand early human expansion throughout the region. They include archaeological sites and landscapes reflecting the settlement and development of island societies.[44] Some such landscapes contain evidence of the settlers' transportation and adaption of systems of agriculture and land tenure. For example, the Kuk Early Agricultural Site in PNG was inscribed on the World Heritage List as a landscape demonstrating the transformation of agricultural practices over time.[45] Other heritage places demonstrate the location and layout of traditional villages, and contain physical features that Pacific Islanders have constructed like burial places, fences, and gardens.[46]

Pacific landscapes may be relics, or they may play an active role in contemporary society because of the continuing living traditions associated with them.[47] The continuity of these traditions is commonly demonstrated through stories, and through customary knowledge and practices.[48] Intertwined with these traditions are the customary legal systems (including land tenure systems) of the sites' owners, which also form part of the region's heritage. Indigenous customary law is itself a critical element of Indigenous culture.[49] Thus, a place may gain its heritage significance from the traditions, customary laws, and governance systems that are associated with it. For example, Chief Roi Mata's Domain in Vanuatu was eligible for

---

[41] See, for example, Smith, above n 10, 24.

[42] See, for example, Michiko Intoh, 'Human Dispersal into Micronesia' (1997) 105 *Anthropological Science* 15.

[43] See, for example, Smith, above n 10, 22.

[44] Lilley (ed), above n 5

[45] Government of Papua New Guinea, *Kuk Early Agricultural Site Cultural Landscape – A Nomination for Consideration as World Heritage Site* (2007).

[46] Smith, above n 10, 32–45.

[47] Smith, above n 10, 58.

[48] Ibid.

[49] S. James Anaya, 'International Human Rights and Indigenous Peoples: The Move Toward the Multicultural State' (2004) 21(1) *Arizona Journal of International and Comparative Law* 13, 49.

World Heritage listing in part because of its association with oral traditions connected to Chief Roi Mata, who lived around 1600 AD.[50] These characteristics distinguish the cultural heritage of the Pacific from many other regions, which has two key implications. Firstly, the World Heritage Committee's early focus on the preservation of the types of heritage prevalent in Western States for many years hampered the recognition of Pacific landscapes on the World Heritage List (see Sect. 3.4) Secondly, the protection of Pacific landscapes will often require different approaches to those employed in other regions. For example, the ongoing management of the Kuk Early Agricultural Site in PNG requires continued occupation and cultivation by the site's customary owners (the Kawelka) because 'they provide a connection between archaeological and contemporary practices through which the site gets its significance'.[51] As such, in the Pacific, conservation measures must often accommodate and support the continued ownership, occupation and use of the site by its customary owners (see Chap. 4).

2.2.3   Sites Reflecting European and American Contact with the Pacific Islands

Pacific Island heritage also comprises sites and landscapes reflecting contact made by Europeans and Americans with Pacific Islanders. Evidence of events such as the conversion of Pacific Islanders to Christianity, colonisation, and activities associated with World War II contribute to the diverse heritage of the region.

The first European contact with the Pacific occurred around 500 years ago, when the Portuguese arrived at the west of the region and the Spanish arrived at the east.[52] In the early nineteenth century, Europeans and Americans began to travel to the Pacific to exploit resources like

---

[50] ICOMOS, *Evaluations of Nominations of Cultural and Mixed Properties to the World Heritage List*, WHC 32nd sess, UN Doc WHC-08/32.COM/INF/8B1 (2008) 92 (Chief Roi Mata's Domain, Vanuatu, Advisory Body Evaluation 1280) 94.

[51] Tim Denham, '*Traim Tasol*... Cultural Heritage Management in Papua New Guinea' in Sue O'Connor, Denis Byrne and Sally Brockwell (eds), *Transcending the Culture-Nature Divide in Cultural Heritage: Views from the Asia-Pacific Region* (ANU E Press, 2012) 117, 120.

[52] See, for example, David A Chappell, 'The Postcontact Period' in Moshe Rapaport (ed), *The Pacific Islands: Environment and Society* (University of Hawai'i Press, 2013) 138, 138.

sandalwood, beche de mer,[53] pearl shell, and whale oil.[54] However, these activities did not require large permanent settlements, so they did not leave a legacy of heritage places.[55] Greater changes to Pacific Island societies were caused by missionaries, who visited the Pacific from around 1800 and quickly converted much of the population to Christianity.[56] The work of missionaries is evidenced in the region's architecturally distinct and diverse churches, and the location and layout of villages[57] (as people were often moved from their traditional communities to larger settlements based around a church). Missionaries also influenced Pacific Island heritage by prohibiting some customary practices they considered to be pagan[58] (see Sect. 2.4.4).

Although colonisation occurred relatively late in the Pacific, by 1900 all Pacific Islands except Tonga[59] were controlled by foreign States,[60] including Solomon Islands, which became a British protectorate in 1893.[61] Some heritage places in the region reflect the process of colonisation in an island environment, and interactions between the colonisers and the population.[62] For example, colonisation was accompanied by the establishment of large-scale industries such as mining and plantations,[63] which impacted Pacific landscapes, including through the disruption of customary practices and tenure systems. The Levuka Historical Port Town in Fiji is an example of this type of heritage site. It was inscribed on the World Heritage List as an example of European settlement in the Pacific Islands, reflecting the contact and interchange of values between colonisers and the Pacific Islanders.[64]

---

[53] Beche de mer is processed from *holothurians*, commonly known as sea cucumbers.

[54] Smith, above n 10, 25.

[55] Ibid., 26.

[56] See, for example, John Barker, 'Religion' in Moshe Rapaport (ed), *The Pacific Islands: Environment and Society* (University of Hawai'i Press, 2013) 214.

[57] Smith and Jones (eds), above n 4, 56.

[58] Fischer, above n 35, 109.

[59] Tonga was a protectorate of the United Kingdom between 1900 and 1970, but even during this period, Tonga maintained its sovereignty.

[60] See generally Fischer, above n 35, 125–174.

[61] *Pacific Order in Council 1893* (UK).

[62] Smith, above n 10, 54–56.

[63] Ibid., 54.

[64] WHC Res 37 COM 8B.25, WHC 37th sess, UN Doc WHC-13/37.COM/20 (5 July 2013) 186.

Many significant battles of World War II occurred in the Pacific, causing loss of life, the destruction of villages and gardens, and damage to island landscapes.[65] In Vanuatu, Solomon Islands, PNG, and some Micronesian islands, tangible evidence of the war can be seen in sites evidencing key battles, intensive bombing, large-scale construction (such as airfields), and the use of wartime machinery.[66] Nuclear weapons testing carried out by the United States and France forever changed the natural and cultural heritage of some parts of the region.[67] Perhaps the most well-known example of this is the Bikini Atoll in Marshall Islands, which was inscribed on the World Heritage List as a place bearing testimony to the birth of the Cold War and the nuclear era.[68] Sites reflecting these important global events form part of the rich heritage of the region. However, as discussed in Sect. 3.6, the global and local significance of such a site may be very different, which can impact conservation efforts.

2.3   THREATS TO PACIFIC ISLAND HERITAGE

The region's biodiversity is vulnerable, as many islands are small and host unique species.[69] While some environmental change in the Pacific Islands was caused by early settlers, the rate of change accelerated with the arrival of Europeans and Americans.[70] Agricultural expansion, plantations, and extractive industries are continuing to damage Pacific habitats, driven by forces such as population growth, urbanisation, and increasing consumption.[71] Marine biodiversity is also being affected by over-exploitation, a shift from subsistence to commercial operations and destructive fishing methods,[72] as well as land based activities that damage coastal vegetation

---

[65] Smith, above n 10, 51–54.

[66] Ibid.

[67] Ibid.

[68] WHC Res 34 COM 8B.20, WHC 34th sess, UN Doc WHC-10/34.COM/20 (3 September 2010) 206.

[69] See, for example, Jupiter, Manguhai and Kingsford, above n 23, 206; Catherine Giraud-Kinley, 'The Effectiveness of International Law: Sustainable Development in the South Pacific Region' (1999–2000) 12 *Georgetown Environmental Law Review* 125, 133.

[70] See, for example, Jupiter, Manguhai and Kingsford, above n 23, 207, 210.

[71] See, for example, P Gerbeaux et al, *Shaping a Sustainable Future in the Pacific: IUCN Regional Programme for Oceania 2007–2012* (IUCN, 2007) 3–5.

[72] See, for example, Michael King et al, *Strategic Plan for Fisheries Management and Sustainable Coastal Fisheries in the Pacific Islands* (Secretariat of the Pacific Community, 2003) 1.

and cause sedimentation and marine pollution.[73] Further threatening heritage, social, and economic changes are contributing to the loss of traditional knowledge and the weakening of customary governance.

Compounding these threats are the effects of climate change, which are likely to be profound in the Pacific. Sea level rise will cause the loss of habitable land on many islands, and increasingly frequent and intense storms may affect biodiversity, fisheries, and crops.[74] These changes will affect Pacific landscapes, as well as national economies and the livelihoods of many people. Pacific Island governments already face the difficult task of balancing development with heritage protection (see Sect. 2.5.1), and climate change is likely to increase that challenge.

Some activities that threaten heritage are driven by Pacific Island governments and multi-national companies seeking to benefit from development, whilst others are undertaken (or at least authorised) by Pacific Islanders themselves. Traditionally, people in the region relied on subsistence agriculture supplemented by fishing, gathering, and hunting for their livelihoods.[75] Today, most subsistence-based economics are increasingly becoming commercialised,[76] and the food security of many islanders is being compromised by urbanisation, population growth, and declining crop yields.[77] In addition, globalisation and modernisation have influenced food preferences and livelihood choices, and Pacific Islanders increasingly want to participate in the cash economy. Limited opportunities for paid work[78] lead some to authorise tourism, agriculture, extractive industries, and other developments on their land in return for cash and in-kind pay-

---

[73] See, for example, Vina Ram-Bidesi, 'Ocean Resources' in Moshe Rapaport (ed), *The Pacific Islands: Environment and Society* (University of Hawai'i Press, 2013) 364, 375.

[74] See, for example, Lai Murari, 'Implications of Climate Change in Small Island Developing Countries of the South Pacific' (2004) 2(1) *Fijian Studies* 15; United Nations Office of the High Representative for the Least Developed Countries, Landlocked Developing Countries and Small Island Developing States (UN-OHRLLS), *Small Island Developing States: Small Islands Big(ger) Stakes* (UN, 2011).

[75] See, for example, Anette Reenberg et al, 'Adaption of Human Coping Strategies in a Small Island Society in the SW Pacific: 50 Years of Change in the Coupled Human-Environment system on Bellona, Solomon Islands' (2008) 3(6) *Human Ecology* 807, 807.

[76] United Nations Environment Programme (UNEP) *Pacific Islands Environment Outlook* (UNEP, 1999) xi.

[77] See, for example, Reenberg et al, above n 75, 808; Donovan Storey and David Abbott, 'Development Prospects' in Moshe Rapaport (ed), *The Pacific Islands: Environment and Society* (University of Hawai'i Press, 2013) 417, 420.

[78] See, for example, Storey and Abbott, above n 77, 421.

ments, which can damage heritage places. Consequently, heritage protection in the Pacific is often intimately related to both national and local economic development.

As Pacific heritage sites are diverse and face a range of threats, different approaches will be required to secure their protection. However, as explained in the next section, all Pacific Island States exhibit legal pluralism and most have high rates of customary land ownership. These characteristics provide a common link between many heritage places in the region.

2.4   PACIFIC ISLAND LEGAL SYSTEMS

'Legal pluralism' is commonly referred to as the existence of two or more legal orders in the same social field.[79] It is therefore not a characteristic of a law or legal system, but of a social field (e.g. a nation, region, or community).[80] As explained below, Pacific Island States are legally plural, in part because their Indigenous and colonial histories have created both customary and State legal systems.

2.4.1   *The Concept of Legal Pluralism and Its Application in the Pacific Islands*

Legal pluralism gained attention during the 1970s as legal analysis of governance arrangements in former colonies became more common.[81] Due to its origins, the early focus of legal pluralism was on the relationship between customary and State legal norms and institutions.[82] This field of study has been described as 'classic legal pluralism'.[83]

Since the 1970s, the concept has expanded to encompass other forms of non-State law in both colonised and non-colonised societies. This

---

[79] Sally Engle Merry, 'Legal Pluralism' (1988) 22 *Law and Society Review* 869, 870; John Griffiths, 'What is Legal Pluralism?' (1986) 24 *Journal of Legal Pluralism* 1, 12.

[80] Griffiths, above n 79, 38.

[81] Simon Roberts, 'Against Legal Pluralism: Some Reflections on the Contemporary Enlargement of the Legal Domain' (1998) 42 *Journal of Legal Pluralism and Unofficial Law* 95, 97. For a comprehensive analysis of the development of concept, see, for example, Miranda Forsyth, *A Bird That Flies with Two Wings: Kastom and State Justice Systems in Vanuatu* (ANU E Press, 2009), ch 2; Brian Z Tamanaha, 'Understanding Legal Pluralism: Past to Present, Local to Global' (2008) 30 *Sydney Law Review* 375, 377–390.

[82] Tamanaha, above n 81, 390.

[83] Merry, above n 79, 872.

broader definition (described as 'new legal pluralism') considers the 'complex and interactive relationship between official and unofficial forms of ordering'.[84] In addition to local, national, regional, and international legal systems, new legal pluralism facilitates consideration of customary, religious, economic, community, and other non-State systems.[85] Pursuant to this broader definition, most, if not all, societies exhibit legal pluralism.[86] However, it is often experienced more intensely in developing countries (such as the Pacific Island States) because of the diversity of legal systems that operate there, the qualitative differences between them, and the lack of an effective overarching framework for regulating their interactions.[87]

As explained further in the sections below, in the Pacific, customary legal systems were developed by islanders over time to regulate their daily commerce, civil life, and land tenure.[88] When the islands became colonies and protectorates, new laws enacted by the colonial legislature or the controlling country were introduced, but customary systems continued to operate, often with the sanction of the controlling nations.[89] At independence, the States adopted systems of law and governance reflecting the outgoing colonial governments, but customary systems remained highly relevant to most Pacific Islanders. Independence also led to the States becoming subject to international legal norms (such as the *World Heritage Convention*) and other forms of law, further enriching their legal pluralism.

Legal pluralism is contrary to the theory of legal centralism, which posits that 'law is and should be the law of the state, uniform for all persons, exclusive of all other law, and administered by a single set of state institutions'.[90] Forsyth (who has comprehensively analysed the development of legal pluralism) notes that while some commentators contend that the

[84] Ibid., 873.

[85] See, for example, Tamanaha, above n 81, 397–399.

[86] Merry, above n 79, 873, 879; Tamanaha, above n 81, 375.

[87] Caroline Sage and Michael Woolcock, 'Introduction' in Brian Z Tamanaha, Caroline Sage and Michael Woolcock (eds), *Legal Pluralism and Development: Scholars and Practitioners in Dialogue* (Cambridge University Press, 2012) 1, 9.

[88] See, for example, Stephan Klingelhofer and David Robinson, *The Rule of Law, Custom and Civil Society in the South Pacific: An Overview* (International Center for Not-for-Profit Law, 2001) 10.

[89] See, for example, Jennifer Corrin, 'Customary Land and the Language of the Common Law' (2008) 37 *Common Law World Review* 305, 309.

[90] Griffiths, above n 79, 3.

concept enjoys wide support,[91] others have challenged that proposition.[92] As Forsyth has said, this may be because the concept is subject to several theoretical debates, including how to define the concept of 'law'.[93] Non-State norms exist on a spectrum, ranging from prohibitions that non-State officials may enforce through sanctions, to norms that constitute mere etiquette or good manners.[94] This raises the question of 'where do we stop speaking of law and find ourselves simply describing social life?'[95] In the Pacific context, this question is most acutely seen in a consideration of when 'customs' may be considered law.

Custom can be described as the 'social norms and practices that make up local approaches to dispute management and everyday social regulation in communities'.[96] It is therefore a broad term, encompassing things like traditional leadership systems, conflict mediation, ceremonial exchange, beliefs, and rights to land, sea, and resources.[97] 'Customary law' is a component of the broader concept of custom.[98] However, this begs the question of how to distinguish customary laws from other customs. It is commonly argued that a custom becomes law through uniform practice and the peoples' subjective belief that the norm must be complied with,[99]

---

[91] See, for example, John Griffiths, 'Legal Pluralism and the Theory of Legislation – With Special Reference to the Regulation of Euthanasia' in Hanne Petersen and Henrik Zahle (eds), *Legal Polycentricity: Consequences of Pluralism in Law* (Hanne Peterson, 1995) 210, cited in Gordon Woodman, 'Why There Can be No Map of Law', *Legal Pluralism and Unofficial Law in Social, Economic and Political Development: Papers of the XIIIth International Congress of the Commission on Folk Law and Legal Pluralism* (Chiangmai, Thailand, 7–10 April, 2002) 383; cf Alan Watson, 'An Approach to Customary Law' (1984) 3 *University of Illinois Law Review* 561. Watson argues that custom only becomes law through recognition by the State: at 576.

[92] Forsyth, above n 81, 38.

[93] Ibid.

[94] Gordon R Woodman, 'Ideological Combat and Social Observation: Recent Debate About Legal Pluralism' (1998) 42 *Journal of Legal Pluralism and Unofficial Law* 21, 44.

[95] Merry, above n 79, 878.

[96] Matthew Allen et al, *Justice Delivered Locally: Systems, Challenges and Innovations in Solomon Islands* (World Bank, 2013) 34.

[97] Ton Otto, 'Transformations of Cultural Heritage in Melanesia: From *Kastam* to *Kalsa*' (2015) 21(2) *International Journal of Heritage Studies* 117, 122.

[98] Miranda Forsyth, 'Beyond Case Law: *Kastom* and Courts in Vanuatu' (2004) 35 *Victoria University of Wellington Law Review* 427, 429.

[99] T W Bennett and T Vermeulen, 'Codification of Customary Law' (1980) 24(2) *Journal of African Law* 206, 215; Francesco Parisi, 'The Formation of Customary Law' (Paper presented at the 96th Annual Conference of the American Political Science Association, Washington DC, August 31–September 3, 2000) 4.

but in practice determining whether a custom has reached that threshold is difficult.[100] Issues that complicate the analysis include how widespread customary rules must be before they can be classified as laws,[101] and how long it takes for a custom to transform into a law.[102]

Tamanaha contends that the lack of any clear definition of what constitutes a 'law' places the concept of legal pluralism on tenuous footing.[103] Others contend there is little utility in attempting to formulate such a definition.[104] For example, Twining proposes that the distinctions between legal and other norms are largely unnecessary because 'in most contexts not much turns on where, or even, whether the line is drawn'.[105] This argument is particularly strong in the Pacific, where the customary system is central to the lives of many and the State often only has marginal significance (see Sect. 2.4.4). No attempt is made here to further this debate, and the terms 'custom' and 'customary law' are used interchangeably.

2.4.2   The Development of Customary Legal Systems in the Pacific Islands

There is no single customary legal system in the Pacific region, or within any Pacific Island State. Rather, numerous distinct customary legal systems developed, as the traditional settlers transported and adapted laws and governance models to suit their island environments. Cultural diversity is greatest in Melanesia, because it contains the region's largest landmass (New Guinea), settlement began up to 50,000 years ago, and a mixing of Papuan and Austronesian cultures occurred there.[106] Consequently, Melanesian customary legal systems vary from island to island, and sometimes even from

---

[100] Sue Farran, 'Is Legal Pluralism an Obstacle to Human Rights? Considerations from the South Pacific' (2006) 52 *Journal of Legal Pluralism and Unofficial Law* 77, 93; Jennifer Corrin Care, 'Wisdom and Worthy Customs: Customary Law in the South Pacific' (2002) 80 *Reform* 31, 32.

[101] Corrin Care, above n 100, 32.

[102] Parisi, above n 99, 5.

[103] Tamanaha, above n 81, 392.

[104] Woodman, above n 94, 45; Merry, above n 79, 889; William Twining, 'Legal Pluralism 101' in Brian Z Tamanaha, Caroline Sage and Michael Woolcock (eds), *Legal Pluralism and Development: Scholars and Practitioners in Dialogue* (Cambridge University Press, 2012) 112, 114.

[105] Twining, above n 104, 114.

[106] Fischer, above n 35, 25–27. See generally Ann Gibbons, 'Genes Point to a New Identity for Pacific Pioneers' (1994) 263(5143) *Science* 32.

village to village,[107] often coinciding with different linguistic and ethnic groups.[108] In contrast, cultures developed much later in Polynesia and Micronesia.[109] Polynesian countries like Samoa, Tonga, and Tuvalu comprise one dominant cultural group and little linguistic diversity, and are thus among the most ethnically homogeneous societies in the world today.[110] Micronesian culture is also relatively homogenous.[111] In some countries in these regions, a customary law may apply country-wide.[112]

Customary laws originally developed when Pacific Islanders had no knowledge of writing or printing, so they were communicated orally and by actions.[113] While Pacific Islanders are increasingly documenting their laws,[114] most remain unwritten. The oral nature of custom allows it to be applied flexibly and adapted to suit new situations,[115] and has facilitated its continuing evolution (see Sect. 2.4.4).

Traditionally, customary laws gained their legitimacy from some form of customary authority within a governance arrangement.[116] Although governance systems varied, most Pacific Islanders lived in separate communities controlled by one or more chiefs or other leaders, who regulated peoples' lives based on the community's customary laws.[117] Sahlins developed the

[107] Jennifer Corrin Care and Jean G Zorn, 'Legislating pluralism: Statutory "Developments" in Melanesian Customary Law' (2001) 46 *Journal of Legal Pluralism* 49, 53, 71; Jennifer Corrin Care and Jean G Zorn, 'Legislating for the Application of Customary Law in Solomon Islands' (2005) 34 *Common Law World Review* 144, 145; Jennifer Corrin, 'A Question of Identity: Complexities of State Law Pluralism in the South Pacific' (2010) 61 *Journal of Legal Pluralism and Unofficial Law* 145, 147; Edvard Hviding, 'Contextual Flexibility: Present Status and Future of Customary Marine Tenure in Solomon Islands' (1998) 40 *Ocean and Coastal Management* 253, 256.

[108] Nicholas Menzies, *Legal Pluralism and the Post-Conflict Transition in the Solomon Islands* (Hertie School of Governance, Berlin, 2007) 4.

[109] Fischer, above n 35, 28–42.

[110] Benjamin Reilly, 'State Functioning and State Failure in the South Pacific' (2004) 58(4) *Australian Journal of International Affairs* 479, 480.

[111] Andrew Pawley, 'Language' in Moshe Rapaport (ed), *The Pacific Islands: Environment and Society* (University of Hawai'i Press, 2013) 159, 160.

[112] Jennifer Corrin and Don Paterson, *Introduction to South Pacific Law* (Palgrave Macmillan, 3rd ed, 2011) 40.

[113] See, for example, Corrin, above n 89, 309.

[114] Allen et al, above n 96, 72; Miranda Forsyth, *The Writing of Community By-Laws and Constitutions in Melanesia: Who? Why? Where? How?* State, Society and Governance in Melanesia in Brief (The Australian National University, 2014).

[115] See, for example, Hviding, above n 107, 255.

[116] Allen et al, above n 96, 34.

[117] Corrin and Paterson, above n 112, 1.

much-cited classic model of Pacific Islander governance.[118] This model describes Polynesian chiefs as gaining their rank through inheritance, with power residing in the position of 'chief', rather than an individual person. The limited scholarship on Micronesian governance suggests that this model applied in that region as well.[119] In contrast, the model describes Melanesia as comprising 'big-man' societies, where leaders achieved their status, rather than inheriting it.[120] For example, a leader might gain status through their skills and their involvement with the community, which allowed them to achieve wealth and distribute it, thus gaining favour among the community members.[121] These leaders tended to exert authority over smaller political units than Polynesian chiefs, so Melanesia has traditionally been characterised by greater social fragmentation than other parts of the Pacific.[122]

While Sahlins' classic model demonstrates basic variations in leadership types in the Pacific, it has been criticised as an over-simplification.[123] For example, while Solomon Islands was primarily characterised by 'big-man' systems, there were also hereditary systems, and systems where status and hereditary title coexisted.[124] Today, the term 'chief' is used to refer to many different types of local leaders in Solomon Islands.[125] Regardless of the type of traditional leadership that existed in Pacific Island societies, as explained in the next sections, all customary legal systems were substantially changed by outside contact.

---

[118] Marshall D Sahlins, 'Poor Man, Rich Man, Big-Man, Chief: Political Types in Melanesia and Polynesia' (1963) 5(3) *Comparative Studies in Society and History* 285.

[119] Abby McLeod, *Leadership Models in the Pacific*, State, Society and Governance Discussion Paper (The Australian National University, 2008) 10–11.

[120] Sahlins, above n 118.

[121] Ibid.

[122] McLeod, above n 119, 7.

[123] See, for example, McLeod, above n 119, 4; B Douglas, 'Rank, Power, Authority; A Reassessment of Traditional Leadership in South Pacific Societies' (1979) 14 *Journal of Pacific History* 2; Christophe Sand, 'Melanesian Tribes vs Polynesian Chiefdoms: Recent Archaeological Assessment of a Classic Model of Socio-Political Types in Oceania' (2002) 41(2) *Asian Perspectives* 284.

[124] Roger M Keesing, 'Killers, Big Men, and Priests on Malaita: Reflections on a Melanesian Troika System' (1985) 24(4) *Ethnology* 237.

[125] Geoffrey White, *Indigenous Governance in Melanesia*, State, Society and Governance in Melanesia Discussion Paper (The Australian National University, 2007).

2.4.3   Colonisation and Independence

During the nineteenth century, some Polynesian leaders developed laws that applied across all or most of the country, to expand their control over that area.[126] However, in most parts of the Pacific, no laws applied at the national scale until the islands fell under the control of outside nations. This began in the late 1800s, and by 1900, all Pacific Islands were under some form of European or American control. Following colonisation, new laws were enacted by the legislature of the controlling country or that of the island colony, imposing a new form of governance on Pacific Islanders.[127]

While the colonising nations imposed new systems of law, they had limited resources to govern their colonies, so to maintain social control they allowed and/or encouraged customary legal systems to continue.[128] Initially, customary and 'formal' legal systems operated independently, except in disputes about customary land where colonial courts were authorised to apply custom.[129] However, over time, customary law was given a greater role within the formal system.[130]

When the Pacific Island States achieved independence, the governments of the new States reflected those of the colonising nations. For example, when Solomon Islands obtained independence from Great Britain in 1978, it adopted the Westminster Parliamentary system, with a unicameral Parliament and the British monarch as the head of state.[131] In some States, traditional leaders were given a formal role within the government.[132] This has not occurred in Solomon Islands, so customary gov-

---

[126] Corrin and Paterson, above n 112, 2.

[127] Ibid., 2–3.

[128] Terence Wesley-Smith, 'Changing Patterns of Power' in Moshe Rapaport (ed), *The Pacific Islands: Environment and Society* (University of Hawai'i Press, 2013) 147; Corrin, above n 89, 309–310.

[129] Corrin Care and Zorn, 'Legislating for the Application of Customary Law', above n 107, 145.

[130] Ibid. For example, in the British Solomon Islands Protectorate, native courts were established and authorised to apply native customs: *Native Courts Ordinance 1942* s 10.

[131] *Solomon Islands Independence Order 1978*, sch (*Constitution of Solomon Islands*) s 1(2), 46.

[132] McLeod, above n 119, 8–11; Richard Scaglion, 'Law' in Moshe Rapaport (ed), *The Pacific Islands: Environment and Society* (University of Hawai'i Press, 2013) 202, 205–207. For example, Vanuatu has a National Council of Chiefs known as Malfatu Mauri (*Constitution of the Republic of Vanuatu* ch 5).

ernance bodies there have no legislative backing.[133] Consequently, the strength of such bodies (including their ability to enforce customs relating to World Heritage protection) is highly dependent on their legitimacy within the relevant local communities.

As well as the governance structures of the controlling nations, introduced laws that were in force before independence were retained, to 'fill the gap' until the Pacific Island legislatures amended or replaced them.[134] Solomon Islands retained laws of the United Kingdom as in force on 1 January 1961, and the rules of common law and equity.[135]

While independence was not used by Pacific Islanders to revert to their customary systems of law and governance, it did provide an opportunity to formalise and strengthen the position of custom within the State system. The Constitutions and legislation of all independent Pacific Island States except Tonga now recognise customary law as a source of law either generally or in the determination of certain disputes.[136] For example, in Solomon Islands, customary law is recognised as a valid source of law, to the extent that it is not inconsistent with any Act of Parliament or certain other written laws.[137] Although complex issues remain regarding the application of customary law by State institutions, its formal recognition in Constitutions and legislation has given it status in modern Pacific Island States.[138] It also continues to be applied independently of the State systems by customary leaders, and indeed, it is the most relevant form of law for many Pacific Islanders.[139] Consequently, although legally custom is

[133] In the lead up to Solomon Islands' independence, the idea that a Council of Elders (comprising an elected group of chiefs) would constitute an upper house was discussed. However, the idea was ultimately not accepted. See Clive Moore, *Decolonising the Solomon Islands: British Theory and Melanesian Practice*, Working Paper 8 (Alfred Deakin Research Institute, Deakin University, 2010) 17–18. The only role of chiefs recognised under Solomon Islands' legislation is in the resolution of disputes over rights to customary land (*Local Courts Act (Cap. 19)* s 12(1)).

[134] See generally, Corrin and Paterson, above n 112, 16–19.

[135] *Constitution of Solomon Islands* s 3

[136] For detailed explanation of the extent to which State laws provide for the recognition of customary law in the Pacific Island countries, see Corrin and Paterson, above n 112, 41–51.

[137] *Constitution of Solomon Islands* sch 3, para 3.

[138] Corrin Care and Zorn, 'Legislating for the Application of Customary Law', above n 107, 147.

[139] Klingelhofer and Robinson above n, 88, 10; Corrin Care and Zorn, 'Legislating for the Application of Customary Law', above n 107, 148; Forsyth, above n 81, 251; Allen et al, above n 96, xi, 34; Hviding, above n 107, 266; Jennifer Corrin, 'Moving Beyond the Hierarchical Approach to Legal Pluralism in the South Pacific' (2009) 59 *Journal of Legal*

only valid to the extent that it is consistent with State legislation,[140] it is somewhat of a fallacy to consider State law as being at the apex of the hierarchy of laws.

As explained in the next section, while customary legal systems remain relevant to many Pacific Islanders, contemporary systems rarely reflect those that existed before European contact.

2.4.4   Contemporary 'Customary' Legal Systems

Customary legal systems are often erroneously interpreted as being the systems that existed in the pre-contact period.[141] That interpretation fails to consider the profound changes to Pacific Island societies and legal systems caused by European and American contact with the islands.[142]

Early impacts on traditional societies included the introduction of foreign diseases[143] and the movement of men away from their communities to work on plantations,[144] both of which caused substantial population losses. Missionaries were another early influence, through prohibiting some traditional practices[145] and moving communities to larger villages, which changed leadership structures.[146] In Solomon Islands, missionaries also changed customary legal systems by installing local leaders and introducing systems of punishment and reconciliation.[147] In some places

Pluralism and Unofficial Law 29, 31; Matthew Zurstrassen, *Customary Dispute Resolution Research Project: Final Report to the Regional PJDP Meetings in Samoa in March 2012*, Pacific Judicial Development Programme (2012) 3.

[140] See, for example, *Constitution of Solomon Islands* sch 3, para 3. For analysis of the status of customary laws pursuant to the Constitutions of Pacific Island States, see Katrina Cuskelly, *Customs and Constitutions: State Recognition of Customary Law Around the World* (IUCN, 2011) 18–21.

[141] Allen et al, above n 96, 9.

[142] See, for example, Smith, above n 10, 25–27; Sand, above n 123, 291; Corrin Care and Zorn, 'Legislating Pluralism', above n 107, 51; Forsyth, above n 98, 429.

[143] See, for example, Smith, above n 10, 26; Sand, above n 123, 291.

[144] See, for example, Forsyth, above n 81, 61; Judith Bennett, *Roots of Conflict in Solomon Islands – Though Much is Taken, Much Abides: Legacies of Tradition and Colonialism*, State, Society and Governance in Melanesia Discussion Paper (The Australian National University, 2002) 3.

[145] See, for example, Michael Goddard, *Justice Delivered Locally, Solomon Islands, Literature Review* (World Bank, 2010) 8, 29; Erika J Techera, 'Samoa: Law, Custom and Conservation' (2006) 10 *New Zealand Journal of Environmental Law* 361, 363.

[146] See, for example, Corrin Care and Zorn, 'Legislating Pluralism', above n 107, 53–54.

[147] Goddard, above n 145, 27, 29.

(including Solomon Islands), customary beliefs and practices were influenced by their integration with Christianity, so there is now significant
overlap between customary and local Christian rules and governance.[148]
Customary legal systems were influenced by the introduction of new
colonial rules, some of which limited the jurisdiction of customary laws.[149]
Colonisation also introduced a new system of law that people could access if
they were not satisfied with the customary system, effectively demoting custom within the legal hierarchy.[150] Pacific Islanders themselves also changed
customary laws, often modelling them on formal laws,[151] particularly when
customary laws were being applied as part of the State system.[152]

Colonisation also affected traditional governance structures, although
the impact varied significantly throughout the region. As colonisers found
the centralised, unified governance structures of Polynesia easier to work
with than the more disparate governance arrangements in Melanesia, they
often sought to adapt the latter to meet their needs. In Vanuatu, for
example, except in a few places where chiefs already existed, the British
introduced the concept of 'chiefs' to help them negotiate with the
natives,[153] and now chiefs are central to 'customary' governance systems in
that country.[154]

The modern notion of 'chiefs' is also a product of colonisation and its
aftermath in Solomon Islands. The pacification of the Solomon Islander
population by the protectorate government (including through the sup-
pression of head-hunting and the slave trade) had destroyed the source of
wealth of many big-men, thus undermining their power base.[155] The protectorate government also affected traditional leadership by appointing some
Solomon Islander men as leaders, many of whom were later given promi-

---

[148] White, above n 125, 4; Allen et al, above n 96, 65; Anne M Brown, 'Custom and
Identity: Reflections on and Representations of Violence in Melanesia' in Nikki Slocum-
Bradley (ed), *Promoting Conflict or Peace through Identity* (Ashgate, 2008) 183, 190.

[149] Corrin Care and Zorn, 'Legislating Pluralism', above n 107, 51.

[150] Ibid.

[151] Ibid.

[152] Jean G Zorn, 'Customary Law in the Papua New Guinea Village Courts' (1990) 2(2)
The Contemporary Pacific 279, 306.

[153] Lissant Bolton, 'Chief Willie Bongmatur Maldo and the Role of Chiefs in Vanuatu'
(1998) 33 *Journal of Pacific History* 179, 180.

[154] Forsyth, above n 98, 430.

[155] Judith Bennett, *Wealth of the Solomons: A History of a Pacific Archipelago, 1800–1978*
(University of Hawaii Press, 1988) 112–114.

nent roles in native tribunals, further elevating their status.[156] However, the population was divided as to the legitimacy of these leaders. In the 1940s and 1950s, during protests against colonial rule, Solomon Islanders themselves led efforts to install invented forms of Indigenous governance, which ultimately led to the development of the universal notion of 'chiefs' in the country. As such, it has been said that in Solomon Islands:

> The dispute-management and governance systems and processes established at … [the time of colonisation] continue to have significant repercussions in the way in which justice and governance are presently observed and practiced at the local level.[157]

Contemporary 'customary' legal systems are therefore the product of several influences that have shaped both norms and governance institutions. These systems continue to play a significant role in most parts of the Pacific, regardless of whether they are truly traditional or not.[158] Their contemporary relevance does however vary between and within countries.[159] They also continue to change and adapt, influenced by forces such as Western education, the cash economy, globalisation, migration, and intermarriage, which in some areas is causing them to weaken.[160] Respect for some traditional leaders is also diminishing, leading to a breakdown in customary governance.[161] As will be explored later in relation to East Rennell, the contemporary relevance of a customary legal system influences the effectiveness of customary protection of World Heritage (see Sect. 6.3).

[156] Allen et al, above n 96, 8–9.
[157] Ibid., 7.
[158] White, above n 125, 2.
[159] Zurstrassen, above n 139, 3.
[160] See, for example, Brown, above n 148, 190; Corrin, above n 107, 147.
[161] See, for example, Joeli Veitayaki et al 'On Cultural Factors and Marine Managed Areas in Fiji' in Jolie Liston, Geoffrey Clark and Dwight Alexander (eds), *Pacific Island Heritage: Archaeology, Identity and Community* (ANU E Press, 2011) 37, 38; Shankar Aswani, 'Customary Sea Tenure in Oceania as a Case of Rights-Based Fishery Management: Does it Work?' (2005) 15 *Reviews in Fish Biology and Fisheries* 285, 304; Pepe Clarke and Stacy D Jupiter, 'Law, Custom and Community-Based Natural Resource Management in Kubulau District (Fiji)' (2010) 37(1) *Environmental Conservation* 98, 104; Marjo Vierros et al, *Traditional Marine Management Areas of the Pacific in the Context of National and International Law and Policy* (United Nations University, 2010) 7; Simon Foale et al 'Tenure and Taboos: Origins and Implications for Fisheries in the Pacific' (2011) 12 *Fish and Fisheries* 357, 364; Jan McDonald, *Marine Resource Management and Conservation in Solomon Islands: Roles, Responsibilities and Opportunities* (Griffith Law School, 2010) 2.

2.4.5   Customary Land Tenure

Land is fundamental to the lives and livelihoods of many Pacific Islanders. Most still have access to their customary land, which forms the basis of islander communities.[162] Land in the Pacific has therefore been described as a 'basic element of human security in the region'.[163] As explained below, rights to customary land are principally governed through the applicable customary tenure system, but State laws are also relevant.

Soon after outsiders made contact with the islands, alienation of customary land began in most Pacific Island States[164] (e.g. to missionaries, traders, and planters).[165] Once the islands became colonies, most colonial administrators enacted laws to restrict alienation on the grounds that it might remove the basis of subsistence for the Indigenous populations.[166] Some States still have laws restricting alienation,[167] and the proportion of land under customary tenure in the region remains high (see Table 2.1). Nearshore marine areas in some States may also be customarily owned.[168] Therefore, most heritage sites in the Pacific (whether they be terrestrial or marine) will include areas under customary ownership.[169] In Solomon

[162] Jim Fingleton, *Pacific 2020 Background Paper: Land* (Commonwealth of Australia, 2005) 5.

[163] Brown, above n 148, 191.

[164] See, for example, Peter Larmour, 'Sharing the Benefits: Customary Landowners and Natural Resource Projects in Melanesia' (1989) 36 *Pacific Viewpoint* 56, 57.

[165] Sue Farran, 'Navigating Between Traditional Land Tenure and Introduced Land Laws in Pacific Island States' (2011) 64 *Journal of Legal Pluralism and Unofficial Law* 65, 67.

[166] Corrin and Paterson, above n 112, 285. For example, in the British protectorate of Solomon Islands, the grant of perpetual estate to foreigners was initially permitted, but later prohibited under *Land Regulation 1914 (King's Regulation No. 3)* (UK) s 3.

[167] For example, in Solomon Islands, only a Solomon Islander can hold an interest in customary land (*Land and Titles Act (Cap. 133)* s 241(1)).

[168] Ron Crocombe, 'Tenure' in Moshe Rapaport (ed), *The Pacific Islands: Environment and Society* (University of Hawai'i Press, 2013) 192, 193. In Solomon Islands, there are conflicting High Court decisions concerning customary ownership of land below the high-water mark. In *Allardyce Lumber Company Ltd v Laore* [1990] SBHC 46, the Court ruled that the foreshore could be customary land but the seabed could not. In *Combined Fera Group v Attorney General* [1997] SBHC 55 the Court found that the seabed could also potentially be under customary tenure. For further discussion see Stephanie Price et al, *Environmental Law in Solomon Islands* (Public Solicitor's Office, 2015) 31–32.

[169] Of the eight World Heritage Sites in the Pacific Island States, seven either partly or entirely comprise customary land (Nan Madol: Ceremonial Centre of Eastern Micronesia, Levuka Historical Port Town, Bikini Atoll Nuclear Site, Rock Islands Southern Lagoon, Kuk Early Agricultural Site, East Rennell and Chief Roi Mata's Domain). See Table 1.2.

Table 2.1 Distribution of land by system of tenure in the independent Pacific Island States

	Public[a]	Freehold[b]	Customary
Cook Islands	Some	Little	95%
Federated States of Micronesia	35%	<1%	65%
Fiji	4%	8%	88%
Kiribati	50%	<5%	>45%
Marshall Islands	<1%	0%	>99%
Nauru	<10%	0%	>90%
Niue	1.5%	0%	98.5%
Palau	Most	Some	Some
Papua New Guinea	2.5%	0.50%	97%
Samoa	15%	4%	81%
Solomon Islands	8%	5%	87%
Tonga	100%	0%	0%
Tuvalu	5%	<0.1%	95%
Vanuatu	2%	0%	98%

Source: Adapted from AusAid, *Making Land Work: Reconciling Customary Land and Development in the Pacific* (Australian Agency for International Development, vol 1, 2008) 4
[a]Includes Crown land and land owned by provincial and local governments
[b]Includes land that is not strictly freehold, but similar in characteristics, such as 'perpetual estates' in Solomon Islands

Islands, alienation was permitted until 1914,[170] so only 87% of land there is customary land.[171]

Except in Tonga (which has no customary land), in the Pacific Island States, rights to customary land are determined according to the applicable customary laws.[172] These laws are often underpinned by closely guarded local knowledge of genealogies and histories,[173] and oral rules and histories that can be easily confused and manipulated by people seeking to rely on the laws to exercise their rights or enforce them against others.[174] As a

[170]Alienation was prohibited by *Land Regulation 1914 (King's Regulation No. 3)* (UK).

[171]AusAid, *Making Land Work: Reconciling Customary Land and Development in the Pacific* (Australian Agency for International Development, vol 1, 2008) 4.

[172]Corrin and Paterson, above n 112. In Solomon Islands, for example, this is provided for in the *Land and Titles Act (Cap. 133)* s 239(1).

[173]White, above n 125, 10.

[174]John McKinnon, 'Resource Management under Traditional Tenure: The Political Ecology of a Contemporary Problem, New Georgia Islands, Solomon Islands' (1993) 14(1) *South Pacific Study* 95, 95.

result, local people sometimes hold different opinions concerning applicable customary rules. Consequently, in Solomon Islands, for example, disputes over rights to customary land are the most common form of dispute,[175] which can hinder heritage conservation efforts.[176]

Customary land tenure laws vary significantly throughout the Pacific, because they were developed and adapted by settlers to suit their island environments, and because of the diverse impacts of colonisation.[177] Consequently, it is difficult to make generalisations about such laws. Some features are however common to many places:

(1) Much literature (including this book) and legislation refer to customary 'ownership' and customary 'owners'. However, those terms over-simplify Pacific land tenure. Under customary laws, people generally do not 'own' land in the Western sense of that word, but rather have rights to it vis-à-vis other people.[178] Rights and obligations are overlapping, as rights of individuals or small groups may be nested in rights of broader groups.[179] Thus, customary tenure is better thought of as a complex and flexible system of rights and obligations, rather than a system of ownership.[180]

(2) Customary land is commonly owned by a group, but the size of landowning unit varies. For example, in Solomon Islands, the landowning unit is often quite large, like a line, clan, or tribe. In other countries, the unit may be smaller such as a family or extended family, or even an individual in some places.[181] On Rennell, where people are of Polynesian descent, land is held individually by male members of the lineage,[182] which differs from most other parts of Solomon Islands.

[175] Allen et al, above n 96, 18.

[176] For example, in Solomon Islands, the Minister for Environment cannot declare an area to be a 'protected area' under the *Protected Areas Act 2010* if there is a dispute over the ownership of the land (*Protected Area Regulations 2012* reg 14(3)). See Sect. 7.2 for analysis of this legislation.

[177] Smith, above n 10, 24, 41.

[178] Jean Guiart, 'Land Tenure and Hierarchies in Eastern Melanesia' (1996) 19(1) *Pacific Studies* 1, 7; Crocombe, above n 168, 192.

[179] Crocombe, above n 168, 192.

[180] Jim Fingleton (ed), *Privatising Land in the Pacific: A Defence of Customary Tenures*, Discussion paper 80 (The Australia Institute, 2005) ix.

[181] Corrin and Paterson, above n 112, 269, 274–275.

[182] Samuel H Elbert and Torben Monberg, *From the Two Canoes: Oral Traditions of Rennell and Bellona Islands* (Danish National Museum and University of Hawaii Press, 1965) 10.

(3) Some people within a landowning group may possess stronger claims than others. For example, the rights of males are generally superior to females. People who have worked the land, stayed in the vicinity, and/or contributed to the community may also have stronger rights.[183] In addition, although rights to land are generally held by a group, group members do not necessarily have equal say over what happens on their land. Key decisions are often made by the senior members (e.g. the chiefs).[184]

(4) Customary land is usually acquired by inheritance, either through the matrilineal or through the patrilineal line.[185] Again, the use of the term 'inheritance' here varies from the Western understanding of that term. In the Pacific, land rights arise at birth but cease upon death.[186]

(5) In some places rights to land are flexible.[187] When the laws were developed, community life was often unsettled, making it difficult for fixed land laws to emerge.[188] In addition, land boundaries were commonly natural features, which could be shifted by nature or people.[189] Tenure laws were also often adjusted to take into account new circumstances, and the need to redistribute land.[190]

(6) Like other customary laws, in many places laws regulating land tenure have changed significantly since pre-colonial times.[191]

In addition to customary laws, State laws regulate the ownership of and dealings in customary land. In Solomon Islands, for example, these laws

---

[183] Ron Crocombe, 'Overview' in *Customary Land Tenure and Sustainable Development: Complementary or Conflict* (South Pacific Commission, 1995) 5, 10–11; Crocombe, above n 168, 192.

[184] Fingleton, above n 162, 7.

[185] Crocombe, 'Overview', above n 183, 10.

[186] Fingleton, above n 162, 7.

[187] See, for example, Donald Denoon, 'Pacific Edens? Myths and Realities of Primitive Affluence' in Donald Denoon, Malama Meleisea, Stewart Firth, Jocelyn Linnekin and Karen Nero (eds), *The Cambridge History of Pacific Islanders* (Cambridge University Press, 2008) 80, 94.

[188] Corrin and Paterson, above n 112, 289.

[189] Ibid.

[190] Fingleton, above n 162, 8.

[191] See, for example, Denoon, above n 187, 90.

cover issues such as land acquisition,[192] land disputes,[193] resource development,[194] and conservation.[195] Understanding the interactions between State laws and rules governing customary land tenure is critical to an assessment of World Heritage protection. As discussed further in Part III of this book, in Solomon Islands, some challenges associated with implementing relevant legislation stem from the laws' failure to appropriately accommodate the variety of customary land tenure systems that exist in that country (see Sects. 7.2.2 and 7.3.1).

2.5   PROTECTION OF WORLD HERITAGE THROUGH PACIFIC ISLAND LEGAL SYSTEMS

2.5.1   *World Heritage Protection Under State Legal Systems*

The governments of the Pacific Island States have comprehensive law-making powers, which could be exercised to enact laws for the protection of heritage. For example, the National Parliament of Solomon Islands has the power to make 'laws for the peace order and good government' of the country.[196] Some Pacific Island States also have sub-national levels of government, whose legislative powers are more limited but may encompass heritage protection. For example, Solomon Islands' nine provincial assemblies[197] have the power to enact ordinances dealing with issues such as 'cultural and environmental matters', 'land and land use', and 'rivers and water'.[198] Furthermore, under the *Solomon Islands Constitution*, national and provincial legislation overrides customary law to the extent of any inconsistency,[199] so State laws could be enacted to protect a heritage site notwithstanding its customary ownership. However, as explained below, political, economic, and social considerations affect the willingness and ability of Pacific Island governments to legislate to protect World Heritage.

---

[192] For example, the *Land and Titles Act (Cap. 133)*.

[193] For example, the *Land and Titles Act (Cap. 133)*; *Local Courts Act (Cap. 19)*.

[194] For example, the *Forest Resources and Timber Utilisation Act (Cap. 40)*; *Mines and Minerals Act (Cap. 42)*; *Environment Act 1998*.

[195] For example, the *Protected Areas Act 2010*.

[196] *Constitution of Solomon Islands* s 59(1).

[197] The nine provinces of Solomon Islands are Central, Choiseul, Guadalcanal, Isabel, Makira-Ulawa, Malaita, Rennell-Bellona, Temotu, and Western Province: see Fig. 1.3.

[198] *Provincial Government Act 1997* ss 31, 33, sch 3.

[199] *Constitution of Solomon Islands* sch 3 para 3.

2.5.1.1 Development and Heritage Protection in the Pacific Islands

Heritage protection is not a high priority for Pacific Island governments.[200] As an officer of the Solomon Islands Ministry of Environment has commented, States in the region are 'flooded with international obligations',[201] so the *World Heritage Convention* is just one of the treaties governments are attempting to comply with. In addition, they have limited resources to dedicate to heritage protection. Per capita economic growth rates are generally very low,[202] as economic development has been hampered by the islands' small size, limited resources, geographical dispersion, and isolation from markets.[203] Solomon Islands, Kiribati, Tuvalu, and Vanuatu have particularly weak economies and are categorised as Least Developed Countries.[204] Consequently, economic and social development is often a higher priority than heritage protection for Pacific Island governments.[205] This is reflected in the budgets of the government ministries charged with implementing the *World Heritage Convention* in the Pacific. A 2012 study found that no such ministry had an adequate budget for heritage

[200] See, for example, Smith and Turk, above n 27, 24; Chape, above n 11, 44; Anita Smith, 'The World Heritage Pacific 2009 Programme' in Anita Smith (ed), *World Heritage in a Sea of Islands: Pacific 2009 Programme*, World Heritage Papers 34 (UNESCO, 2012) 2, 9; Denham, above n 207, 101; Salamat Ali Tabbasum, 'Developing the Solomon Islands Tentative List' in Anita Smith (ed), *World Heritage in a Sea of Islands: Pacific 2009 Programme*, World Heritage Papers 34 (UNESCO, 2012) 34, 34.

[201] Interview by the author with a conservation officer in the Ministry of Environment (Honiara, 2 August 2013).

[202] See, for example, AusAid, *Pacific 2020: Challenges and Opportunities for Growth* (Commonwealth of Australia, 2006) 1, 19; Storey and Abbott, above n 77, 417; Geoff Bertram, 'Pacific Island Economies' in Moshe Rapaport (ed), *The Pacific Islands: Environment and Society* (University of Hawai'i, 2013) 325. For example, Solomon Islands' economy grew by 2.9% in 2015 (Central Bank of Solomon Islands, *Annual Report 2015* (Solomon Islands Government, 2016) 1) and 3.5% in 2016 (Central Bank of Solomon Islands, *Annual Report 2016* (Solomon Islands Government, 2017) 3).

[203] See, for example, *Agenda 21, Report of the UNCED, I*, UN Doc. A/CONF.151/26/Rev.1 (1992) para 17.123.

[204] United Nations Committee for Development Policy, *List of Least Developed Countries (as of March 2018)* (2018) https://www.un.org/development/desa/dpad/wp-content/uploads/sites/45/publication/ldc_list.pdf. 'Least Developed Countries' are those that meet certain low-income, human resource weakness and economic vulnerability criterion specified by the Economic and Social Council of the United Nations.

[205] See, for example, Herr, above n 8, 43; Tabbasum, above n 200, 34; Peter Shelley, 'Contracting for Conservation in the Central Pacific: An Overview of the Phoenix Islands Protected Area' (2012) 106 *Proceedings of the Annual Meeting (American Society of International Law* 511, 514.

protection,[206] presenting a significant challenge for the conservation of World Heritage. A lack of resources and institutional capacity is one of the reasons why Pacific Island States have struggled to comply with their *Convention* obligations.[207]

The fact that some Pacific Island governments are economically dependent on activities that can harm heritage is a further challenge. For example, in Solomon Islands, substantial government revenue is earned from the logging industry,[208] which has caused widespread environmental and social damage.[209] In the Pacific (as elsewhere), the tension between heritage protection and economic development can influence the State's willingness to implement the *Convention* (discussed further in Sects. 5.3.3.2 and 7.3.1.1).

2.5.1.2 Governance Issues and the (Ir)relevance of State Legal Systems

Many Pacific Island States are plagued by governance issues, which are barriers to heritage protection.[210] These issues contribute to the lack of relevance and legitimacy afforded to the national level of government by many Pacific Islanders, which limits the effectiveness of the State legal system.

Several factors contribute to these governance problems, beginning with the colonisation process. The boundaries of most Pacific Island States were determined by colonial powers and not based on cultural or geographical logic,[211] so many States had little sense of national unity

---

[206] *Final Report on the Results of the Second Cycle of the Periodic Reporting Exercise for Asia and the Pacific*, WHC 36th sess, UN Doc WHC-12/36.COM/10A (1 June 2012) 22.

[207] See, for example, Tim Denham, 'Building Institutional and Community Capacity for World Heritage in Papua New Guinea: The Kuk Early Agricultural Site and Beyond' in Anita Smith (ed), *World Heritage in a Sea of Islands: Pacific 2009 Programme*, World Heritage Papers 34 (UNESCO, 2012) 98, 101.

[208] See, for example, Daniel Gay (ed), *Solomon Islands Diagnostic Trade Integration Study 2009 Report* (Solomon Islands Government, 2009) 48; Morgan Wairiu, 'History of the Forestry Industry in Solomon Islands: The Case of Guadalcanal' (2007) 42(2) *Journal of Pacific History* 233, 243. See Sect. 7.3.2 for further discussion.

[209] See, for example, Pacific Horizon Consultancy Group, *Solomon Islands State of Environment Report* (Solomon Islands Government, 2008) 81.

[210] *Pacific World Heritage Action Plan 2016–2020* (2016) 3.

[211] Clive Moore, 'Indigenous Participation in Constitutional Development' (2013) 48(2) *The Journal of Pacific History* 162, 163; Sinclair Dinnen, 'State-Building in a Post-Colonial Society: The Case of Solomon Islands' (2008) 9 *Chicago Journal of International Law* 51, 53.

before colonisation. This feature is most prevalent in Melanesia, where islanders did not have a long history of contact and cooperation,[212] but less significant in areas with greater ethnic, linguistic, and cultural homogeneity. In some States, including Solomon Islands, the independence process did not engender nationalist sentiments, as it was initiated from the top down rather than from a struggle by the people.[213] This lack of national unity has made it challenging for many governments to establish a strong presence among their populations, particularly in rural areas. Post-independence State-building in some places has been further impeded by political instability, weak parliaments and executive governments, and corruption,[214] and exacerbated by the lack of many checks and balances found in countries with larger populations.[215]

Melanesian national governments have been particularly unstable. Reflecting the 'big-man' style of leadership characteristic of many traditional societies, politicians in Melanesia often see their role as rewarding the people who voted for them (generally a sub-group of their electorate) rather than implementing policies in the broader public interest[216] (such as protecting World Heritage). Instead of being members of well-established political parties, these politicians tend to form loose coalitions, with affiliations frequently changing, contributing to political instability.[217] Governance issues contributed to the conflicts experienced in some States, such as the secessionist struggle in Bougainville, coups in Fiji, and the ethnic tensions in Solomon Islands.[218]

---

[212] Stephen Levine, 'The Experience of Sovereignty in the Pacific: Island States and Political Autonomy in the Twenty-First Century' (2012) 50(4) *Commonwealth & Comparative Politics* 439, 444.

[213] Fischer, above n 35, 249; Sinclair Dinnen, 'The Solomon Islands Intervention and the Instabilities of the Post-Colonial State' (2008) 20(3) *Global Change, Peace and Security* (formerly *Pacific Review: Peace, Security and Global Change*) 338, 347. An exception to this is Samoa, where from the 1930s there was an indigenous independence movement: see, for example, Crocombe, above n 32, 438.

[214] Cedric Saldanha, *Pacific 2020 Background Paper: Political Governance* (Commonwealth of Australia, 2005) 4.

[215] Ibid.; Ron Duncan, 'An Overview of Decentralisation and Local Governance Structures in the Pacific Region' (Paper presented at the Pacific Regional Symposium 'Making Local Governance Work', Suva, Fiji, 4–8 December 2004) 10.

[216] Reilly, above n 110, 482–483; McLeod, above n 119, 8.

[217] Dinnen, above n 211, 57.

[218] Solomon Islands' ethnic tensions are explained briefly in Sect. 5.3.3.1.

These issues exacerbate the lack of relevance and legitimacy afforded to the national level of governance and law-making by many Pacific Islanders, particularly those living in rural areas. In the ethnically diverse Melanesia, people's main association rests with their clan, tribe, and island, rather than with their State,[219] and the idea of a national government is often viewed as foreign.[220] While this characteristic is less evident in Polynesia, some in that region still view their national government with suspicion.[221]

In Solomon Islands, the relevance of the national government is further diminished by the fact that most people do not rely on the State for their day-to-day needs. They live predominantly subsistence lifestyles, and social services (where they exist) are often provided by non-State entitles such as churches.[222] Disenchantment with the government has been fuelled by limited opportunities for rural development, decreasing provision of government services, and the perceived greed of politicians, many of whom have benefited significantly from the logging industry.[223] In addition, frequently people are not aware of or do not understand State laws, because they rarely translated into local languages or explained to the public.[224] Thus, the State legal system in Solomon Islands is of marginal significance to much of the population,[225] which means people often have little impetus to comply with State heritage protection laws.

2.5.1.3 The (Lack of) Implementation and Enforcement of State Laws
Pacific Island States have historically poor records of compliance with and enforcement of some State laws, including heritage protection legislation.[226] In some cases, this is because the legislation is based on a 'command and control' approach to regulation, which is a poor fit in

---

[219] Reilly, above n 110, 482; Bennett, above n 144, 14; Ian Frazer, 'The Struggle for Control of Solomon Island Forests' (1997) 9(1) *Contemporary Pacific* 39, 44.

[220] Bennett, above n 144, 14.

[221] Wesley-Smith, above n 128, 151.

[222] Jane Turnbull, 'Solomon Islands: Blending Traditional Power and Modern Structures in the State' (2002) 22 *Public Administration and Development* 191, 197.

[223] Dinnen, above n 211, 58.

[224] Klingelhofer and Robinson, above n 88, 9.

[225] Allen et al, above n 96, 45.

[226] Laurence Cordonnery, 'Environmental Law Issues in the South Pacific and the Quest for Sustainable Development and Good Governance' in Anita Jowitt and Tess Newton Cain (eds), *Passage of Change: Law, Society and Governance in the Pacific* (ANU Press, 2010) 233, 238; *Final Report on the Results of the Second Cycle of the Periodic Reporting Exercise for Asia and the Pacific*, WHC 36th sess, UN Doc WHC-12/36.COM/10A (1 June 2012) 43; Ben

the Pacific.[227] Furthermore, government ministries charged with enforcing the legislation are under-resourced, impeding their ability to carry out their statutory duties.[228] This challenge is exacerbated by the geography of some States. Solomon Islands, for example, comprises almost 1000 islands stretching across 1450 km of ocean. Enforcement of State laws in isolated places requires substantial human and financial resources, which are often beyond the capacity of the government. A further challenge is the highly 'Honiara-centric' nature of Solomon Islands' State legal system, with most courts and the bulk of legal services being located in the nation's capital.[229] This makes it extremely difficult for people living on the outer islands to access the court system to enforce their rights under State legislation.[230]

A lack of implementation and enforcement also hampers the effectiveness of many provincial ordinances. Solomon Islands' provincial governments could play an important role in heritage protection, including through monitoring and enforcing compliance with national and provincial laws. However, while some provinces have enacted relevant ordinances, few have been effectively implemented.[231]

Due to the challenges referred to above, in States such as Solomon Islands, heritage protection is unlikely to be achieved in a purely centralised manner. The potential for customary legal systems to contribute to World Heritage protection must therefore be considered.

Boer and Pepe Clarke, *Legal Frameworks for Ecosystem-Based Adaptation to Climate Change in the Pacific Islands* (SPREP, 2012) 25. See generally, Price et al, above n 168.

[227] Govan, et al above n 7, 17.

[228] See above n 200.

[229] Allen et al, above n 96, 44–45.

[230] A promising development in this regard is the publication of the *Solomon Islands Environmental Crime Manual*, which is aimed to assist members of the Royal Solomon Islands Police Force to identify and enforce environmental crimes, including those committed under logging and mining laws, the *Environment Act 1998* and the *Protected Areas Act 2010*: See Katrina Moore, *Solomon Islands Environmental Crime Manual* (Solomon Islands Government, 2015). In time, this may lead to some improvement in the enforcement of such legislation.

[231] Phillip Iro Tagini, *The Search for King Solomon's Gold: An Examination of the Policy and Regulatory Framework for Mining in Solomon Islands* (PhD Thesis, The Australian National University, 2007) 391.

2.5.2   World Heritage Protection Under Customary Legal Systems

Pacific Islanders have repeatedly stressed that the operation of the *World Heritage Convention* in the region 'can only be effected through recognition of local customary and other forms of tenure of land and sea, and traditional custodianship of cultural heritage'.[232] In some cases this is because customary systems form an integral part of the heritage value of the place (see Sect. 2.2.2). Customary laws can also contribute to the protection of other forms of heritage, including the natural environment.

Over time, Pacific Islanders developed management practices to regulate access to and use of land and resources.[233] In some parts of Solomon Islands, for example, customary practices restricted access to important sites, regulated the consumption of certain species, and limited some peoples' harvesting rights.[234] Around the Pacific, practices such as these coevolved with customary laws and tenure systems, and hence all are integrated.[235] Pacific Islanders also developed processes for making decisions and resolving disputes.

The motivation behind Pacific Islanders' development of customs governing rights to land and resources varied. A much-cited paper by Johannes noted that Pacific Islanders understood that their vital fisheries resources could be depleted, so they developed management techniques to guard against this.[236] However, the idea that all Indigenous people

---

[232] *Identification of World Heritage Properties in the Pacific: First World Heritage Global Strategy Meeting for the Pacific Islands Region* (Suva, Fiji, 15–18 July 1997) para 7. See also *Presentation of the World Heritage Programme for the Pacific*, WHC 31st sess, UN Doc WHC-07/31.COM/11C (10 May 2007) annex I (Appeal to the World Heritage Committee from the Pacific Island State Parties).

[233] See, for example, Smith, above n 10, 60; Ballard and Wilson, above n 27, 130; Hugh Govan, 'Achieving the Potential of Locally Managed Marine Areas in the South Pacific' (2009) 25 *SPC Traditional Marine Resource Management and Knowledge Information Bulletin* 16, 17; L M Scherl and A J O'Keefe, *Capacity Development for Protected and Other Conserved Areas in the Pacific Islands Region: Strategy and Action Framework 2015–2020* (IUCN, 2016) 1.

[234] Reuben Sulu, 'Traditional law and the Environment in the Solomon Islands' (2004) 17 *SPC Traditional Marine Resource Management and Knowledge Information Bulletin* 20, 20.

[235] Scherl and O'Keefe, above n 233, 1.

[236] R E Johannes, 'Traditional Marine Conservation Methods in Oceania and their Demise' 9 (1978) *Annual Review of Ecology and Systematics* 349, 350.

lived in harmony with nature is no longer well accepted.[237] For example, a study of customary marine management in Melanesia found that human population densities there were generally too low to generate the population pressures required to stimulate a conservation ethic.[238] Therefore, the reverence that Pacific Islanders have for nature cannot be confused with the possession of a conservation ethic.[239] While resource management in some communities was driven by a desire to conserve resources, other motivations included allocation of resources and customary and religious beliefs.[240]

The existence of these different motivations has implications for the contemporary role of customary systems in World Heritage protection, as it cannot be assumed that Indigenous values are consistent with the conservation of heritage.[241] However, even where customary management was not designed for conservation, it may still provide the basis for good resource stewardship.[242] Thus, the role of customary systems in the protection of natural heritage places is being increasingly recognised.[243]

The limits to customary protection of World Heritage must however be understood. Throughout the Pacific, many systems have been weakened by

---

[237] See, for example, Foale et al, above n 161, 365; Crocombe, above n 32, 25; Simon Foale, 'The Intersection of Scientific and Indigenous Ecological Knowledge in Coastal Melanesia: Implications for Contemporary Marine Resource Management' (2006) 58 (187) *International Social Science Journal* 129, 129; R E Johannes and F R Hickey, *Evolution of Village-Based Marine Resource Management in Vanuatu Between 1993 and 2001*, Coastal Region and Small Island Papers 15 (UNESCO, 2004) 29; K Ruddle, E Hviding and R E Johannes, 'Marine Resources Management in the Context of Customary Tenure' (1992) 7 *Marine Resource Economics* 249, 267; Marianne Pederson, *Conservation Complexities: Conservationists' and Local Landowners' Different Perceptions of Development and Conservation in Dandaun Province, Papua New Guinea*, State, Society and Governance in Melanesia Discussion Paper 7 (Australian National University, 2013) 3.

[238] Foale et al, above n 161, 357.

[239] McNeill, above n 16, 309.

[240] Ruddle, Hiving, Johannes, above n 237, 262.

[241] Giraud-Kinley, above n 69, 157; Foale et al, above n 161, 365; Crocombe, 'Overview', above n 183; Ruddle, Hviding and Johannes, above n 237, 267.

[242] S Aswani et al, 'Customary Management as Precautionary and Adaptive Principles for Protecting Coral Reefs in Oceania' (2007) 26 *Coral Reefs* 1009, 1010.

[243] See, for example, Aswani et al, above n 242; Clark and Jupiter, above n 161; Govan, above n 233; McDonald, above n 161; David Doulman, 'Community-Based Fishery Management: Towards Restoration of Traditional Practices in the South Pacific' (1993) *Marine Policy* 108; R E Johannes and F R Hickey, above n 237, 28; J E Cinner and T R McClanahan, 'Socioeconomic Factors that Lead to Overfishing in Small-Scale Coral Reef Fisheries of Papua New Guinea' (2006) 33 *Environmental Conservation* 73.

colonisation and later influences (see Sect. 2.4.4). Customary laws are today less relevant to some young people, particularly those who have moved from their village and been exposed to other ideas.[244] The availability of the State system as an alternative dispute resolution mechanism[245] and the loss of respect for chiefs and traditional practices and protocols[246] have also reduced the legitimacy of customary legal systems. Thus, customary resource management in some parts of the Pacific is not strong.[247]

Furthermore, customary systems are seldom able to deal with all pressures affecting heritage places.[248] As Crocombe has noted:

No traditional precedents exist for chain saws, bulldozers, hunting rifles, metal traps, power torches, spearguns, scuba gear, filament nets, dynamite, outboard motors or global markets for timber, coral, bird of paradise feathers, sea shells, clams for soup and nautilus shells for tourist mantel pieces.[249]

Similarly, Ballard and Wilson have said:

Community control, in and of itself, is seldom sufficient as a basis for long-term management under novel conditions that include pressure to sell or lease land, to sign contracts for timber, fisheries or oil-palm production, or to enter into agreements for protected natural or cultural areas.[250]

---

[244] Menzies, above n 108, 10; Corrin, above n 139, 30.

[245] Corrin Care and Zorn, 'Legislating for the Application of Customary Law', above n 107, 149.

[246] Forsyth, above n 81, 114–120; Veitayaki et al, above n 161, 40.

[247] See, for example, Foale et al, above n 161, 364; McDonald, above n 161, 2; Govan et al, above n 7, 25; Johannes, above n 236, 356; Francis R Hickey, 'Traditional Marine Resource Management in Vanuatu: Acknowledging, Supporting and Strengthening Indigenous Management Systems' (2006) 20 *SPC Traditional Marine Resource Management and Knowledge Information Bulletin* 11, 11; Tom Graham and Noah Idechong, 'Reconciling Customary and Constitutional Law: Managing Marine Resources in Palau, Micronesia' (1998) 40 *Ocean and Coastal Management* 143, 146–7; Kenneth Ruddle, 'The Context of Policy Design for Existing Community-Based Fisheries Management Systems in the Pacific Islands' (1998) 40 *Ocean and Coastal Management* 105, 108.

[248] Ballard and Wilson, above n 27, 132, 149; Smith, above n 200, 5; Pepe Clarke and Charles Taylor Gillespie, *Legal Mechanisms for the Establishment and Management of Terrestrial Protected Areas in Fiji* (IUCN, 2009) 2.

[249] Crocombe, above n 32, 26.

[250] Ballard and Wilson, above n 27, 132.

Importantly, customary systems often cannot protect a site against activities undertaken by outsiders,[251] or threats arising from beyond the area under the jurisdiction of the relevant customary governance body. Therefore, even if a site is subject to customary protection, other measures (including State legislation) will usually be needed to ensure its long-term conservation.

2.6 Conclusion

Pacific heritage is diverse, encompassing impressive natural landscapes, and sites associated with the development of island societies or later events of global significance. While the law alone cannot ensure the protection of these places, it plays an important role. Pacific Island governments have broad legislative powers, but legal, political, economic, and social issues influence their willingness and ability to develop and implement heritage protection laws. Such laws must be tailored to the nature of Pacific heritage sites, the resource capacities of the governments, and the legal and land tenure systems prevalent in the region.

The growing acceptance of the concept of legal pluralism has given non-State legal systems increased legitimacy in academic discourse.[252] Thus, as Twining has noted, 'a conception of law confined to state law … leaves out too many significant phenomena deserving sustained juristic attention' including customary law.[253] In the context of World Heritage, it is clear that customary legal systems, through their relationship with traditional practices and land tenure, are a key component of the legal framework of Pacific Island States. However, while customary systems can contribute to World Heritage protection, they have been significantly altered (and often weakened) since colonisation, and there are limits to the issues that they can deal with.

Legal pluralism requires consideration not just of the existence of multiple legal systems, but also the relationship between those systems.[254] As Forsyth has noted however, it does not greatly assist in working out how different systems of law can most effectively relate to each other.[255] There

---

[251] Jonathan M Lindsay, *Creating Legal Space for Community-Based Management: Principles and Dilemmas* (Food and Agriculture Organisation of the United Nations, 1998) 3; Veitayaki et al, above n 161, 41.

[252] Forsyth, above n 81, 44.

[253] Twining, above n 104, 114.

[254] Merry, above n 79, 873.

[255] Forsyth, above n 81, 46.

is still a need for greater understanding of how customary and State legal systems can best operate and interact to support World Heritage protection. This book makes an important contribution to knowledge in this area by exploring these issues in relation to Solomon Islands.

Building upon the foundation laid by this chapter, the next two chapters analyse the *World Heritage Convention* regime in the Pacific context. Chapter 3 considers the scope for Pacific Island heritage to be recognised as 'World Heritage', and thus to fall within the ambit of the treaty. Chapter 4 assesses the protection regime established by the *Convention* and its application in the Pacific.

REFERENCES

ARTICLES, BOOKS AND REPORTS

Ballard, Chris and Meredith Wilson, 'Unseen Monuments: Managing Melanesian Cultural Landscapes' in Ken Taylor and Jane L Lennon (eds), *Managing Cultural Landscapes* (Routledge, 2012) 130

Allen, Matthew, Sinclair Dinnen, Daniel Evans, Rebecca Monson, *Justice Delivered Locally: Systems, Challenges and Innovations in Solomon Islands* (World Bank, 2013)

Anaya, S James, 'International Human Rights and Indigenous Peoples: The Move Toward the Multicultural State' (2004) 21(1) *Arizona Journal of International and Comparative Law* 13

Aswani, Shankar, 'Customary Sea Tenure in Oceania as a Case of Rights-Based Fishery Management: Does it Work?' (2005) 15 *Reviews in Fish Biology and Fisheries* 285

Aswani, S, S Albert, A Sabetian, T Furusawa, 'Customary Management as Precautionary and Adaptive Principles for Protecting Coral Reefs in Oceania' (2007) 26 *Coral Reefs* 1009

AusAid, *Pacific 2020: Challenges and Opportunities for Growth* (Commonwealth of Australia, 2006)

AusAid, *Making Land Work: Reconciling Customary Land and Development in the Pacific* (Australian Agency for International Development, vol 1, 2008)

Barker, John, 'Religion' in Moshe Rapaport (ed), *The Pacific Islands: Environment and Society* (University of Hawai'i Press, 2013) 214

Bennett, Judith, *Wealth of the Solomons: A History of a Pacific Archipelago, 1800–1978* (University of Hawaii Press, 1988)

Bennett, Judith, *Roots of Conflict in Solomon Islands – Though Much is Taken, Much Abides: Legacies of Tradition and Colonialism*, State, Society and Governance in Melanesia Discussion Paper (The Australian National University, 2002)

Bennett, T W and T Vermeulen, 'Codification of Customary Law' (1980) 24(2) *Journal of African Law* 206

Bertram, Geoff, 'Pacific Island Economies' in Moshe Rapaport (ed), *The Pacific Islands: Environment and Society* (University of Hawai'i, 2013) 325

Boccardi, Giovanni, 'The World Heritage Pacific 2009 Programme: Addressing the Aims of the Global Strategy in the Pacific Regions' in Anita Smith (ed), *World Heritage in a Sea of Islands: Pacific 2009 Programme*, World Heritage Papers 34 (UNESCO, 2012) 12

Boer, Ben and Pepe Clarke, *Legal Frameworks for Ecosystem-Based Adaptation to Climate Change in the Pacific Islands* (SPREP, 2012)

Boer, Ben and Stefan Gruber, 'Heritage Discourses' in Brad Jessup and Kim Rubenstein (eds), *Environmental Discourses in Public and International Law* (Cambridge University Press, 2012) 375

Bolton, Lissant, 'Chief Willie Bongmatur Maldo and the Role of Chiefs in Vanuatu' (1998) 33 *Journal of Pacific History* 179

Brown, Anne M, 'Custom and Identity: Reflections on and Representations of Violence in Melanesia' in Nikki Slocum-Bradley (ed), *Promoting Conflict or Peace through Identity* (Ashgate, 2008) 183

Central Bank of Solomon Islands, *Annual Report 2015* (Solomon Islands Government, 2016)

Central Bank of Solomon Islands, *Annual Report 2016* (Solomon Islands Government, 2017)

Chape, Stuart, 'Natural World Heritage in Oceania: Challenges and Opportunities' in Anita Smith (ed), *World Heritage in a Sea of Islands: Pacific 2009 Programme*, World Heritage Papers 34 (UNESCO, 2012) 40

Chappell, David A, 'The Postcontact Period' in Moshe Rapaport (ed), *The Pacific Islands: Environment and Society* (University of Hawai'i Press, 2013) 138

Cinner, J E and T R McClanahan, 'Socioeconomic Factors that Lead to Overfishing in Small-Scale Coral Reef Fisheries of Papua New Guinea' (2006) 33 *Environmental Conservation* 73

Clarke, Pepe and Charles Taylor Gillespie, *Legal Mechanisms for the Establishment and Management of Terrestrial Protected Areas in Fiji* (IUCN, 2009)

Clarke, Pepe and Stacy D Jupiter, 'Law, Custom and Community-Based Natural Resource Management in Kubulau District (Fiji)' (2010) 37(1) *Environmental Conservation* 98

Cordonnery, Laurence, 'Environmental Law Issues in the South Pacific and the Quest for Sustainable Development and Good Governance' in Anita Jowitt and Tess Newton Cain (eds), *Passage of Change: Law, Society and Governance in the Pacific* (ANU Press, 2010) 233

Corrin Care, Jennifer, 'Wisdom and Worthy Customs: Customary Law in the South Pacific' (2002) 80 *Reform* 31

Corrin, Jennifer, 'Customary Land and the Language of the Common Law' (2008) 37 *Common Law World Review* 305

Corrin, Jennifer, 'Moving Beyond the Hierarchical Approach to Legal Pluralism in the South Pacific' (2009) 59 *Journal of Legal Pluralism and Unofficial Law* 29

Corrin, Jennifer, 'A Question of Identity: Complexities of State Law Pluralism in the South Pacific' (2010) 61 *Journal of Legal Pluralism and Unofficial Law* 145

Corrin, Jennifer and Don Paterson, *Introduction to South Pacific Law* (Palgrave Macmillan, 3rd ed, 2011)

Corrin Care, Jennifer and Jean G Zorn, 'Legislating pluralism: Statutory 'Developments' in Melanesian Customary Law' (2001) 46 *Journal of Legal Pluralism* 49

Corrin Care, Jennifer and Jean G Zorn, 'Legislating for the Application of Customary Law in Solomon Islands' (2005) 34 *Common Law World Review* 144

Cox, Barry and Peter Moore, *Biogeography: An Ecological and Evolutionary Approach* (Oxford, 1980)

Crocombe, Ron, 'Overview' in *Customary Land Tenure and Sustainable Development: Complementary or Conflict* (South Pacific Commission, 1995) 5

Crocombe, Ron, *The South Pacific* (University of the South Pacific, 2001)

Crocombe, Ron, 'Tenure' in Moshe Rapaport (ed), *The Pacific Islands: Environment and Society* (University of Hawai'i Press, 2013) 192

Cuskelly, Katrina, *Customs and Constitutions: State Recognition of Customary Law Around the World* (IUCN, 2011)

Denham, Tim, 'Building Institutional and Community Capacity for World Heritage in Papua New Guinea: The Kuk Early Agricultural Site and Beyond' in Anita Smith (ed), *World Heritage in a Sea of Islands: Pacific 2009 Programme*, World Heritage Papers 34 (UNESCO, 2012) 98

Denham, Tim, '*Traim Tasol*... Cultural Heritage Management in Papua New Guinea' in Sue O'Connor, Denis Byrne and Sally Brockwell (eds), *Transcending the Culture-Nature Divide in Cultural Heritage: Views from the Asia-Pacific Region* (ANU E Press, 2012) 117

Denoon, Donald, 'Human Settlement' in Donald Denoon, Malama Meleisea, Stewart Firth, Jocelyn Linnekin and Karen Nero (eds), *The Cambridge History of Pacific Islanders* (Cambridge University Press, 2008) 37

Denoon, Donald, 'Pacific Edens? Myths and Realities of Primitive Affluence' in Donald Denoon, Malama Meleisea, Stewart Firth, Jocelyn Linnekin and Karen Nero (eds), *The Cambridge History of Pacific Islanders* (Cambridge University Press, 2008) 80

Dingwall, Paul, 'Pacific Islands World Heritage Tentative Lists' in Anita Smith (ed), *World Heritage in a Sea of Islands: Pacific 2009 Programme*, World Heritage Papers 34 (UNESCO, 2012) 28

Dinnen, Sinclair, 'State-Building in a Post-Colonial Society: The Case of Solomon Islands' (2008) 9 *Chicago Journal of International Law* 51

Dinnen, Sinclair, 'The Solomon Islands Intervention and the Instabilities of the Post-Colonial State' (2008) 20(3) *Global Change, Peace and Security* (formerly *Pacific Review: Peace, Security and Global Change*) 338

Douglas, B, 'Rank, Power, Authority; A Reassessment of Traditional Leadership in South Pacific Societies' (1979) 14 *Journal of Pacific History* 2

Doulman, David, 'Community-Based Fishery Management: Towards Restoration of Traditional Practices in the South Pacific' (1993) *Marine Policy* 108

Duncan, Ron, 'An Overview of Decentralisation and Local Governance Structures in the Pacific Region' (Paper presented at the Pacific Regional Symposium 'Making Local Governance Work', Suva, Fiji, 4–8 December 2004)

Elbert, Samuel H and Torben Monberg, *From the Two Canoes: Oral Traditions of Rennell and Bellona Islands* (Danish National Museum and University of Hawaii Press, 1965)

Farran, Sue, 'Is Legal Pluralism an Obstacle to Human Rights? Considerations from the South Pacific' (2006) 52 *Journal of Legal Pluralism and Unofficial Law* 77

Farran, Sue, 'Navigating Between Traditional Land Tenure and Introduced Land Laws in Pacific Island States' (2011) 64 *Journal of Legal Pluralism and Unofficial Law* 65

Fingleton, Jim, *Pacific 2020 Background Paper: Land* (Commonwealth of Australia, 2005)

Fingleton, Jim (ed), *Privatising Land in the Pacific: A Defence of Customary Tenures*, Discussion Paper 80 (The Australia Institute, 2005)

Fischer, Steven Roger, *A History of the Pacific Islands* (Palgrave Macmillan, 2nd ed, 2013)

Foale, Simon, 'The Intersection of Scientific and Indigenous Ecological Knowledge in Coastal Melanesia: Implications for Contemporary Marine Resource Management' (2006) 58 (187) *International Social Science Journal* 129

Foale, Simon, Phillipa Cohen, Stephanie Januchowski-Hartley, Amelia Wenger and Martha Macintyre, 'Tenure and Taboos: Origins and Implications for Fisheries in the Pacific' (2011) 12 *Fish and Fisheries* 357

Forsyth, Miranda, 'Beyond Case Law: *Kastom* and Courts in Vanuatu' (2004) 35 *Victoria University of Wellington Law Review* 427

Forsyth, Miranda, *A Bird That Flies with Two Wings: Kastom and State Justice Systems in Vanuatu* (ANU E Press, 2009)

Forsyth, Miranda, *The Writing of Community By-Laws and Constitutions in Melanesia: Who? Why? Where? How?* State, Society and Governance in Melanesia In Brief (The Australian National University, 2014)

Frazer, Ian, 'The Struggle for Control of Solomon Island Forests' (1997) 9(1) *Contemporary Pacific* 39

Gay, Daniel (ed), *Solomon Islands Diagnostic Trade Integration Study 2009 Report* (Solomon Islands Government, 2009)

Gerbeaux, P, T Kami, P Clarke and T Gillespie, *Shaping a Sustainable Future in the Pacific: IUCN Regional Programme for Oceania 2007–2012* (IUCN, 2007)

Gibbons, Ann, 'Genes Point to a New Identity for Pacific Pioneers' (1994) 263(5143) *Science* 32

Giraud-Kinley, Catherine, 'The Effectiveness of International Law: Sustainable Development in the South Pacific Region' (1999–2000) 12 *Georgetown Environmental Law Review* 125

Goddard, Michael, *Justice Delivered Locally, Solomon Islands, Literature Review* (World Bank, 2010)

Govan, Hugh, 'Achieving the Potential of Locally Managed Marine Areas in the South Pacific' (2009) 25 *SPC Traditional Marine Resource Management and Knowledge Information Bulletin* 16

Govan, Hugh et al, *Status and Potential of Locally-Managed Marine Areas in the South Pacific: Meeting Nature Conservation and Sustainable Livelihood Targets Through Wide-Spread Implementation of LMMAs* (SPREP/WWF/WorldFish-Reefbase/CRISP, 2009)

Government of Papua New Guinea, *Kuk Early Agricultural Site Cultural Landscape – A Nomination for Consideration as World Heritage Site* (2007)

Graham, Tom and Noah Idechong, 'Reconciling Customary and Constitutional Law: Managing Marine Resources in Palau, Micronesia' (1998) 40 *Ocean and Coastal Management* 143

Griffiths, John, 'What is Legal Pluralism?' (1986) 24 *Journal of Legal Pluralism* 1

Griffiths, John, 'Legal Pluralism and the Theory of Legislation – With Special Reference to the Regulation of Euthanasia' in Hanne Petersen and Henrik Zahle (eds), *Legal Polycentricity: Consequences of Pluralism in Law* (Hanne Peterson, 1995) 210

Guiart, Jean, 'Land Tenure and Hierarchies in Eastern Melanesia' (1996) 19(1) *Pacific Studies* 1

Herr, Richard, 'Environmental Protection in the South Pacific: The Effectiveness of SPREP and its Conventions' in Olav Schram Stokke and Øystein B Thommessen (eds), *Yearbook of International Co-operation on Environment and Development 2002/2003* (Earthscan Publications, 2002)

Hickey, Francis R, 'Traditional Marine Resource Management in Vanuatu: Acknowledging, Supporting and Strengthening Indigenous Management

Systems' (2006) 20 *SPC Traditional Marine Resource Management and Knowledge Information Bulletin* 11

Hviding, Edvard, 'Contextual Flexibility: Present Status and Future of Customary Marine Tenure in Solomon Islands' (1998) 40 *Ocean and Coastal Management* 253

Intoh, Michiko, 'Human Dispersal into Micronesia' (1997) 105 *Anthropological Science* 15

Irwin, Geoffrey, 'Navigation and Seafaring' in Ian Lilley (ed), *Early Human Expansion and Innovation in the Pacific: Thematic Study* (ICOMOS, 2010) 47

Johannes, R E, 'Traditional Marine Conservation Methods in Oceania and their Demise' 9 (1978) *Annual Review of Ecology and Systematics* 349

Johannes, R E and F R Hickey, *Evolution of Village-Based Marine Resource Management in Vanuatu Between 1993 and 2001*, Coastal Region and Small Island Papers 15 (UNESCO, 2004)

Jupiter, Stacy, Sangeeta Manguhai and Richard T Kingsford, 'Conservation of Biodiversity in the Pacific Islands of Oceania: Challenges and Opportunities' (2014) 20(2) *Pacific Conservation Biology* 206

Keesing, Roger M, 'Killers, Big Men, and Priests on Malaita: Reflections on a Melanesian Troika System' (1985) 24(4) *Ethnology* 237

Keppel, Gunnar, Clare Morrison, Jean-Yves Meyer and Hans Juergen Boehmer, 'Isolated and Vulnerable: The History and Future of Pacific Island Terrestrial Biodiversity' (2014) 20(2) *Pacific Conservation Biology* 136

King, Michael, Ueta Fa'asili, Semisi Fakahau and Aliti Vunisea, *Strategic Plan for Fisheries Management and Sustainable Coastal Fisheries in the Pacific Islands* (Secretariat of the Pacific Community, 2003)

Kirch, Patrick V, 'Late Holocene Human-Induced Modifications to a Central Polynesian Island Ecosystem' (1996) 93 *Proceedings of the National Academy of Sciences* 5296

Klingelhofer, Stephan and David Robinson, *The Rule of Law, Custom and Civil Society in the South Pacific: An Overview* (International Center for Not-for-Profit Law, 2001)

Kwa, Eric L, 'Climate Change and Indigenous People in the South Pacific' (Paper presented at IUCN Academy of Environmental Law Conference on 'Climate Law in Developing Countries Post-2012: North and South Perspectives', Ottawa, Canada, 26–28 September 2008)

Larmour, Peter, 'Sharing the Benefits: Customary Landowners and Natural Resource Projects in Melanesia' (1989) 36 *Pacific Viewpoint* 56

Levine, Stephen, 'The Experience of Sovereignty in the Pacific: Island States and Political Autonomy in the Twenty-First Century' (2012) 50(4) *Commonwealth & Comparative Politics* 439

Lilley, Ian (ed), *Early Human Expansion and Innovation in the Pacific: Thematic Study* (ICOMOS, 2010)

Lilley, Ian and Christophe Sand, 'Thematic Frameworks for the Cultural Values of the Pacific' in Anita Smith (ed), *World Heritage in a Sea of Islands: Pacific 2009 Programme*, World Heritage Papers 34 (UNESCO, 2012) 22

Lindsay, Jonathan M, *Creating Legal Space for Community-Based Management: Principles and Dilemmas* (Food and Agriculture Organisation of the United Nations, 1998)

McDonald, Jan, *Marine Resource Management and Conservation in Solomon Islands: Roles, Responsibilities and Opportunities* (Griffith Law School, 2010)

McIntyre, Matt, *Pacific Environment Outlook* (United Nations Environment Programme and the Secretariat of the Pacific Regional Environment Programme, 2005)

McKinnon, John, 'Resource Management under Traditional Tenure: The Political Ecology of a Contemporary Problem, New Georgia Islands, Solomon Islands' (1993) 14(1) *South Pacific Study* 95

McLeod, Abby, *Leadership Models in the Pacific*, State, Society and Governance Discussion Paper (The Australian National University, 2008)

McNeill, John R, 'Of Rats and Men: A Synoptic Environmental History of the Island Pacific' (1994) 5(2) *Journal of World History* 299

Menzies, Nicholas, *Legal Pluralism and the Post-Conflict Transition in the Solomon Islands* (Hertie School of Governance, Berlin, 2007)

Merry, Sally Engle, 'Legal Pluralism' (1988) 22 *Law and Society Review* 869

Moore, Clive, *Decolonising the Solomon Islands: British Theory and Melanesian Practice*, Working Paper 8 (Alfred Deakin Research Institute, Deakin University, 2010)

Moore, Clive, 'Indigenous Participation in Constitutional Development' (2013) 48(2) *The Journal of Pacific History* 162

Moore, Katrina, *Solomon Islands Environmental Crime Manual* (Solomon Islands Government, 2015)

Murari, Lai, 'Implications of Climate Change in Small Island Developing Countries of the South Pacific' (2004) 2(1) *Fijian Studies* 15

Nunn, Patrick D, 'Nature-society interactions in the Pacific Islands' (2013) 85(4) *Geografiska Annaler, Series B, Human Geography* 219

Otto, Ton, 'Transformations of Cultural Heritage in Melanesia: From *Kastam* to *Kalsa*' (2015) 21(2) *International Journal of Heritage Studies* 117

Pacific Horizon Consultancy Group, *Solomon Islands State of Environment Report* (Solomon Islands Government, 2008)

Parisi, Francesco, 'The Formation of Customary Law' (Paper presented at the 96th Annual Conference of the American Political Science Association, Washington DC, August 31–September 3, 2000)

Pawley, Andrew, 'Language' in Moshe Rapaport (ed), *The Pacific Islands: Environment and Society* (University of Hawai'i Press, 2013) 159

Pederson, Marianne, *Conservation Complexities: Conservationists' and Local Landowners' Different Perceptions of Development and Conservation in Dandaun Province, Papua New Guinea*, State, Society and Governance in Melanesia Discussion Paper 7 (Australian National University, 2013)

Posey, Darrell Addison, 'Introduction: Culture and Nature – The Inextricable Link' in Darrell Addison Posey (ed), *Cultural and Spiritual Values of Biodiversity* (UNEP, 1999) 1

Price, Stephanie, Adam Beeson, Joe Fardin and Jennifer Radford, *Environmental Law in Solomon Islands* (Public Solicitor's Office, Solomon Islands Government, 2015)

Ram-Bidesi, Vina, 'Ocean Resources' in Moshe Rapaport (ed), *The Pacific Islands: Environment and Society* (University of Hawai'i Press, 2013) 364

Reenberg, Anette, Torben Birch-Thomsen, Ole Mertz, Bjarne Fog and Sofus Christiansen, 'Adaption of Human Coping Strategies in a Small Island Society in the SW Pacific: 50 Years of Change in the Coupled Human-Environment system on Bellona, Solomon Islands' (2008) 3(6) *Human Ecology* 807

Reilly, Benjamin, 'State Functioning and State Failure in the South Pacific' (2004) 58(4) *Australian Journal of International Affairs* 479

Roberts, Simon, 'Against Legal Pluralism: Some Reflections on the Contemporary Enlargement of the Legal Domain' (1998) 42 *Journal of Legal Pluralism and Unofficial Law* 95

Ruddle, Kenneth, 'The Context of Policy Design for Existing Community-Based Fisheries Management Systems in the Pacific Islands' (1998) 40 *Ocean and Coastal Management* 105

Ruddle, K, E Hviding and R E Johannes, 'Marine Resources Management in the Context of Customary Tenure' (1992) 7 *Marine Resource Economics* 249

Sage, Caroline and Michael Woolcock, 'Introduction' in Brian Z Tamanaha, Caroline Sage and Michael Woolcock (eds), *Legal Pluralism and Development: Scholars and Practitioners in Dialogue* (Cambridge University Press, 2012) 1

Sahlins, Marshall D, 'Poor Man, Rich Man, Big-Man, Chief: Political Types in Melanesia and Polynesia' (1963) 5(3) *Comparative Studies in Society and History* 285

Saldanha, Cedric, *Pacific 2020 Background Paper: Political Governance* (Commonwealth of Australia, 2005)

Sand, Christophe, 'Melanesian Tribes vs Polynesian Chiefdoms: Recent Archaeological Assessment of a Classic Model of Socio-Political Types in Oceania' (2002) 41(2) *Asian Perspectives* 284

Scaglion, Richard, 'Law' in Moshe Rapaport (ed), *The Pacific Islands: Environment and Society* (University of Hawai'i Press, 2013) 202

Scherl, L M and A J O'Keefe, *Capacity Development for Protected and Other Conserved Areas in the Pacific Islands Region: Strategy and Action Framework 2015–2020* (IUCN, 2016)

Shelley, Peter, 'Contracting for Conservation in the Central Pacific: An Overview of the Phoenix Islands Protected Area' (2012) 106 *Proceedings of the Annual Meeting (American Society of International Law* 511

Smith, Anita, 'The Cultural Landscapes of the Pacific Islands' in Anita Smith and Kevin L Jones (eds), *Cultural Landscapes of the Pacific Islands* (ICOMOS, 2007)

Smith, Anita, 'The World Heritage Pacific 2009 Programme' in Anita Smith (ed), *World Heritage in a Sea of Islands: Pacific 2009 Programme*, World Heritage Papers 34 (UNESCO, 2012) 2

Smith, Anita and Kevin L Jones (eds), *Cultural Landscapes of the Pacific Islands* (ICOMOS, 2007b)

Smith, Anita and Cate Turk, 'Customary Systems of Management and World Heritage in the Pacific Islands' in Sue O'Connor, Denis Byrne and Sally Brockwell (eds), *Transcending the Culture-Nature Divide in Cultural Heritage: Views from the Asia-Pacific Region* (ANU E Press, 2012) 22

Steadman, David W, 'Prehistoric Extinctions of Pacific Island Birds: Biodiversity Meets Zooarchaeology' (1995) 267 *Science* 1123

Storey, Donovan and David Abbott, 'Development Prospects' in Moshe Rapaport (ed), *The Pacific Islands: Environment and Society* (University of Hawai'i Press, 2013) 417

Sulu, Reuben, 'Traditional law and the Environment in the Solomon Islands' (2004) 17 *SPC Traditional Marine Resource Management and Knowledge Information Bulletin* 20

Tabbasum, Salamat Ali, 'Developing the Solomon Islands Tentative List' in Anita Smith (ed), *World Heritage in a Sea of Islands: Pacific 2009 Programme*, World Heritage Papers 34 (UNESCO, 2012) 34

Tagini, Phillip Iro, *The Search for King Solomon's Gold: An Examination of the Policy and Regulatory Framework for Mining in Solomon Islands* (PhD Thesis, The Australian National University, 2007)

Tamanaha, Brian Z, 'Understanding Legal Pluralism: Past to Present, Local to Global' (2008) 30 *Sydney Law Review* 375

Techera, Erika J, 'Samoa: Law, Custom and Conservation' (2006) 10 *New Zealand Journal of Environmental Law* 361

Thomas, Frank R, 'The Precontact Period' in Moshe Rapaport (ed), *The Pacific Islands: Environment and Society* (University of Hawai'i Press, 2013) 125

Turnbull, Jane, 'Solomon Islands: Blending Traditional Power and Modern Structures in the State' (2002) 22 *Public Administration and Development* 191

Twining, William, 'Legal Pluralism 101' in Brian Z Tamanaha, Caroline Sage and Michael Woolcock (eds), *Legal Pluralism and Development: Scholars and Practitioners in Dialogue* (Cambridge University Press, 2012) 112

United Nations Environment Programme (UNEP) *Pacific Islands Environment Outlook* (UNEP, 1999)

United Nations Office of the High Representative for the Least Developed Countries, Landlocked Developing Countries and Small Island Developing States (UN-OHRLLS), *Small Island Developing States: Small Islands Big(ger) Stakes* (UN, 2011)

Veitayaki, Joeli, Akosita D R Nakoro, Tareguci Sigarua and Nanise Bulai, 'On Cultural Factors and Marine Managed Areas in Fiji' in Jolie Liston, Geoffrey Clark and Dwight Alexander (eds), *Pacific Island Heritage: Archaeology, Identity and Community* (ANU E Press, 2011) 37

Vierros, Marjo, Alifereti Tawake, Francis Hickey, Ana Tiraa and Rahera Noa, *Traditional Marine Management Areas of the Pacific in the Context of National and International Law and Policy* (United Nations University, 2010)

Wairiu, Morgan, 'History of the Forestry Industry in Solomon Islands: The Case of Guadalcanal' (2007) 42(2) *Journal of Pacific History* 233

Watson, Alan, 'An Approach to Customary Law' (1984) 3 *University of Illinois Law Review* 561

Wesley-Smith, Terence, 'Changing Patterns of Power' in Moshe Rapaport (ed), *The Pacific Islands: Environment and Society* (University of Hawai'i Press, 2013) 147

West, Paige and Dan Brockington, 'An Anthropological Perspective on Some Unexpected Consequences of Protected Areas' (2006) 20(3) *Conservation Biology* 609

White, Geoffrey, *Indigenous Governance in Melanesia*, State, Society and Governance in Melanesia Discussion Paper (The Australian National University, 2007)

Wingham, Elspeth J, *Nomination of East Rennell, Solomon Islands by the Government of Solomon Islands for Inclusion in the World Heritage List Natural Sites* (1997)

Woodman, Gordon R, 'Ideological Combat and Social Observation: Recent Debate About Legal Pluralism' (1998) 42 *Journal of Legal Pluralism and Unofficial Law* 21

Woodman, Gordon, 'Why There Can be No Map of Law', *Legal Pluralism and Unofficial Law in Social, Economic and Political Development: Papers of the XIIIth International Congress of the Commission on Folk Law and Legal Pluralism* (Chiangmai, Thailand, 7–10 April, 2002)

Zorn, Jean G, 'Customary Law in the Papua New Guinea Village Courts' (1990) 2(2) *The Contemporary Pacific* 279

Zurstrassen, Matthew, *Customary Dispute Resolution Research Project: Final Report to the Regional PJDP Meetings in Samoa in March 2012*, Pacific Judicial Development Programme (2012)

CASES

Allardyce Lumber Company Ltd v Laore [1990] SBHC 46
Combined Fera Group v Attorney General [1997] SBHC 55

LEGISLATION AND BILLS: SOLOMON ISLANDS

Environment Act 1998
Forest Resources and Timber Utilisation Act (Cap. 40)
Land and Titles Act (Cap. 133)
Land Regulation 1914 (King's Regulation No. 3) (UK)
Local Courts Act (Cap. 19)
Mines and Minerals Act (Cap. 42)
Native Courts Ordinance 1942
Pacific Order in Council 1893 (UK)
Protected Areas Act 2010
Protected Area Regulations 2012
Provincial Government Act 1997
Solomon Islands Independence Order 1978, sch (*Constitution of Solomon Islands*)

CONVENTIONS

Convention Concerning the Protection of the World Cultural and Natural Heritage,
    opened for signature 16 November 1972, 1037 UNTS 151 (entered into force
    17 December 1975)

UNITED NATIONS DOCUMENTS

Agenda 21, Report of the UNCED, I, UN Doc. A/CONF.151/26/Rev.1 (1992)
*Final Report on the Results of the Second Cycle of the Periodic Reporting Exercise for
    Asia and the Pacific,* WHC 36th sess, UN Doc WHC-12/36.COM/10A (1
    June 2012)
ICOMOS, *Evaluations of Nominations of Cultural and Mixed Properties to the
    World Heritage List,* WHC 32nd sess, UN Doc WHC-08/32.COM/INF/8B1
    (2008) 92 (Chief Roi Mata's Domain, Vanuatu, Advisory Body Evaluation
    1280) 94
Presentation of the World Heritage Programme for the Pacific, WHC 31st sess, UN
    Doc WHC-07/31.COM/11C (10 May 2007) annex I (Appeal to the World
    Heritage Committee from the Pacific Island State Parties)
WHC Res 34 COM 8B.20, WHC 34th sess, UN Doc WHC-10/34.COM/20 (3
    September 2010) 206
WHC Res 37 COM 8B.25, WHC 37th sess, UN Doc WHC-13/37.COM/20 (5
    July 2013) 186

INTERNET MATERIALS

United Nations Committee for Development Policy, *List of Least Developed Countries (as of March 2018)* (2018) https://www.un.org/development/desa/dpad/wp-content/uploads/sites/45/publication/ldc_list.pdf

INTERVIEWS

Interview by the author with a conservation officer in the Ministry of Environment (Honiara, 2 August 2013)

OTHER

Identification of World Heritage Properties in the Pacific: First World Heritage Global Strategy Meeting for the Pacific Islands Region (Suva, Fiji, 15–18 July 1997)

Identification of World Heritage Properties in the Pacific: Second World Heritage Global Strategy Meeting for the Pacific Islands Region (Port Vila, Vanuatu, 24–27 August 1999)

Pacific World Heritage Action Plan 2016–2020 (2016)

CHAPTER 3

The Concept of 'World Heritage' and Its Application in the Pacific

3.1 Introduction

Chapter 2 discussed the nature of Pacific Island heritage and legal systems, and identified key issues concerning the protection of heritage places. This chapter builds upon that analysis by exploring how Pacific Island heritage 'fits' within the concept of 'World Heritage'.[1]

The term 'World Heritage' is not defined in the *World Heritage Convention*, and in fact only appears in the treaty's preamble.[2] Instead, sites that fall within the scope of the *Convention* are those that meet the definitions of 'cultural heritage' and 'natural heritage' in Articles 1 and 2, respectively. 'Cultural heritage' is defined as monuments and groups of buildings that have outstanding universal value (OUV) from the point of view of history, art or science; as well as sites that have OUV from an historical, aesthetic, ethnological, or anthropological point of view.[3] 'Natural heritage' is defined as natural features, geological, and physiographical formations and natural areas of OUV from the point of view of science,

---

[1] See Sect. 1.6.1 for discussion of the use of the term 'World Heritage' in this book.

[2] *Convention Concerning the Protection of the World Cultural and Natural Heritage*, opened for signature 16 November 1972, 1037 UNTS 151 (entered into force 17 December 1975) ('*World Heritage Convention*') preamble para 6.

[3] Ibid., art 1.

© The Author(s) 2018
S. C. Price, *World Heritage Conservation in the Pacific*,
Palgrave Series in Asia and Pacific Studies,
https://doi.org/10.1007/978-981-13-0602-0_3

conservation, or aesthetics.[4] Therefore, 'World Heritage' is essentially a site that expresses cultural and/or natural heritage values, and has OUV.

This chapter explores the concept of 'World Heritage', explaining how the scope of the *World Heritage Convention* reflects developments such as the growing recognition of the interrelationship between humankind and the environment, and the notion of intergenerational equity (Sect. 3.2). The term 'OUV' was introduced into the *Convention* to restrict the treaty's scope to sites of global significance. The *Convention* does not define the term, but rather gives the World Heritage Committee the power to prescribe the criteria to be applied when determining whether a site meets that threshold (Sect. 3.3). As will be explained, while the criteria initially set by the Committee were relatively narrow, emerging views concerning cultural diversity led to them being broadened (Sect. 3.4).

This chapter explains that while significant impediments to the nomination of Pacific sites exist, the Committee's broadened approach to the concept of OUV has increased the potential for the *World Heritage Convention* to be successfully implemented in the Pacific (Sect. 3.5). Many Pacific places could qualify for World Heritage listing, including sites of value because of their association with continuing living traditions and customs. The implications of listing sites which possess markedly different global and local significance do however warrant careful consideration, including the challenges this presents for the sites' conservation. In particular, issues that may stem from recognising a Pacific place as a natural World Heritage site should be taken into account before such a site is nominated.

3.2   The Concept of 'World Heritage'

The scope of the concept of 'World Heritage' reflects the era in which the *World Heritage Convention* was developed. As explained below, the *Convention* was a product of growing awareness among the international community of the need for broader international laws to protect cultural and natural places from the impacts of human activities, as well as increasing recognition of the interrelationship between people and the environment and the concept of intergenerational equity. Reflecting these developments, it was the first international agreement to protect 'heritage', as well as the first to cover both cultural and natural places.

[4] Ibid., art 2.

3.2.1 A Brief History of the Development of the World Heritage Convention[5]

Laws to protect cultural properties and objects have a long history.[6] Their progressive development cannot be described in a linear or logical fashion[7] because each used different terminology to describe the items or places that fell within its scope, and defined such terms for the purposes of that instrument alone.[8] In general, however, the law evolved from focusing on the physical manifestations of culture (such as objects, individual monuments, and buildings) to the more holistic notion of 'cultural heritage'.[9]

Laws for the protection of monuments and art work began to be enacted in Europe in the fifteenth century, but were initially narrow in scope.[10] As cultural monuments and objects have long been a 'victim of war',[11] the first international legal principles and rules applying to such properties emerged through the development of the laws of war and international humanitarian law.[12] The progressive codification of the international laws of war provided some protections to cultural properties in the

---

[5] For detailed discussion of the history of the *World Heritage Convention*, see, for example, Sarah M Titchen, *On the Construction of Outstanding Universal Value: UNESCO's World Heritage Convention (Convention Concerning the Protection of the World Cultural and Natural Heritage, 1972) and the Identification and Assessment of Cultural Places for Inclusion in the World Heritage List* (PhD Thesis, The Australian National University, 1995) chs 2, 3; Francesco Francioni (ed), *The 1972 World Heritage Convention: A Commentary* (Oxford University Press, 2008).

[6] Ben Boer and Graeme Wiffen, *Heritage Law in Australia* (Oxford University Press, 2006) 9. For discussion of the history of such laws, see generally Francesco Francioni, 'A Dynamic Evolution of Concept and Scope: From Cultural Property to Cultural Heritage' in Yusuf A Abdulqawi (ed), *Standard-Setting in UNESCO Volume 1: Normative Action in Education, Science and Culture* (Martinus Nijoff and UNESCO Publishing, 2007) 221; David Lowenthal, 'Natural and Cultural Heritage' (2005) 11(1) *International Journal of Heritage Studies* 81; Craig Forrest, *International Law and the Protection of Cultural Heritage* (Routledge, 2011).

[7] Seong-Yong Park, *On Intangible Heritage Safeguarding Governance: An Asia-Pacific Context* (Cambridge Scholars Publishing, 2013) 9.

[8] Lyndel V Prott and P J O'Keefe, *Law and the Cultural Heritage – Volume I* (Professional Books, 1984) 8.

[9] See, for example, Forrest, above n 6, xxi.

[10] Prott and O'Keefe, above n 8, 34.

[11] Forrest, above n 6, 56.

[12] Francioni, above n 6, 223. For detailed discussion of the history of war and cultural heritage, see Forrest, above n 6, ch 3.

event of armed conflicts.[13] However, it was the immense destruction of cultural properties during World War II and the subsequent establishment of the United Nations Educational, Scientific and Cultural Organisation (UNESCO) that provided the impetus needed for the first comprehensive multi-lateral treaty to protect cultural properties.[14]

The resulting agreement, the *Convention for the Protection of Cultural Property During Armed Conflict 1954*, affords protection to 'cultural property'.[15] It defines that term to include monuments and objects that are worthy of protection because of their importance 'to the cultural heritage of every people'.[16] The term 'cultural property' was also used in the *Convention on the Means of Prohibiting and Preventing the Illicit Import, Export and Transfer of Ownership of Cultural Property 1970*.[17] Reflecting the subject matter of that treaty, it confined the term to moveable cultural objects.[18] Like the 1954 treaty, it referred to heritage, stating that the illicit import, export, and transfer of ownership of these objects is 'one of the main causes of the impoverishment of the cultural heritage of the countries of origin of such property'.[19] Therefore, although these laws did not establish an agreed definition of 'cultural property',[20] they did introduce the concept of 'cultural heritage' into international law.

Unlike the concept of 'cultural heritage', the term 'natural heritage' was not used in an international law before the *World Heritage Convention*. A conservation movement began in the late 1800s in the United States, and gained traction in the 1960s when several landmark publications highlighted the impact of humans on the environment.[21] With increasing

---

[13] Guido Carducci, 'The 1972 World Heritage Convention in the Framework of other UNESCO Conventions on Cultural Heritage' in Francesco Francioni (ed), *The 1972 World Heritage Convention: A Commentary* (Oxford University Press, 2008) 363, 365.

[14] Forrest, above n 6, 78.

[15] *Convention for the Protection of Cultural Property during Armed Conflict*, opened for signature 14 May 1954, 249 UNTS 240 (entered into force 7 August 1956) arts 2–3.

[16] Article 1.

[17] *Convention on the Means of Prohibiting and Preventing the Illicit Import, Export and Transfer of Ownership of Cultural Property*, opened for signature 14 November 1970, 823 UNTS 231 (entered into force 24 April 1972).

[18] Article 1.

[19] Article 2.

[20] Prott and O'Keefe, above n 8, 8.

[21] See, for example, Rachel Carson, *Silent Spring* (1962); Paul Ehrlich, *The Population Bomb* (Sierra Club and Ballantine Books, 1968); Donella H Meadows et al, *Limits to Growth* (Universe Books, 1972).

evidence that rapid industrialisation and urbanisation were threatening natural areas,[22] awareness about the impact of human activities on the natural environment turned to concern by the early 1970s.[23] This led to the convening of the United Nations Conference on the Human Environment (UNCHE) in Stockholm in 1972, at which State parties adopted the now famous *Stockholm Declaration*.[24] That conference 'marked the emergence of international environment law as a separate branch of international law'[25] and led to a proliferation of treaties on the subject. While the *World Heritage Convention* was not adopted at the UNCHE, its negotiation and drafting were intertwined with preparations for that conference, so it reflects many of the principles underlying the *Stockholm Declaration*.

In the years leading up to the UNCHE, UNESCO and the International Council on Monuments and Sites (ICOMOS) sought to expand international legal protection of cultural properties by drafting a treaty for the conservation of monuments, buildings and sites of universal value.[26] At the same time, the International Union for the Conservation of Nature (IUCN) was preparing a draft *Convention for the Conservation of the World's Heritage* which would protect significant natural areas.[27] When a working group established to assist with preparations for the UNCHE was asked to review these draft treaties, it recommended that they be combined into one agreement.[28]

The working group's recommendation built upon an idea raised by the United States in 1965 for the creation of a World Heritage Trust that would preserve natural and scenic areas and historic sites.[29] More gener-

---

[22] Francesco Francioni, 'The Preamble' in Francesco Francioni (ed), *The 1972 World Heritage Convention: A Commentary* (Oxford University Press, 2008) 11, 12; Lowenthal, above n 6, 84.

[23] Douglas Pocock, 'Some Reflections on World Heritage' (1997) 29(3) *Area* 260, 260.

[24] *Report of the United Nations Conference on the Human Environment*, UN Doc A/CONF.48/14/Rev.1 (5–16 June 1972) ch 1 (Declaration of the United Nations Conference on the Human Environment) ('*Stockholm Declaration*').

[25] Edith Brown Weiss, Daniel B Magraw and Paul C Szasz, *International Environmental Law: Basic Instruments and References* (Transnational Publishers, 1992) 171.

[26] *International Instruments for the Protection of Monuments, Groups of Buildings and Sites*, UN Doc SHC/MD/17 (30 June 1971) annex II.

[27] Barbara J Lausche, *Weaving a Web of Environmental Law: Contributions of the IUCN Environmental Law Programme* (IUCN/ICEL, 2008) 89.

[28] Francioni, above n 22, 14.

[29] Titchen, above n 5, 52, 62; Francesco Bandarin, *World Heritage: Challenges for the Millennium* (UNESCO, 2007) 28; Catherine Redgwell, 'Article 2 Definition of Natural

ally, it reflected growing recognition of links between humans and the environment,[30] which was increasingly being reflected in international agreements. The first such agreement was the 1971 *Man and the Biosphere Program*, which sought to promote conservation and sustainable use of reserves.[31] The link also underpinned the *Stockholm Declaration*, which begins with the bold declaration that '[m]an is both creature and moulder of his environment'.[32]

Consistent with this trend, the working group's recommendation was accepted, and the UNESCO/ICOMOS draft treaty on the protection of cultural properties was broadened to include natural areas.[33] This expanded treaty became the *World Heritage Convention*, which was adopted by the UNESCO General Assembly in November 1972.

3.2.2   The Scope of the Concept of 'World Heritage'

The *World Heritage Convention* was the first international agreement designed to protect 'heritage'. As noted above, previous international laws dealing with places of cultural significance had referred to 'cultural heritage', but had sought to protect the narrower concept of 'cultural property'.[34] International environmental laws had addressed the conservation of 'nature' or specific flora and fauna, but not natural heritage.[35]

The shift in language from 'cultural property' to 'cultural heritage' in international law, which was solidified by the *World Heritage Convention*, was partly a response to the need to accommodate cultural and natural sites under one agreement.[36] 'Property' is a key concept under Western law, which implies control by the owner and the right to alienate and exclude.[37] Because of the connotations associated with that term, it would

Heritage' in Francesco Francioni (ed), *The 1972 World Heritage Convention: A Commentary* (Oxford University Press, 2008) 63, 64.

[30] Ralph O Slatyer, 'The Origin and Development of the World Heritage Convention' (1984) *Monumentum* 3, 4.

[31] *Records of the General Conference*, 16th sess, UNESCO Res 2.313 (1970) 35 (Intergovernmental Programme on Man and the Biosphere).

[32] *Stockholm Declaration*, UN Doc A/CONF.48/14/Rev.1, art 1.

[33] Titchen, above n 5, 40.

[34] Forrest, above n 6, xxi.

[35] Redgwell, above n 29, 64.

[36] Francioni, above n 12, 229.

[37] Lyndel V Prott and Patrick J O'Keefe, 'Cultural Heritage' or 'Cultural Property' (1992) 1(2) *International Journal of Cultural Property* 307, 310.

have been inappropriate to use it to describe features of the natural environment worthy of protection.[38] In addition, the term 'heritage' was more consistent with another view emerging at the time, namely the need for intergenerational equity.

Before 1972, international environmental laws were limited in scope and based on the idea that the environment should be conserved for the benefit of present (rather than future) generations.[39] In the 1960s and early 1970s, it was increasingly accepted that certain cultural properties and natural areas are non-renewable resources that should be preserved for future generations.[40] Thus, a principle underpinning the *Stockholm Declaration* was that humans should 'protect and improve the environment for present and future generations'.[41] This principle would become known as 'intergenerational equity', and is now firmly enshrined in international environmental law.[42] As the policy underlying the concept of 'property' is the protection of the rights of the possessor, that term does not fit well with the principle of intergenerational equity. In contrast, 'heritage' is more consistent with the idea that some sites must be conserved for future generations,[43] so was a more appropriate term for use in the *World Heritage Convention*.

Heritage may be defined as 'those valuable features of our environment which we seek to conserve from the ravages of development and decay'.[44] However, it is a term of art, so can have many meanings, in part because it is used in a variety of fields, including law, architecture, art, and archaeology.[45] The meaning of the term will not be explored in detail here,

---

[38] Janet Blake, 'On Defining the Cultural Heritage' (2000) 49(1) *The International and Comparative Law Quarterly* 61, 67.

[39] Redgwell, above n 29, 64.

[40] Blake, above n 38, 67; Francioni, above n 12, 229.

[41] *Stockholm Declaration*, UN Doc A/CONF.48/14/Rev.1, art 1.

[42] See, for example, *Convention on Biological Diversity*, opened for signature 5 June 1992, 1760 UNTS 79 (entered into force 29 December 1993) art 22.

[43] Francioni, above n 12, 229. For example, Aplin contends that 'heritage' implies a gift for future generations and benefits for the community: Graeme Aplin, *Heritage: Identification, Conservation and Management* (Oxford University Press, 2002) 13. Lowenthal defines 'heritage' as 'everything we suppose has been handed down to us from the past': Lowenthal, above n 6, 81.

[44] Graeme Davison, 'The Meanings of "Heritage"' in Graeme Davison and Chris McConville (eds), *A Heritage Handbook* (Allen and Unwin, 1991) 1, 1.

[45] Josephine Suzanne Gillespie, *Monumental Challenges: Local Perspectives on World Heritage Landscape Regulation at Angkor Archaeological Park, Cambodia* (PhD Thesis, The University of Sydney, 2010) 67.

as that has been done extensively elsewhere.[46] However, three aspects of the concept of 'World Heritage' of particular relevance to the Pacific Islands are highlighted below.

Firstly, World Heritage is limited to immoveable heritage. Heritage can encompass many elements, including cultural, natural, Indigenous, moveable, immoveable, tangible, and intangible aspects.[47] While World Heritage may reflect some of these attributes, it does not include moveable heritage. As such, objects that may be of significance to Pacific Islanders such as handicrafts, ceramics, or other artefacts are not directly covered under the *Convention*. In addition, while a place that is related to or expresses intangible heritage values may be considered World Heritage, purely intangible heritage may not. This limitation can be explained by the fact that the *World Heritage Convention* was 'conceived, supported and nurtured by the industrially developed societies' and thus it reflects 'concern for a type of heritage that was highly valued in those countries'.[48] It has therefore been said that the *Convention* regime is 'not really appropriate for the kinds of heritage most common in regions where cultural energies have been concentrated in other forms of expression such as artefacts, dance or oral traditions'[49] such as the Pacific. There is some scope for sites associated with intangible values to be considered World Heritage (see Sect. 3.4). However, much of the intangible heritage of Pacific Islanders (including their traditional knowledge, customs, songs, stories, and dances) is not directly protected under the *World Heritage Convention*, which limits the treaty's relevance in the region.[50]

---

[46] See, for example, Aplin, above n 43, ch 1; Boer and Wiffen, above n 6, ch 1; Davison, above n 44; Lowenthal, above n 6; Maurice Evans, *Principles of Environmental Heritage* (Prospect Media, 2000) ch 2; Rodney Harrison, *Heritage: Critical Approaches* (Routledge, 2013) chs 2, 3.

[47] Boer and Wiffen, above n 6, 7.

[48] World Commission on Culture and Development, *Our Creative Diversity* (2nd ed, 1996) quoted in Ian Strasser 'Putting Reform into Action: Thirty Years of the World Heritage Convention: How to Reform a Convention without Changing its Regulations' (2002) 11(2) *International Journal of Cultural Property* 215, 224.

[49] Ibid.

[50] Intangible cultural heritage is now covered by the *Convention for the Safeguarding of the Intangible Cultural Heritage*, opened for signature 17 October 2003, 2368 UNTS 3 (entered into force 20 April 2006). See UNESCO, *Intangible Cultural Heritage in the Pacific* (UNESCO, 2011) for discussion of the application of this Convention in the Pacific.

Secondly, World Heritage encompasses both cultural and natural heritage, reflecting the origins of the *Convention*. This creates potential for the *Convention* to be usefully applied in the Pacific, where the distinction between sites of cultural and natural significance is often blurred (see Sect. 2.2.1). However, the dichotomy between natural and cultural World Heritage sites under the *Convention* regime continues to present challenges for the protection of such places (discussed in Sect. 3.3.1).

Thirdly, World Heritage is heritage that has OUV and thus has value to 'mankind as a whole'.[51] Heritage is an inherently subjective concept, as a site's value depends on who makes that judgement.[52] In the World Heritage context, in practice, the decision as to whether a particular site has OUV is made by the people who represent State parties on the World Heritage Committee,[53] so their views influence the scope of the concept of World Heritage. As explained below, the Committee's relatively narrow interpretation of the term OUV for many years limited the extent to which Pacific heritage could be considered as World Heritage.

3.3   THE ASSESSMENT OF OUTSTANDING UNIVERSAL VALUE

The *World Heritage Convention* does not attempt to protect all natural and cultural heritage, only that which is exceptional and thus has value for 'mankind as a whole'. The term OUV was introduced into the *Convention* to limit its scope to such places, rather than sites of purely local, regional, or national significance.[54] As the term had not been used in international law prior to its inclusion in the *Convention*,[55] it had no clear legal definition. The *Convention* also does not define the term or instruct the World

---

[51] *World Heritage Convention* preamble para 6, arts 1–2.

[52] Forrest, above n 6, 7–8; Ben Boer and Stefan Gruber, 'Heritage Discourses' in Brad Jessup and Kim Rubenstein (eds), *Environmental Discourses in Public and International Law* (Cambridge University Press, 2012) 375, 383.

[53] *World Heritage Convention* art 12. A site may have OUV but not be included on the World Heritage List, for example, if the relevant State party has not nominated it. However, because the focus of the *World Heritage Convention* regime is on sites inscribed on the World Heritage List, in practice the Committee's decision as to whether a site has OUV (and therefore whether it should be listed) is central to the operation of the regime.

[54] For detailed analysis of the origins of the term 'outstanding universal value', see Titchen, above n 5.

[55] Sarah M Titchen, 'On the Construction of "Outstanding Universal Value": Some Comments on the Implementation of the 1972 UNESCO World Heritage Convention' (1996) 1 *Conservation and Management of Archaeological Sites* 235, 236.

Heritage Committee to formulate a definition. However, the Committee is charged with determining whether a site should be inscribed on the World Heritage List,[56] and for defining the criteria by which a site may be listed,[57] which has allowed it to give further meaning to the concept of OUV.

Early nominations for World Heritage listing were assessed by the Committee and the Advisory Bodies in a fairly ad hoc manner.[58] However, their decision-making became more standardised when the Committee included provisions to guide the assessment of a site's value in the *Operational Guidelines for the Implementation of the World Heritage Convention*. The *Operational Guidelines* define OUV to mean 'cultural and/or natural significance which is so exceptional as to transcend national boundaries and to be of common importance for present and future generations of all humanity'.[59] They state that to be considered to have OUV, a site must

(1) meet one or more of the prescribed criteria[60];
(2) meet the conditions of integrity and authenticity[61]; and
(3) have adequate protection and management.[62]

Requirements (1) and (2) are discussed below. Requirement (3) is explored in Sect. 4.3.3.

3.3.1   The Criteria for the Assessment of Outstanding Universal Value

Pursuant to paragraph 77 of the 2016 version of the *Operational Guidelines*, to be considered to have OUV a property must

(i) represent a masterpiece of human creative genius;
(ii) exhibit an important interchange of human values on developments in architecture or technology, monumental arts, town planning, or landscape design;

---

[56] *World Heritage Convention* art 11(2).

[57] Ibid., art 11(5).

[58] Lasse Steiner and Bruno S Frey, 'Correcting the Imbalance of the World Heritage List: Did the UNESCO Strategy Work?' (2012) 3 *Journal of International Organisation Studies* 25, 27.

[59] UNESCO, *Operational Guidelines for the Implementation of the World Heritage Convention*, UN Doc WHC.15/01 (8 July 2015) ('*Operational Guidelines 2016*') para 49.

[60] Ibid., paras 77–78.

[61] Ibid., paras 79–95.

[62] Ibid., paras 96–115.

(iii) bear an exceptional testimony to a cultural tradition or civilisation which may be living or historical;

(iv) be an outstanding example of a type of building, architectural or technological ensemble, or landscape which illustrates a significant stage in human history;

(v) be an outstanding example of a traditional human settlement, land use, or sea use which is representative of a culture or human interaction with the environment;

(vi) be associated with events or living traditions, with ideas, or with beliefs, with artistic and literary works of outstanding universal significance;

(vii) contain superlative natural phenomena or areas of exceptional natural beauty and aesthetic importance;

(viii) be an outstanding example representing major stages of earth's history, significant ongoing geological processes in the development of landforms, or significant geomorphic or physiographic features;

(ix) be an outstanding example representing significant ongoing processes in the evolution and development of ecosystems and communities of plants and animals; and/or

(x) contain the most important natural habitats for in-situ conservation of biological diversity.

Two issues concerning these criteria warrant particular mention here: firstly, the dichotomy between cultural and natural World Heritage sites; and secondly, the importance of the selected criteria to the ongoing protection of the site.

The *Operational Guidelines* previously contained two separate lists of criteria, one for cultural sites and the other for natural sites, reflecting the separate definitions of 'cultural heritage' and 'natural heritage' in the *World Heritage Convention*.[63] In 2003, the World Heritage Committee resolved to merge these lists.[64] However, in practice, sites meeting criteria (i)–(vi) in paragraph 77 of the *Operational Guidelines* are considered to be cultural sites, and those meeting criteria (vii)–(x) are considered to be

---

[63] See, for example, UNESCO, *Operational Guidelines for the Implementation of the World Heritage Convention*, UN Doc CC-77/CONF.008 (30 July 1977) para 5(ii); UNESCO, *Operational Guidelines for the Implementation of the World Heritage Convention*, UN Doc WHC/2 (1978) 7, 10.

[64] WHC Res 6 EXT.COM 5.1, 6th extraordinary WHC sess, UN Doc WHC-03/6 EXT. COM/8 (27 May 2003) 5.

natural sites. Sites meeting a criterion in each group are referred to as 'mixed sites'.[65] This dichotomy is reinforced by the existence of different Advisory Bodies[66] for cultural and natural sites.[67] In addition, UNESCO has published different guidance documents for the management of cultural[68] and natural sites,[69] and the World Heritage Committee still refers in its documents and decisions to natural, cultural, and mixed sites.

The Advisory Bodies are working to better coordinate their work, particularly in relation to mixed sites.[70] In addition, the Committee's recognition of 'cultural landscapes' as a category of World Heritage site helped reinforce the link between culture and nature (see Sect. 3.4). However, a clear distinction remains between the treatment of cultural and natural sites under the regime, even though a founding principle of the *Convention* was the intrinsic link between culture and nature.

The practical importance of the criteria for World Heritage listing must also be recognised. Although the criteria are located in the *Operational Guidelines* not the *Convention*, they are critical for two key reasons. Firstly, a nomination for World Heritage listing will be deemed incomplete unless it demonstrates how the site complies with the criteria.[71] Thus, they impact the composition of the World Heritage List.

Secondly, they potentially influence the ongoing management and protection of the site. The *Operational Guidelines* state that a World Heritage property should be protected to ensure that its OUV is sustained or enhanced over time.[72] Consequently, the attributes of the site that give it OUV often become the focus of the Committee's concerns regarding the

[65] *Operational Guidelines 2016*, UN Doc WHC.16/01, para 46.

[66] The three Advisory Bodies are the International Centre for the Study of the Preservation and Restoration of Cultural Property (ICCROM), the International Council of Monuments and Sites (ICOMOS), and the International Union for Conservation of Nature and Natural Resources (IUCN). See Table 1.1 for description of their roles.

[67] ICOMOS and ICCROM are the Advisory Bodies for cultural sites. IUCN is the Advisory Body for natural sites: *Operational Guidelines 2016*, UN Doc WHC.16/01, paras 32–37.

[68] UNESCO et al, *Managing Cultural World Heritage*, World Heritage Resource Manual (UNESCO, 2013).

[69] UNESCO et al, *Managing Natural World Heritage*, World Heritage Resource Manual (UNESCO, 2012).

[70] See, for example, *Reports of the Advisory Bodies*, WHC 39th sess, UN Doc WHC-15/39. COM/5B (15 May 2015) para 23; *Progress Report on the Reflection on Processes for Mixed Nominations*, WHC 39th sess, UN Doc WHC-15/39.COM/9B (15 May 2015).

[71] *Operational Guidelines 2016*, UN Doc WHC.16/01, para 132(3).

[72] Ibid., para 96.

site's protection. The Committee does of course recognise that the conservation of OUV cannot be considered in isolation to other issues. Indeed, UNESCO's World Heritage management manuals state that a site should be managed to conserve *all* its heritage values.[73] In addition, it is now widely recognised that heritage conservation is a component of sustainable development[74] (discussed further in Sect. 4.3.1). However, because of its mandate, the Committee is often most concerned to ensure that a World Heritage site retains its OUV.

For example, East Rennell was inscribed on the World Heritage List based on the criterion now found in paragraph 77 (ix) of the *Operational Guidelines*, so it is considered to be a natural World Heritage site. Reflecting this, the Committee's resolutions concerning the protection of East Rennell have centred on threats to the natural environment such as resource development and over-harvesting. The preservation of the area's cultural values does not fall directly within the remit of the *Convention* regime, even though from a local perspective nature and culture are intrinsically linked. At East Rennell, this exacerbates the disconnect between the global and local perceptions of the site's value, which is a challenge for its protection (see Sect. 5.2.1.1).

3.3.2   *The Conditions of Integrity and Authenticity*

In addition to the criteria discussed above, the Committee considers that all sites must meet the condition of integrity to be eligible for World Heritage listing.[75] An assessment of a site's integrity considers the wholeness and intactness of the property. Among other things, it requires the Committee to consider whether the property contains all elements necessary to express its OUV, and whether it suffers from adverse effects of development or neglect.[76]

Cultural sites must also meet the condition of authenticity.[77] A property will be found to meet this requirement if its value is credibly and truthfully expressed rather than being a copy or replica.[78] The issue of whether a

---

[73] See, for example, UNESCO et al, above n 69, 37.
[74] See, for example, UNESCO et al, above n 68, 2.
[75] *Operational Guidelines 2016*, UN Doc WHC.16/01, para 87.
[76] Ibid., para 88. See also paras 89–95.
[77] Ibid., para 79.
[78] Ibid., para 80. See also paras 81–86.

heritage site meets this condition may arise, for example, when a site comprises structures that have been reconstructed.

As the criteria and requirements for OUV are contained in the *Operational Guidelines* not in the *Convention*, the Committee has been able to amend them to accommodate changing perceptions concerning heritage and its protection. As explained in the next section, over time the Committee has broadened the criteria and requirements for cultural sites so that a greater range of heritage places are now eligible for World Heritage listing.[79] The criteria for natural World Heritage sites have also been amended, reflecting developments in international environmental law.

3.4   THE WORLD HERITAGE COMMITTEE'S APPROACH TO THE ASSESSMENT OF OUTSTANDING UNIVERSAL VALUE

The feasibility of prescribing criteria for assessing whether a site has OUV was debated by delegates at the first World Heritage Committee meeting in 1977.[80] Heritage is an inherently subjective concept, with the value of a piece of heritage depending on who is making that assessment. As a result, 'the value of heritage may be skewed in favour of the current fashions favoured by those in the heritage industry' as opposed to 'reflect[ing] the views of those who "own" the heritage'.[81] At the 1977 Committee meeting, delegates expressed concern over how criteria would be applied given the subjectivity of an evaluation of heritage values, the potential impact of Western views on that evaluation, and the fact that heritage may be perceived differently by those within a culture as compared to those on the outside.[82] Despite these concerns, criteria were prescribed in the 1978 version of the *Operational Guidelines*,[83] and have been retained (albeit in a revised form) in all subsequent revisions.

---

[79] For comprehensive analysis of the criteria for cultural heritage sites, see, for example, Titchen, above n 5, in particular chs 5, 8; Jukka Jokilehto, *What is OUV? Defining the Outstanding Universal Value of Cultural World Heritage Properties* (ICOMOS, 2008), in particular chs 3–4.

[80] *Report of the World Heritage Committee*, WHC 1st sess, UN Doc CC-77/CONF.001/9 (17 October 1977) para 19.

[81] Gillespie, above n 45, 67.

[82] *Report of the World Heritage Committee*, WHC 1st sess, UN Doc CC-77/CONF.001/9 (17 October 1977) para 19.

[83] UNESCO, *Operational Guidelines for the Implementation of the World Heritage Convention*, UN Doc WHC/2 (1978) 7, 10.

The concerns described above have been played out. In the early years of the implementation of the *Convention*, the Committee (which was dominated by Europeans) tended to be most concerned about the protection of ancient structures and the monumental heritage of Europe.[84] This was reflected in its drafting of the cultural criteria, which until 1992 favoured sites of value because of their architectural or artistic characteristics, rather than places with less tangible heritage values.[85]

Soon after sites began to be inscribed on the World Heritage List, the influence of the criteria on the List's composition became a topic of discussion among the Committee and the Advisory Bodies. However, the Committee did not have a formal plan to address the imbalances that were emerging in the List until 1994, when it adopted the *Global Strategy for a Representative, Balanced and Credible World Heritage List*.[86] The meetings and studies that preceded the adoption of that strategy highlighted the need for the Committee to reconsider what constitutes heritage of OUV so that the List better reflects the diversity of heritage places around the world.[87]

Discussions concerning the imbalances in the World Heritage List also raised questions about the potential for sites demonstrating the interactions between people and the environment to be listed.[88] Of particular concern was the absence of 'cultural landscapes' on the List, being places that illustrate the evolution of human society and settlement over time, as influenced by the natural environment, social, economic, and cultural forces.[89]

---

[84] Sophia Labadi, 'A Review of the Global Strategy for a Balanced, Representative and Credible World Heritage List 1994–2004' (2005) 7(2) *Conservation and Management of Archaeological Sites* 89, 89–90.

[85] UNESCO, *World Heritage: Challenges for the Millennium* (UNESCO, 2007) 39.

[86] WHC Res CONF 003 X.10, WHC 18th sess, UN Doc WHC-94/CONF.003/16 (31 January 1995) 41–44. See Sect. 1.3 for discussion of the *Global Strategy*.

[87] Abdulqawi A Yusuf, 'Article 1 Definition of Cultural Heritage' in Francesco Francioni (ed), *The 1972 World Heritage Convention: A Commentary* (Oxford University Press, 2008) 23, 31–40.

[88] Mechtild Rössler, 'Managing World Heritage Cultural Landscapes and Sacred Sites' in Eléonore de Merode, Rieks Smeets and Carol Westrik (eds), *Linking Universal and Local Values: Managing a Sustainable Future for World Heritage*, World Heritage Papers 13 (UNESCO, 2004) 45, 45; Nora Mitchell, Mechtild Rössler and Pierre-Marie Tricaud, *World Heritage Cultural Landscapes: A Handbook for Conservation and Management*, World Heritage Papers 26 (UNESCO, 2009) 3.

[89] For discussion of the recognition of cultural landscapes under the *World Heritage Convention* regime in the Asia Pacific context, see, for example, Ken Taylor and Kirsty Altenburg, 'Cultural Landscapes in Asia-Pacific: Potential for Filling World Heritage Gaps' (2006) 12(3) *International Journal of Heritage Studies* 267; Natsuko Akagawa and Tiamsoon

Before 1992, there was some scope for cultural landscapes to be listed pursuant to the natural criteria, which referred to sites representing 'man's interaction with his natural environment' and 'exceptional combinations of natural and cultural elements'.[90] However, there was confusion as to how these criteria should be applied, given that the definition of 'natural heritage' in Article 2 of the *Convention* is not sufficiently broad to encompass sites of that type.[91] In contrast, 'cultural heritage' as defined under the *Convention* is clearly able to encompass cultural landscapes. Article 1 defines 'cultural heritage' to include sites that represent the 'combined works of nature and man', as well as buildings of OUV because of their place in the landscape.

In recognition of these issues, in 1994 the Committee significantly amended the criteria for World Heritage listing in the *Operational Guidelines*.[92] It removed references to interactions between culture and nature from the natural criteria, and broadened the cultural criteria by moving from a 'purely architectural view of the cultural heritage of humanity towards one which [is] much more anthropological, multi-functional and universal'.[93] Among other things, this involved amending the cultural criteria so they now encompass sites associated with living cultures[94] and places evidencing human interaction with the environment.[95] The Committee also formally recognised cultural landscapes as a category of World Heritage site, and included guidance principles for the listing of such sites in the *Operational Guidelines*.[96]

The Committee's broadening of the criteria for World Heritage listing corresponded with changing views concerning authenticity. The *Operational Guidelines* now state that an assessment of a site's authenticity should be

Sirisrisak, 'Cultural Landscapes in Asia and the Pacific: Implications of the World Heritage Convention' (2008) 14(2) *International Journal of Heritage Studies* 176.

[90] See, for example, UNESCO, *Operational Guidelines for the Implementation of the World Heritage Convention*, UN Doc WHC/2/Revised (December 1988) para 36(a) (ii)–(iii).

[91] Titchen, above n 5, 209.

[92] WHC Res CONF 003 XIV.3, WHC 18th sess, UN Doc WHC-94/CONF.003/16 (31 January 1995) 64–68.

[93] *Report on the Expert Meeting on the 'Global Strategy' and Thematic Studies for a Representative World Heritage List*, WHC 18th sess, UN Doc WHC-94/CONF.003/INF.6 (13 October 1994) 4.

[94] *Operational Guidelines 2016*, UN Doc WHC.16/01, para 77(iii), (vi).

[95] Ibid., para 77(v).

[96] Ibid., para 47, annex 3.

based on the *Nara Document on Authenticity*,[97] which was adopted by participants at the 1994 'Nara Conference on Authenticity in Relation to the World Heritage Convention'.[98] The *Nara Document* acknowledges that values attributed to cultural properties may differ from culture to culture, and within cultures, so judgements about authenticity cannot be based on fixed criteria.[99] Rather, heritage properties must be judged within their cultural context.[100] Importantly, the Committee now recognises that authenticity may be expressed through a variety of attributes, including traditions, techniques, and management systems; language and other forms of intangible heritage; and spirit and feeling.[101] This has made assessments of authenticity more applicable to a range of cultural contexts.[102]

The natural criteria (as now found in (vii)–(x) of paragraph 77 in the 2016 *Operational Guidelines*) have also been amended over time,[103] but the changes have been less contentious than those made to the cultural criteria. The most significant amendments were made in 1994, reflecting the substantial developments in international environmental law that occurred in 1992.[104] The United Nations Conference on Environment and Development held in that year led to the adoption of several instruments that introduced new concepts to international law, including 'ecosystems' and 'biodiversity conservation'.[105] Those concepts are now referred to in criteria (ix) and (x). No substantial changes have been made to the natural criteria since 1994; however, they were renumbered following the Committee's decision in 2003 to merge the cultural and natural criteria into one list.

---

[97] *Nara Document on Authenticity* (1994); *Operational Guidelines 2016*, UN Doc WHC.16/01, para 79, annex 4.

[98] For history of the *Nara Document*, see, for example, Christina Cameron and Nobuko Inaba, 'The Making of the Nara Document on Authenticity' (2015) 46(4) *APT Bulletin* 30.

[99] *Nara Document on Authenticity* (1994) para 11.

[100] Ibid.

[101] Ibid., para 13.

[102] Naomi Deegan, 'The Local-Global Nexus in the Politics of World Heritage: Space for Community Development?' in Marie-Theres Albert, Marielle Richon, Marie José Viñals and Andrea Witcomb (eds), *Community Development through World Heritage*, World Heritage Papers 31 (UNESCO, 2012) 77, 79. For detailed analysis of the changing concept of authenticity, see, Christina Cameron, 'From Warsaw to Mostar: The World Heritage Committee and Authenticity' (2008) 39(2/3) *APT Bulletin* 19.

[103] For analysis of the development of the criteria for natural sites, see Redgwell, above n 29; Titchen, above n 5, ch 5.

[104] Redgwell, above n 29, 67.

[105] Ibid., 75.

3.5   THE RECOGNITION OF PACIFIC ISLAND HERITAGE AS 'WORLD HERITAGE'

3.5.1   Cultural World Heritage Sites

The relatively narrow scope of the cultural criteria prior to 1992 may have contributed to the under-representation of Pacific Island heritage on the World Heritage List. However, notwithstanding the expansion of the criteria, few sites in the independent Pacific Island States have been listed. Furthermore, only two of these represent the living cultures of Pacific Islanders. This raises the question of whether the cultural criteria still present impediments to the listing of Pacific sites.

This question has been explored by Anita Smith, who concluded that the concept of OUV (as framed in the current criteria), and the arguments and evidence required to demonstrate that a site meets that threshold, can accommodate sites of value to Pacific Islanders.[106] Her conclusion was based on several sites that were being considered for nomination or had been listed, including Chief Roi Mata's Domain in Vanuatu.[107]

Chief Roi Mata's Domain (a cultural landscape) was listed in 2008[108] on the basis of criteria (iii), (v), and (vi) in the 2008 version of the *Operational Guidelines*.[109] The site comprises areas associated with the life and death of Chief Roi Mata, who died in around 1600 AD and is credited with initiating important social reforms.[110] Criterion (iii) was previously limited to sites

[106] Anita Smith, 'World Heritage and Outstanding Universal Value in the Pacific Islands' (2015) 21(2) *International Journal of Heritage Studies* 177.

[107] In addition to Chief Roi Mata's Domain, in support of her argument, Smith refers to the listing of Papahānaumokuākea, in Hawaii, as a mixed site, and two sites being considered for nomination: the 'Yapese Stone Money' site (a proposed transnational serial site from Palau and Yap in the Federated Sites of Micronesia) and the Sacred Site of Taputapuatea/Te Po and the Opoa Valley (in French Polynesia).

[108] WHC Res 32 COM 8B.27, WHC 32nd sess, UN Doc WHC-08/32.COM/24Rev (31 March 2009) 170.

[109] UNESCO, *Operational Guidelines for the Implementation of the World Heritage Convention*, UN Doc WHC.08/01 (January 2008) ('*Operational Guidelines 2008*'). For discussion of the heritage values of the site, see, for example, Meredith Wilson, Chris Ballard and Douglas Kalotiti, 'Chief Roi Mata's Domain: Challenges for a World Heritage Property in Vanuatu' (2011) 23(2) *Historic Environment* 5.

[110] Republic of Vanuatu, *Chief Roi Mata's Domain – Nomination by the Republic of Vanuatu for Inscription on the World Heritage List* (2007); WHC Res 32 COM 8B.27, WHC 32nd sess, UN Doc WHC-08/32.COM/24Rev (31 March 2009) 170 para 3.

that bore testimony to an *extinct* civilisation,[111] but in 1994 was expanded to also apply to *living* cultural traditions and civilisations.[112] While substantial archaeological research provides some evidence of the heritage value of Chief Roi Mata's Domain, the site has OUV because of the local communities' continuing customary knowledge of and respect for the place.[113] Consequently, the expansion of criterion (iii) facilitated the site's listing based on its association with the living traditions of its customary owners.

When Smith conducted her analysis, Chief Roi Mata's Domain was the only listed World Heritage site within the independent Pacific Island States inscribed because of its association with living cultures. Since that time, a site in the Federated States of Micronesia referred to as 'Nan Madol: Ceremonial Centre of Eastern Micronesia' has been listed.[114] That site contains remains of stone palaces, temples, mortuaries, and residential domains bearing testimony to the development of chiefly societies.[115] The continuing association of the site with social and ceremonial traditions and systems of customary governance was also recognised in the site's listing.[116] The inscription of Nan Madol therefore reinforces Smith's finding that the expansion of the cultural criteria has opened the door for the listing of sites associated with the living customs of Pacific Islanders.

The Kuk Early Agricultural Site is another Pacific cultural landscape on the World Heritage List. That site, in the western highlands of Papua New Guinea (PNG), contains archaeological remains demonstrating a transformation of agricultural practices that occurred around 6500 years ago.[117] It was found to have OUV on the basis of criteria (iii) and (iv) in the 2008

---

[111] UNESCO, *Operational Guidelines for the Implementation of the World Heritage Convention*, UN Doc WHC/2/Revised (27 March 1992) para 24(a) (iii) ('*Operational Guidelines 1992*').

[112] WHC Res CONF 003 XIV.3, WHC 18th sess, UN Doc WHC-94/CONF.003/16 (31 January 1995) 64–68. See *Operational Guidelines 2008*, UN Doc WHC.08/01, para 77(iii).

[113] Smith, above n 106, 182.

[114] WHC Res 40 COM 8B.22, WHC 40th sess, UN Doc WHC/16/40.COM/19 (15 November 2016) 217.

[115] ICOMOS, *Evaluations of Nominations of Cultural and Mixed Properties to the World Heritage List*, WHC 40th sess, UN Doc WHC/16/40.COM/INF.8B1 (July 2016) 103 (Nan Madol, Federated States of Micronesia, Advisory Body Evaluation 1503) 106.

[116] Ibid.

[117] Government of Papua New Guinea, *Kuk Early Agricultural Site Cultural Landscape – A Nomination for Consideration as World Heritage Site* (2007).

version of the *Operational Guidelines*.[118] Criterion (iv) applies to sites that illustrate a significant stage in history, and was expanded in 1994 from *buildings* and *architectural ensembles*[119] to also encompass *landscapes*.[120] This amendment made the criterion applicable to Kuk, and potentially more relevant to other Pacific heritage places.

While the inclusion of Kuk in the World Heritage List was important in terms of the recognition of Pacific landscapes, the site is what Smith describes as an 'Oceanic or island expression of a global narrative',[121] rather than one representing the living traditions of Pacific Islanders. Other cultural sites on the World Heritage List in the region have also been found to have OUV because of their interpretation through global narratives: the Levuka Historical Port Town in Fiji is an example of European settlement in the Pacific Islands, which reflects the contact and interchange of values between colonisers and the Pacific Islanders[122]; the Bikini Atoll Nuclear Test Site in Marshall Islands bears testimony to the birth of the Cold War and the nuclear era[123]; and the Rock Islands Southern Lagoon in Palau is a mixed site, gaining its OUV from the remains of stone villages, rock art, cave deposits, and burials, which evidence the development of Pacific Island societies, as well as its exceptional marine environment and biodiversity.[124]

---

[118]WHC Res 32 COM 8B.26, WHC 32nd sess, UN Doc WHC-08/32.COM/24Rev (31 March 2009) 168; *Operational Guidelines 2008*, UN Doc WHC.08/01, para 77(iii)–(iv). For discussion of the heritage values of the site, see, for example, John Denham, Tim Muke and Vagi Genorupa, 'Nominating and Managing a World Heritage Site in the Highlands of Papua New Guinea' (2007) 39(3) *World Archaeology* 324, 331.

[119]*Operational Guidelines 1992*, UN Doc WHC/2/Revised, para 24(a) (iv).

[120]WHC Res CONF 003 XIV.3, WHC 18th sess, UN Doc WHC-94/CONF.003/16 (31 January 1995) 64–68. See *Operational Guidelines 2008*, UN Doc WHC.08/01, para 77(iv).

[121]Smith, above n 106, 181.

[122]WHC Res 37 COM 8B.25, WHC 37th sess, UN Doc WHC-13/37.COM/20 (5 July 2013) 186, 186 para 3. For discussion of the heritage values of the site, see, for example, David Harrison, 'Levuka, Fiji: Contested Heritage?' (2004) 7(4) *Current Issues in Tourism* 346.

[123]WHC Res 34 COM 8B.20, WHC 34th sess, UN Doc WHC-10/34.COM/20 (3 September 2010) 206, 207 para 3. For discussion of the heritage values of the site, see, for example, Steve Brown, 'Poetics and Politics: Bikini Atoll and World Heritage Listing' in Sue O'Connor, Denis Byrne and Sally Brockwell (eds), *Transcending the Culture-Nature Divide in Cultural Heritage: Views from the Asia-Pacific Region* (ANU E Press, 2012) 35.

[124]WHC Res 36 COM 8B.12, WHC 36th sess, UN Doc WHC-12/36.COM/19 (June–July 2012) 165, 165 para 3. For discussion of the heritage values of the site, see, for example, Christian Reepmeyer et al, 'Selecting Cultural Sites for the UNESCO World Heritage List:

Documents such as the *Pacific Appeal* suggest that Pacific Islanders are most concerned to ensure the protection of their 'spiritually-valued natural features and cultural places', which are related to the 'origins of peoples, the land and sea, and other sacred stories'.[125] While the cultural World Heritage sites referred to in the paragraph above are significant, they are not examples of the types of places most valued by Pacific Islanders. Consequently, although the criteria for World Heritage listing are now broad enough to accommodate such sites, impediments to their nomination and listing (which were noted in Sect. 1.3) remain to be addressed. A key issue is likely to be the lack of heritage inventories and research documenting the heritage values of places of significance to Pacific Islanders. Furthermore, the implications of listing sites that possess very different global and local significance warrant consideration (see further discussion in Sect. 3.6).

3.5.2 Natural World Heritage Sites

Three sites in the independent Pacific Island States have been inscribed on the World Heritage List on the basis of natural criteria. The first was East Rennell in Solomon Islands, which was listed in 1998.[126] It was found to meet criterion (ix) because of the evolutionary and speciation processes that have happened on the island, particularly in relation to bird life.[127] The Phoenix Islands Protected Area in Kiribati was listed in 2008 based on criteria (xii) and (ix).[128] It is considered to have OUV as an 'oceanscape' exhibiting exceptional natural beauty, and because of its contribution to

Recent Work in the Rock Islands – Southern Lagoon Area, Republic of Palau' in Jolie Liston, Geoffrey Clark and Dwight Alexander (eds), *Pacific Island Heritage: Archaeology, Identity and Community* (ANU E Press, 2011) 85.

[125] *Presentation of the World Heritage Programme for the Pacific*, WHC 31st sess, UN Doc WHC-07/31.COM/11C (10 May 2007) annex I (Appeal to the World Heritage Committee from the Pacific Island State Parties) para 11. The *Pacific Appeal* is discussed in Sect. 1.5.

[126] WHC Res CONF 203 VIII.A.1, WHC 22nd sess, UN Doc WHC-98/CONF/203/18 (29 January 1999) 25.

[127] *Adoption of Retrospective Statements of Outstanding Universal Value*, WHC 36th sess, UN Doc WHC-12/36.COM/8E (15 June 2012) 55–6 (East Rennell, Solomon Islands); WHC Res 36 COM 8E, WHC 36th sess, UN Doc WHC-12/36.COM/19 (June–July 2012) 225.

[128] WHC Res 34 COM 8B.2, WHC 34th sess, UN Doc WHC-10/34.COM/20 (3 September 2010) 165.

evolutionary processes and the development of global marine ecosystems.[129] In 2012, Rock Islands Southern Lagoon in Palau was listed as a mixed site. In addition to some cultural criteria, it was found to meet criteria (vii), (iv), and (x) due to its exceptional marine environment and biodiversity.[130]

Studies undertaken by IUCN as part of the implementation of the *Global Strategy* suggest there is scope for the listing of further natural World Heritage sites in the Pacific. IUCN has noted that natural heritage of OUV is not evenly distributed around the world, and therefore regional balance of listed natural sites is neither desirable nor achievable.[131] Consequently, most IUCN studies focus on the global distribution of World Heritage in terms of biogeographic realms, biomes, and habitats, or themes such as wetlands, coastal areas, mountains, forests, and geological sites, rather than their regional distribution. IUCN's work did however identify several sites in the Pacific Islands worthy of inscription on the World Heritage List as natural sites.[132] Hazen and Anthamatten's analysis of the ecological representativeness of the World Heritage List also highlighted that determining an optimal definition of 'representation' in the context of the natural listed World Heritage sites is controversial, because of the diverse ways that site distribution can be assessed.[133] However, they too identified some ecological realms that were clearly under-represented on the World Heritage List, including the Pacific Islands.[134] These studies do not suggest that the natural criteria in the *Operational Guidelines* present any barrier to the recognition of Pacific

[129] *Nominations to the World Heritage List*, WHC 35th sess, UN Doc WHC-11/35. COM/8B.Add (27 May 2011) 10–11 (Statement of Outstanding Universal Value: Phoenix Islands Protected Area, Kiribati); WHC Res 35 COM 8B.60, WHC 35th sess, UN Doc WHC-11/35.COM/20 (7 July 2011) 249.

[130] WHC Res 36 COM 8B.12, WHC 36th sess, UN Doc WHC-12/36.COM/19 (June–July 2012) 165, 165 para 3.

[131] IUCN, *The World Heritage List: Future Priorities for a Credible and Complete List of Natural and Mixed Sites* (IUCN, 2004) 3.

[132] See, for example, Jim Thorsell and Todd Sigaty, *A Global Overview of Forest Protected Areas on the World Heritage List* (IUCN, 1997) 21; Jim Thorsell, Renée Ferster and Todd Sigarty, *A Global Overview of Wetland and Marine Protected Areas on the World Heritage List* (IUCN, 1997) 21; Tim Wong et al (eds), *Proceedings of the Asia-Pacific Forum on Karst Ecosystems and World Heritage* (Gunung Mulu National Park World Heritage Area, Malaysia, 26–30 May 2001) 45.

[133] Helen Hazen and Peter Anthamatten, 'Unnatural Selection: An Analysis of the Ecological Representativeness of Natural World Heritage Sites' (2007) 59(2) *The Professional Geographer* 256, 256.

[134] Ibid., 264.

places as World Heritage sites. As discussed further in the next section, the more pertinent question is whether it is appropriate to list such places purely based on natural criteria.

3.6 Local and Global Values at Pacific World Heritage Sites

A variation between the global and local significance of a place can exist at any type of World Heritage site.[135] Indeed, as noted in Sect. 3.5.1, most cultural World Heritage sites in the Pacific have been found to have OUV because of their interpretation through global narratives rather than being representative of Pacific Islander values, as articulated in documents such as the *Pacific Appeal*.[136] For example, the Bikini Atoll Nuclear Test Site in Marshall Islands was inscribed on the World Heritage List as a tangible testimony to the birth of the nuclear era and as a source of globally significant symbols and icons.[137] As Brown notes, this is 'not strictly the history and heritage of the Bikinian people', who were forced to move from the island to make way for the United States' nuclear programme.[138] The site's listing therefore privileged 12 years of the area's history (1946–1958) and 'reduced the Bikinian people's story to a subplot'.[139]

Another example is the Levuka Historical Port Town in Fiji, which was found to have OUV in part because the urban landscape 'exhibits the important interchange of human values and cultural contact' associated with colonisation.[140] That site includes the land in Levuka Town boundary, within which there are numerous commercial, residential, and civic buildings. Fisher notes there is a distinction between how Indigenous Fijians and people of European origin perceive 'heritage' at Levuka. For

[135] For discussion of this issue more generally, see David Harrison, 'Introduction: Contested narratives in the domain of World Heritage' (2004) 7:4–5 *Current Issues in Tourism* 281. Harrison notes that 'what is defined as heritage is linked to *power*: the power to impose a view of the world, especially of the past, on others' (at 287).

[136] Smith, above n 106, 181.

[137] WHC Res 34 COM 8B.20, WHC 34th sess, UN Doc WHC-10/34.COM/20 (3 September 2010) 206.

[138] Brown, above n 123, 36.

[139] Steve Brown, 'Archaeology of Brutal Encounter: Heritage and Bomb Testing on Bikini Atoll, Republic of the Marshall Islands' (2013) 48 *Archaeology in Oceania* 26, 36.

[140] ICOMOS, *Evaluations of Nominations of Cultural and Mixed Properties to the World Heritage List*, WHC 37th sess, UN Doc WHC-13/37.COM/INF.8B1 (June 2013) 87 (Levuka Historical Port Town, Fiji, Advisory Body Evaluation 1399) 95.

the former, 'buildings and artefacts are of little consequence—mere adjectives in a flowery piece of prose'. For the latter, 'buildings are the substance and structure of the writing, without which the meaning becomes obscured'.[141] Consequently, most Indigenous Fijians consider Levuka 'at best, an irrelevance to modern Fiji, while some resent the attention it periodically receives from government and the international community'.[142]

A variation between the global and local significance of a site will almost certainly exist at natural World Heritage sites in the Pacific, where most land and inshore areas are under customary tenure, and where many people possess strong spiritual connections to their land. As Ballard and Wilson have said, '[C]lassifying any Melanesian landscape as natural, whether under a national conservation programme or as a World Heritage site, effectively obscures a series of claims to cultural knowledge and ownership by local communities.'[143] As such, the listing of natural World Heritage sites in the Pacific will often create a situation where the values that make the site eligible for inscription are very different to those that local communities attach to the property.

The question of whether Pacific places should be listed as natural World Heritage sites should be considered on a case-by-case basis. It must be remembered that while the concepts of nature and culture are closely linked in the Pacific, not all significant sites will qualify for World Heritage listing as cultural landscapes or mixed sites. It is likely that many such places will only meet the OUV threshold based on their natural values. Indeed, there are several sites on the Tentative Lists of Pacific Island States which are identified as meeting natural heritage criteria only.[144] If the listing of natural World Heritage sites in the Pacific was ruled out, such places could not be listed at all (under the current *Operational Guidelines*). Whether that would be a better outcome than listing them purely based on natural criteria needs to be considered on an individual basis.

Regardless of the criteria upon which a site is nominated, the implications of any disconnect between the global and local significance of a

---

[141] David Fisher, *The Socio-Economic Consequences of Tourism in Levuka, Fiji* (PhD Thesis, Lincoln University, 2000) 134–135.

[142] Harrison, above n 122, 365.

[143] Chris Ballard and Meredith Wilson, 'Unseen Monuments: Managing Melanesian Cultural Landscapes' in Ken Taylor and Jane L Lennon (eds), *Managing Cultural Landscapes* (Routledge, 2012) 130, 134.

[144] For example, a site referred to as 'Tropical Rainforest Heritage of Solomon Islands' (which is on Solomon Islands' Tentative List) is proposed as a natural World Heritage site.

World Heritage site should be explored at the nomination stage. The analysis of East Rennell in Part III of this book demonstrates that substantial variation between the local and international significance of a site can create or at least exacerbate challenges associated with the site's protection. Strategies to safeguard the OUV of such a place will often need to try to bridge any variation between the site's global and local significance,[145] but achieving that can be difficult in practice.

3.7 CONCLUSION

This chapter highlighted three key features of the concept of 'World Heritage'. Firstly, it is limited to immoveable heritage, which limits the extent to which the *World Heritage Convention* can be used to protect heritage of value to Pacific Islanders. Secondly, it encompasses both natural and cultural heritage, reflecting the era in which the *Convention* was drafted. While this creates significant potential for the regime to apply to Pacific sites, a dichotomy remains between natural and cultural World Heritage, which poses challenges for the recognition and protection of such places. Finally, World Heritage is an inherently subjective concept, and the World Heritage Committee's assessment of the concept of OUV essentially dictates the scope of the regime. Over time, the Committee has broadened its interpretation of the concept, and has recognised 'cultural landscapes' as a category of World Heritage site. This has allowed a greater range of Pacific places to meet the threshold for World Heritage listing.

Smith's work demonstrates that the criteria for World Heritage listing are now sufficiently broad to encompass many heritage places in the Pacific. However, such sites (particularly those associated with the living cultures of Pacific Islanders) remain barely represented on the World Heritage List, so barriers to their nomination still need to be addressed. The implications of listing sites which possess markedly different global and local significance also warrant further consideration, including the challenges this presents for conservation.

As the boundaries of the concept of 'World Heritage' are broadening, there is a corresponding need to also expand our thinking concerning how

---

[145] Eric L Edroma, 'Linking Universal and Local Values for the Sustainable Management of World Heritage Sites' in Eléonore de Merode, Rieks Smeets and Carol Westrik (eds), *Linking Universal and Local Values: Managing a Sustainable Future for World Heritage* (UNESCO, 2004) 36, 40.

heritage places should be protected. Therefore, the next chapter explains the protection regime established by the *Convention*, and the Committee's changing approach to World Heritage conservation.

REFERENCES

ARTICLES, BOOKS AND REPORTS

Akagawa, Natsuko and Tiamsoon Sirisrisak, 'Cultural Landscapes in Asia and the Pacific: Implications of the World Heritage Convention' (2008) 14(2) *International Journal of Heritage Studies* 176

Aplin, Graeme, *Heritage: Identification, Conservation and Management* (Oxford University Press, 2002)

Ballard, Chris and Meredith Wilson, 'Unseen Monuments: Managing Melanesian Cultural Landscapes' in Ken Taylor and Jane L Lennon (eds), *Managing Cultural Landscapes* (Routledge, 2012) 130

Bandarin, Francesco, *World Heritage: Challenges for the Millennium* (UNESCO, 2007)

Blake, Janet, 'On Defining the Cultural Heritage' (2000) 49(1) *The International and Comparative Law Quarterly* 61

Boer, Ben and Stefan Gruber, 'Heritage Discourses' in Brad Jessup and Kim Rubenstein (eds), *Environmental Discourses in Public and International Law* (Cambridge University Press, 2012) 375

Boer, Ben and Graeme Wiffen, *Heritage Law in Australia* (Oxford University Press, 2006)

Brown, Steve, 'Poetics and Politics: Bikini Atoll and World Heritage Listing' in Sue O'Connor, Byrne, Denis and Sally Brockwell (eds), *Transcending the Culture-Nature Divide in Cultural Heritage: Views from the Asia-Pacific Region* (ANU E Press, 2012) 35

Brown, Steve, 'Archaeology of Brutal Encounter: Heritage and Bomb Testing on Bikini Atoll, Republic of the Marshall Islands' (2013) 48 *Archaeology in Oceania* 26

Cameron, Christina, 'From Warsaw to Mostar: The World Heritage Committee and Authenticity' (2008) 39(2/3) *APT Bulletin* 19

Cameron, Christina and Nobuko Inaba, 'The Making of the Nara Document on Authenticity' (2015) 46(4) *APT Bulletin* 30

Carducci, Guido, 'The 1972 World Heritage Convention in the Framework of other UNESCO Conventions on Cultural Heritage' in Francesco Francioni (ed), *The 1972 World Heritage Convention: A Commentary* (Oxford University Press, 2008) 363

Carson, Rachel, *Silent Spring* (1962)

Davison, Graeme, 'The Meanings of 'Heritage'' in Graeme Davison and Chris McConville (eds), *A Heritage Handbook* (Allen and Unwin, 1991) 1

Deegan, Naomi, 'The Local-Global Nexus in the Politics of World Heritage: Space for Community Development?' in Marie-Theres Albert, Marielle Richon, Marie José Viñals and Andrea Witcomb (eds), *Community Development through World Heritage*, World Heritage Papers 31 (UNESCO, 2012) 77

Denham, John, Tim Muke and Vagi Genorupa, 'Nominating and Managing a World Heritage Site in the Highlands of Papua New Guinea' (2007) 39(3) *World Archaeology* 324

Edroma, Eric L, 'Linking Universal and Local Values for the Sustainable Management of World Heritage Sites' in Eléonore de Merode, Rieks Smeets and Carol Westrik (eds), *Linking Universal and Local Values: Managing a Sustainable Future for World Heritage* (UNESCO, 2004) 36

Ehrlich, Paul, *The Population Bomb* (Sierra Club and Ballantine Books, 1968)

Evans, Maurice, *Principles of Environmental Heritage* (Prospect Media, 2000)

Fisher, David, *The Socio-Economic Consequences of Tourism in Levuka, Fiji* (PhD Thesis, Lincoln University, 2000)

Forrest, Craig, *International Law and the Protection of Cultural Heritage* (Routledge, 2011)

Francioni, Francesco, 'A Dynamic Evolution of Concept and Scope: From Cultural Property to Cultural Heritage' in Yusuf A Abdulqawi (ed), *Standard-Setting in UNESCO Volume 1: Normative Action in Education, Science and Culture* (Martinus Nijoff and UNESCO Publishing, 2007) 221

Francioni, Francesco (ed), *The 1972 World Heritage Convention: A Commentary* (Oxford University Press, 2008)

Francioni, Francesco, 'The Preamble' in Francesco Francioni (ed), *The 1972 World Heritage Convention: A Commentary* (Oxford University Press, 2008) 11

Government of Papua New Guinea, *Kuk Early Agricultural Site Cultural Landscape – A Nomination for Consideration as World Heritage Site* (2007)

Harrison, David, 'Introduction: Contested narratives in the domain of World Heritage' (2004) 7:4–5 *Current Issues in Tourism* 281

Harrison, David, 'Levuka, Fiji: Contested Heritage' (2004) 7:4–5 *Current Issues in Tourism* 346

Harrison, Rodney, *Heritage: Critical Approaches* (Routledge, 2013)

Hazen, Helen and Peter Anthamatten, 'Unnatural Selection: An Analysis of the Ecological Representativeness of Natural World Heritage Sites' (2007) 59(2) *The Professional Geographer* 256

IUCN, *The World Heritage List: Future Priorities for a Credible and Complete List of Natural and Mixed Sites* (IUCN, 2004)

Jokilehto, Jukka, *What is OUV? Defining the Outstanding Universal Value of Cultural World Heritage Properties* (ICOMOS, 2008)

Labadi, Sophia, 'A Review of the Global Strategy for a Balanced, Representative and Credible World Heritage List 1994–2004' (2005) 7(2) *Conservation and Management of Archaeological Sites* 89

Lausche, Barbara J, *Weaving a Web of Environmental Law: Contributions of the IUCN Environmental Law Programme* (IUCN/ICEL, 2008)

Lowenthal, David, 'Natural and Cultural Heritage' (2005) 11(1) *International Journal of Heritage Studies* 81

Meadows, Donella H, Dennis L Meadows, Jørgen Randers and William W Behrens III, *Limits to Growth* (Universe Books, 1972)

Mitchell, Nora, Mechtild Rössler and Pierre-Marie Tricaud, *World Heritage Cultural Landscapes: A Handbook for Conservation and Management*, World Heritage Papers 26

Park, Seong-Yong, *On Intangible Heritage Safeguarding Governance: An Asia-Pacific Context* (Cambridge Scholars Publishing, 2013)

Pocock, Douglas, 'Some Reflections on World Heritage' (1997) 29(3) *Area* 260

Prott, Lyndel V and P J O'Keefe, *Law and the Cultural Heritage – Volume I* (Professional Books, 1984)

Prott, Lyndel V and Patrick J O'Keefe, 'Cultural Heritage' or 'Cultural Property' (1992) 1(2) *International Journal of Cultural Property* 307

Redgwell, Catherine, 'Article 2 Definition of Natural Heritage' in Francesco Francioni (ed), *The 1972 World Heritage Convention: A Commentary* (Oxford University Press, 2008) 63

Reepmeyer, Christian, Geoffrey Clark, Dwight Alexander, Ilebrang U Olkeriil, Jolie Liston and Ann Hillmann Kitalong, 'Selecting Cultural Sites for the UNESCO World Heritage List: Recent Work in the Rock Islands – Southern Lagoon Area, Republic of Palau' in Jolie Liston, Geoffrey Clark and Dwight Alexander (eds), *Pacific Island Heritage: Archaeology, Identity and Community* (ANU E Press, 2011) 85

Republic of Vanuatu, *Chief Roi Mata's Domain – Nomination by the Republic of Vanuatu for Inscription on the World Heritage List* (2007)

Rössler, Mechtild, 'Managing World Heritage Cultural Landscapes and Sacred Sites' in Eléonore de Merode, Rieks Smeets and Carol Westrik (eds), *Linking Universal and Local Values: Managing a Sustainable Future for World Heritage*, World Heritage Papers 13 (UNESCO, 2004) 45

Slatyer, Ralph O, 'The Origin and Development of the World Heritage Convention' (1984) *Monumentum* 3

Smith, Anita, 'World Heritage and Outstanding Universal Value in the Pacific Islands' (2015) 21(2) *International Journal of Heritage Studies* 177

Steiner, Lasse and Bruno S Frey, 'Correcting the Imbalance of the World Heritage List: Did the UNESCO Strategy Work?' (2012) 3 *Journal of International Organisation Studies* 25

Strasser, Ian, 'Putting Reform into Action: Thirty Years of the World Heritage Convention: How to Reform a Convention without Changing its Regulations' (2002) 11(2) *International Journal of Cultural Property* 215

Taylor, Ken and Kirsty Altenburg, 'Cultural Landscapes in Asia-Pacific: Potential for Filling World Heritage Gaps' (2006) 12(3) *International Journal of Heritage Studies* 267

Thorsell, Jim and Todd Sigaty, *A Global Overview of Forest Protected Areas on the World Heritage List* (IUCN, 1997)

Thorsell, Jim, Renée Ferster and Todd Sigarty, *A Global Overview of Wetland and Marine Protected Areas on the World Heritage List* (IUCN, 1997)

Titchen, Sarah M, 'On the Construction of 'Outstanding Universal Value': Some Comments on the Implementation of the 1972 UNESCO World Heritage Convention' (1996) 1 *Conservation and Management of Archaeological Sites* 235

UNESCO/ICCROM/ICOMOS/IUCN, *Managing Cultural World Heritage*, World Heritage Resource Manual (UNESCO, 2013)

UNESCO/ICCROM/ICOMOS/IUCN, *Managing Natural World Heritage*, World Heritage Resource Manual (UNESCO, 2012)

UNESCO, *Intangible Cultural Heritage in the Pacific* (UNESCO, 2011)

UNESCO, *World Heritage: Challenges for the Millennium* (UNESCO, 2007)

Weiss, Edith Brown, Daniel B Magraw and Paul C Szasz, *International Environmental Law: Basic Instruments and References* (Transnational Publishers, 1992) 171

Wilson, Meredith, Chris Ballard and Douglas Kalotiti, 'Chief Roi Mata's Domain: Challenges for a World Heritage Property in Vanuatu' (2011) 23(2) *Historic Environment* 5

Wong, Tim, Elery Hamilton-Smith, Stuart Chape and Hans Friederich (eds), *Proceedings of the Asia-Pacific Forum on Karst Ecosystems and World Heritage* (Gunung Mulu National Park World Heritage Area, Malaysia, 26–30 May 2001)

World Commission on Culture and Development, *Our Creative Diversity* (2nd ed, 1996)

Yusuf, Abdulqawi A, 'Article 1 Definition of Cultural Heritage' in Francesco Francioni (ed), *The 1972 World Heritage Convention: A Commentary* (Oxford University Press, 2008) 23, 31–40

CONVENTIONS

Convention Concerning the Protection of the World Cultural and Natural Heritage, opened for signature 16 November 1972, 1037 UNTS 151 (entered into force 17 December 1975)

Convention for the Protection of Cultural Property during Armed Conflict, opened for signature 14 May 1954, 249 UNTS 240 (entered into force 7 August 1956)

Convention on Biological Diversity, opened for signature 5 June 1992, 1760 UNTS 79 (entered into force 29 December 1993)

Convention on the Means of Prohibiting and Preventing the Illicit Import, Export and Transfer of Ownership of Cultural Property, opened for signature 14 November 1970, 823 UNTS 231 (entered into force 24 April 1972)

Convention for the Safeguarding of the Intangible Cultural Heritage, opened for signature 17 October 2003, 2368 UNTS 3 (entered into force 20 April 2006)

UNITED NATIONS DOCUMENTS

Adoption of Retrospective Statements of Outstanding Universal Value, WHC 36th sess, UN Doc WHC-12/36.COM/8E (15 June 2012) 55 (East Rennell, Solomon Islands)

ICOMOS, *Evaluations of Nominations of Cultural and Mixed Properties to the World Heritage List*, WHC 37th sess, UN Doc WHC-13/37.COM/INF.8B1 (June 2013) 87 (Levuka Historical Port Town, Fiji, Advisory Body Evaluation 1399)

ICOMOS, *Evaluations of Nominations of Cultural and Mixed Properties to the World Heritage List*, WHC 40th sess, UN Doc WHC/16/40.COM/INF.8B1 (July 2016) 103 (Nan Madol, Federated States of Micronesia, Advisory Body Evaluation 1503)

International Instruments for the Protection of Monuments, Groups of Buildings and Sites, UN Doc SHC/MD/17 (30 June 1971) annex II

Nominations to the World Heritage List, WHC 35th sess, UN Doc WHC-11/35. COM/8B.Add (27 May 2011) 10 (Statement of Outstanding Universal Value: Phoenix Islands Protected Area, Kiribati)

Presentation of the World Heritage Programme for the Pacific, WHC 31st sess, UN Doc WHC-07/31.COM/11C (10 May 2007) annex I (Appeal to the World Heritage Committee from the Pacific Island State Parties)

Progress Report on the Reflection on Processes for Mixed Nominations, WHC 39th sess, UN Doc WHC-15/39.COM/9B (15 May 2015)

Records of the General Conference, 16th sess, UNESCO Res 2.313 (1970) 35 (Intergovernmental Programme on Man and the Biosphere)

Report of the United Nations Conference on the Human Environment, UN Doc A/ CONF.48/14/Rev.1 (5–16 June 1972) ch 1 (Declaration of the United Nations Conference on the Human Environment)

Report of the World Heritage Committee, WHC 1st sess, UN Doc CC-77/ CONF.001/9 (17 October 1977)

Report on the Expert Meeting on the 'Global Strategy' and Thematic Studies for a Representative World Heritage List, WHC 18th sess, UN Doc WHC-94/ CONF.003/INF.6 (13 October 1994)

Reports of the Advisory Bodies, WHC 39th sess, UN Doc WHC-15/39.COM/5B (15 May 2015)

UNESCO, *Operational Guidelines for the Implementation of the World Heritage Convention*, UN Doc CC-77/CONF.008 (30 July 1977)

UNESCO, *Operational Guidelines for the Implementation of the World Heritage Convention*, UN Doc WHC/2 (1978)

UNESCO, *Operational Guidelines for the Implementation of the World Heritage Convention*, UN Doc WHC/2/Revised (December 1988)

UNESCO, *Operational Guidelines for the Implementation of the World Heritage Convention*, UN Doc WHC/2/Revised (27 March 1992)

UNESCO, *Operational Guidelines for the Implementation of the World Heritage Convention*, UN Doc WHC.08/01 (January 2008)

UNESCO, *Operational Guidelines for the Implementation of the World Heritage Convention*, UN Doc WHC.15/01 (8 July 2015)

WHC Res CONF 003 X.10, WHC 18th sess, UN Doc WHC-94/CONF.003/16 (31 January 1995a) 41

WHC Res CONF 003 XIV.3, WHC 18th sess, UN Doc WHC-94/CONF.003/16 (31 January 1995b) 64

WHC Res CONF 203 VIII.A.1, WHC 22nd sess, UN Doc WHC-98/ CONF/203/18 (29 January 1999) 25

WHC Res 6 EXT.COM 5.1, 6th extraordinary WHC sess, UN Doc WHC-03/6 EXT.COM/8 (27 May 2003) 5

WHC Res 32 COM 8B.26, WHC 32nd sess, UN Doc WHC-08/32.COM/24Rev (31 March 2009a) 168

WHC Res 32 COM 8B.27, WHC 32nd sess, UN Doc WHC-08/32.COM/24Rev (31 March 2009b) 170

WHC Res 34 COM 8B.2, WHC 34th sess, UN Doc WHC-10/34.COM/20 (3 September 2010a) 165

WHC Res 34 COM 8B.20, WHC 34th sess, UN Doc WHC-10/34.COM/20 (3 September 2010b) 206

WHC Res 35 COM 8B.60, WHC 35th sess, UN Doc WHC-11/35.COM/20 (7 July 2011) 249

WHC Res 36 COM 8B.12, WHC 36th sess, UN Doc WHC-12/36.COM/19 (June–July 2012a) 165

WHC Res 36 COM 8E, WHC 36th sess, UN Doc WHC-12/36.COM/19 (June–July 2012b) 225

WHC Res 37 COM 8B.25, WHC 37th sess, UN Doc WHC-13/37.COM/20 (5 July 2013) 186

WHC Res 40 COM 8B.22, WHC 40th sess, UN Doc WHC/16/40.COM/19 (15 November 2016) 217

THESES

Gillespie, Josephine Suzanne, *Monumental Challenges: Local Perspectives on World Heritage Landscape Regulation at Angkor Archaeological Park, Cambodia* (PhD Thesis, The University of Sydney, 2010)

Titchen, Sarah M, *On the Construction of Outstanding Universal Value: UNESCO's World Heritage Convention (Convention Concerning the Protection of the World Cultural and Natural Heritage, 1972) and the Identification and Assessment of Cultural Places for Inclusion in the World Heritage List* (PhD Thesis, The Australian National University, 1995)

OTHER

Nara Document on Authenticity (1994)

CHAPTER 4

The Protection of Pacific Island Heritage Through the *World Heritage Convention* Regime

4.1 Introduction

The last chapter explored the concept of 'World Heritage' and concluded that the World Heritage Committee's broadened interpretation of the notion of 'outstanding universal value' (OUV) has increased the potential for Pacific sites to qualify for World Heritage listing. This chapter considers the protection of such places, by analysing the *World Heritage Convention*[1] text and the Committee's changing approach to heritage conservation. From this, key opportunities and challenges concerning the protection of World Heritage in the independent Pacific Island States are identified.

As will be explained, the *World Heritage Convention* regime is a product of its time. The obligations the *Convention* imposes on State parties and the international community, and the structural elements it creates, attempt to balance national sovereignty over heritage sites with the international community's interest in the preservation of such places. Also reflecting the era in which the law was drafted, the *Convention* does not mention the role of non-State actors in heritage conservation, other than the three international Advisory Bodies. Each of these features influence the scope for the *Convention* to be used to protect Pacific heritage (Sect. 4.2).

---

[1] *Convention Concerning the Protection of the World Cultural and Natural Heritage*, opened for signature 16 November 1972, 1037 UNTS 151 (entered into force 17 December 1975) ('*World Heritage Convention*').

© The Author(s) 2018   119
S. C. Price, *World Heritage Conservation in the Pacific*,
Palgrave Series in Asia and Pacific Studies,
https://doi.org/10.1007/978-981-13-0602-0_4

The *World Heritage Convention* regime is, however, evolving. This is possible because the *Convention* text establishes a framework only, allowing it to be implemented in accordance with contemporary views. This chapter thus considers the World Heritage Committee's changing approach to heritage conservation (Sect. 4.3). It focuses on three issues of particular relevance in the Pacific: the Committee's recognition of the relationship between the protection of World Heritage and sustainable development, its growing appreciation of the rights and roles of local communities in heritage protection, and its decision to allow sites protected through customary systems to be inscribed on the World Heritage List. This chapter shows that while these changes have made the *Convention* regime a better fit for the Pacific, significant challenges remain. Many of these arise from inherent features of the *Convention*, provisions of the *Operational Guidelines for the Implementation of the World Heritage Convention*,[2] and the nature of Pacific Island States. Key challenges identified here are explored further in the Solomon Islands context in Part III of this book.

4.2   THE PROTECTION REGIME ESTABLISHED BY THE *WORLD HERITAGE CONVENTION*

4.2.1   *Balancing National Sovereignty and the International Community's Interest in World Heritage Protection*

Before the *World Heritage Convention* was adopted, most States maintained that State sovereignty was paramount, and should only be 'pierced' in relation to the most important of issues, such as human rights.[3] Therefore, States tended to view heritage sites as being wholly subject to their sovereignty.[4] Developments such as the 1954 *Convention for the Protection of Cultural Property during Armed Conflicts*, which declared

---

[2] UNESCO, *Operational Guidelines for the Implementation of the World Heritage Convention*, UN Doc WHC.16/01 (26 October 2016) ('*Operational Guidelines 2016*').

[3] Craig Forrest, *International Law and the Protection of Cultural Heritage* (Routledge, 2011) 390; Francesco Francioni and Federico Lenzerini, 'The Future of the World Heritage Convention: Problems and Prospects' in Francesco Francioni (ed), *The 1972 World Heritage Convention: A Commentary* (Oxford University Press, 2008) 401, 404.

[4] Guido Carducci, 'Articles 4–7 National and International Protection of the Cultural and Natural Heritage' in Francesco Francioni (ed), *The 1972 World Heritage Convention: A Commentary* (Oxford University Press, 2008) 103, 115.

that certain properties form part of the 'cultural heritage of mankind',[5] reflected a growing view that the international community had an interest in heritage protection, notwithstanding State sovereignty.[6]

In the years leading up to the adoption of the *World Heritage Convention*, it also became increasingly evident that the international community could play a valuable role in heritage protection. This was highlighted by successful campaigns in the 1960s to rescue important heritage sites, led by the United Nations Educational, Scientific and Cultural Organisation (UNESCO).[7] The most notable campaign aimed to save the Abu Simbel temples from rising waters of the Nile, caused by the Egyptian government's construction of the Aswan Dam. In a demonstration of international commitment and cooperation, over 50 nations donated half of the $80 million required to relocate the temples.[8] Campaigns to save cultural objects in Venice and Florence from flooding were similarly successful,[9] making it clear to UNESCO that the *World Heritage Convention* should promote cooperative efforts to protect heritage. Furthermore, during this era, many States were achieving independence, and it was evident that they would need help to protect their heritage whilst also striving for economic development.[10]

Due to these views, the *Convention* regime was designed to encourage international cooperation for the protection of World Heritage, whilst not unduly intruding on State sovereignty.[11] This is reflected in Articles 4–7 of

[5] *Convention for the Protection of Cultural Property during Armed Conflict*, opened for signature 14 May 1954, 249 UNTS 240 (entered into force 7 August 1956) preamble para 2.
[6] Francioni and Lenzerini, above n 3, 404.
[7] See, for example, Edward J Goodwin, 'The World Heritage Convention, the Environment and Compliance' (2008–2009) 20 *Colorado Journal of International Environmental Law and Policy* 157, 158–159; Forrest, above n 3, 227.
[8] Allan Galis, 'UNESCO Documents and Procedure: The Need to Account for Political Conflict When Designating World Heritage Sites' (2009–2010) 38 *Georgia Journal of International and Comparative Law* 205, 208.
[9] Francesco Francioni, 'The Preamble' in Francesco Francioni (ed), *The 1972 World Heritage Convention: A Commentary* (Oxford University Press, 2008) 11.
[10] Forrest, above n 3, 229.
[11] Gionata P Buzzini and Luigi Condorelli, 'Article 11 List of World Heritage in Danger and Deletion of a Property from the World Heritage List' in Francesco Francioni (ed), *The 1972 World Heritage Convention: A Commentary* (Oxford University Press, 2008) 175, 179; Susan Shearing, 'Here Today, Gone Tomorrow? Climate Change and World Heritage' (2008) 12(2) *The Australasian Journal of Natural Resources Law and Policy* 161, 164; Ian Strasser, 'Putting Reform into Action: Thirty Years of the World Heritage Convention: How

the *Convention*, which set out the respective obligations of State parties and the international community in the protection of World Heritage.

Articles 4 and 5 contain the principal obligations of State parties regarding the protection of World Heritage. Article 4 states:

> Each State Party to this Convention recognizes that the duty of ensuring the identification, protection, conservation, presentation and transmission to future generations of the cultural and natural heritage referred to in Articles 1 and 2 and situated on its territory, belongs primarily to that State. It will do all it can to this end, to the utmost of its own resources and, where appropriate, with any international assistance and co-operation, in particular, financial, artistic, scientific and technical, which it may be able to obtain.

Article 5 (discussed further in Sect. 4.2.3) then lists some broad measures that a State party must take to comply with its Article 4 duties.

Although the *Convention* imposes the primary duty to protect World Heritage on State parties, it acknowledges that such sites have value for humankind as a whole, and that State action may be insufficient to effectively protect heritage.[12] Thus, pursuant to Articles 6 and 7, the international community also has obligations concerning World Heritage conservation. Article 6 states that the international community has a duty to cooperate for the protection of World Heritage, and as such, each State party undertakes to help others comply with their *Convention* duties, when requested to do so. Article 7 then says that 'international protection' of World Heritage means 'the establishment of a system of international cooperation and assistance designed to support State parties to the Convention in their efforts to conserve and identify that heritage'. Read together, these articles confirm that the international community's role is 'secondary and auxiliary',[13] designed to supplement not supplant the role of the State party. This is confirmed by the *Convention's* Preamble, which notes that although it is incumbent on the international community to participate in the protection of World Heritage, collective action shall not take the place of action by the State concerned.[14]

to Reform a Convention without Changing its Regulations' (2002) 11(2) *International Journal of Cultural Property* 215, 216–217.

[12] *World Heritage Convention* preamble paras 3, 6.

[13] Stefano Battini, 'The Procedural Side of Legal Globalisation: The Case of the World Heritage Convention' (2011) 9(2) *International Journal of Constitutional Law* 340, 351.

[14] *World Heritage Convention* preamble para 7. See also *World Heritage Convention* art 25; *Operational Guidelines 2016*, UN Doc WHC.16/01, para 233.

Through its delineation of the roles of State parties and the international community, the *Convention* seeks a 'delicate balance between national sovereignty and international intervention'.[15] This can also be seen in the structural elements established by the *Convention* (the World Heritage Committee, the World Heritage List, and the World Heritage Fund), which are discussed in Sect. 4.2.4. However, there remains a degree of tension between State sovereignty over heritage sites and the international community's interest in their preservation,[16] which has been a concern for some involved with implementing the *Convention* since it was first adopted.[17] It is particularly evident when the *Convention* bodies (i.e. the Committee and the Advisory Bodies) and the relevant State party hold different views about a site. As will be explored in Part III of this book, to some extent, this is the case in relation to East Rennell. This book argues that the gap between the positions of the *Convention* bodies and the Solomon Islands government (SIG) concerning the protection of East Rennell must be narrowed if all parties are to work cooperatively, as envisaged by the *Convention*.

4.2.2   The Role of Non-State Actors in the World Heritage Convention *Regime*

While the *World Heritage Convention* addresses the roles of State parties and the international community in the protection of World Heritage, it makes little reference to non-State actors, other than the three international Advisory Bodies.[18] These bodies hold significant positions within the regime, which is not surprising given they were involved with the development of the *Convention*.[19] Their role includes making recommen-

---

[15] Christina Cameron, 'The Strengths and Weaknesses of the World Heritage Convention' (1992) 28(3) *Nature and Resources* 18, 18.

[16] Natasha Affolder, 'Democratising or Demonising the World Heritage Convention?' (2007) 39 *University of Wellington Law Review* (2007) 341, 342.

[17] Francesco Francioni, 'The 1972 World Heritage Convention: An Introduction' in Francesco Francioni (ed), *The 1972 World Heritage Convention: A Commentary* (Oxford University Press, 2008) 3, 5–6.

[18] The three Advisory Bodies are the International Centre for the Study of the Preservation and Restoration of Cultural Property (ICCROM), the International Council of Monuments and Sites (ICOMOS), and the International Union for Conservation of Nature and Natural Resources (IUCN).

[19] Ana Filipa Vrdoljak, 'Article 14 The Secretariat and Support of the World Heritage Committee' in Francesco Francioni (ed), *The 1972 World Heritage Convention: A Commentary* (Oxford University Press, 2008) 243, 260.

dations to the Committee on World Heritage List nominations and applications for international assistance, and participating in Committee meetings, albeit in an advisory capacity.[20]

The lack of references to other non-State actors in the *Convention* reflects the approach to the protection of heritage that was most common in industrialised countries when the treaty was drafted. That approach (often referred to as 'fortress conservation') arose from the conservation movement of the late 1800s, and is characterised by centralised State ownership, control, and management.[21] It reflects the Judeo-Christian philosophy that humans are set apart from nature[22] and the belief that the purpose of conservation is to protect nature from people.[23] When the *Convention* was adopted, fortress conservation was widely accepted by governments and protected area managers as being appropriate for the preservation of wilderness areas. That approach did not take into account the fact that humans have impacted 'natural' areas for millennia, or the practical need for collaborative approaches to conservation efforts.[24] If regard was paid to local communities, it was generally only because they were viewed as a threat to the environment.[25]

The traditional model for the protection of cultural properties was similarly based on State control. When the *World Heritage Convention* was drafted, most places recognised as having cultural value were individual historic monuments or buildings, or other places under public ownership.[26] The goal of conservation efforts was often to prolong the life of the

---

[20] *World Heritage Convention* art 8(3); *Operational Guidelines 2016*, UN Doc WHC.16/01, para 31.

[21] Barbara Lausche, *Guidelines for Protected Area Legislation* (IUCN, 2011) 79.

[22] Phillipe Bourdeau, 'The Man-Nature Relationship and Environmental Ethics' (2004) 72 *Journal of Environmental Radioactivity* 9, 9. See also Grazia Borrini-Feyerabend, Ashish Kothari and Gonzalo Oviedo, *Indigenous and Local Communities and Protected Areas: Towards Equity and Enhanced Conservation*, Best Practice Protected Area Guidelines 11 (World Conservation Union, 2004) xiv.

[23] Jeremy Carew-Reid, 'Conservation and Protected Areas on South-Pacific Islands: The Importance of Tradition' (1990) 17(1) *Environmental Conservation* 29, 34.

[24] Adrian Phillips, 'Cultural Landscapes: IUCN's Changing Vision of Protected Areas' in *Cultural Landscapes: The Challenges of Conservation*, World Heritage Papers 7 (UNESCO, 2003) 40, 41.

[25] Adrian Phillips, 'Turning Ideas on their Head: The New Paradigm for Protected Areas' (2003) 20(2) *The George Wright Forum* 8, 14.

[26] UNESCO et al, *Managing Cultural World Heritage*, World Heritage Resource Manual (UNESCO, 2013) 12.

physical fabric of such structures.[27] Little attention was paid to the relationship between the structures and their surroundings, or the associations between the places and local communities.[28]

Reflecting these approaches, the *Convention* imposes responsibility for the protection of World Heritage on State parties, and contains little recognition of the role or interests of non-State actors operating at the regional, national, or local level.[29] This feature can be contrasted with later treaties, which recognise the involvement of a broader range of groups.[30] Importantly, unlike later treaties, the *World Heritage Convention* does not require or even encourage State parties to involve local communities in the identification of heritage places[31] or their protection.[32]

The impacts of the designation and protection of World Heritage sites on local communities received little attention for many years.[33] However, as will be explored later in this chapter, since the *Convention* was adopted, the international community's approach to heritage protection has changed. Although the *Convention* has not been amended to reflect these views, the Committee now encourages State parties to ensure that the rights and roles of local communities are respected in the identification and conservation of heritage places, which has helped make the *Convention* regime a better fit for Pacific Island States.

---

[27] Ibid., 24.

[28] Ibid., 12.

[29] An exception to this statement is *World Heritage Convention* Article 13(7), which requires the Committee to cooperate with international and national NGOs with similar objectives to the *Convention*. This article states that the Committee may call upon public and private bodies and individuals to assist with the implementation of its programmes.

[30] See, for example, *Convention on the Protection of the Underwater Cultural Heritage*, opened for signature 2 November 2001, 2562 UNTS 48 (entered into force 2 January 2009) preamble para 10 ('*Underwater Heritage Convention*'); *Convention on Biological Diversity*, opened for signature 5 June 1992, 1760 UNTS 79 (entered into force 29 December 1993) preamble paras 12–14 ('*Convention on Biological Diversity*'); *Convention for the Safeguarding of the Intangible Cultural Heritage*, opened for signature 17 October 2003, 2368 UNTS 3 (entered into force 20 April 2006) art 11(b) ('*Intangible Cultural Heritage Convention*') preamble para 6.

[31] Cf *Intangible Cultural Heritage Convention* art 11(b).

[32] Cf *Convention on Biological Diversity* art 8(j); *Intangible Cultural Heritage Convention* art 15.

[33] Josephine Gillespie, 'Legal Pluralism and World Heritage Management at Angkor, Cambodia' (2012) 14(1&2) *Asia Pacific Journal of Environmental Law* 1, 12.

4.2.3   State Parties' Duty to Protect World Heritage

4.2.3.1   The Duty to Protect, Conserve, Present, and Transmit World Heritage to Future Generations

Article 4 of the *World Heritage Convention* refers to State parties having a duty to identify, protect, conserve, present, and transmit World Heritage to future generations. However, the *Convention* does not define those terms. As such, State parties and the Committee are entitled to interpret them according to their ordinary meaning, in light of the purpose of the *Convention*.[34]

'Protection' is a term commonly used in international heritage laws, but it is not defined consistently or with precision in those laws.[35] While the Committee does not define the term in the *Operational Guidelines*, it does specify that protection must ensure the safeguarding of the site's OUV.[36] This is one of the reasons why the OUV criterion upon which a site is inscribed is critical. The criterion not only signifies its eligibility to be included in the World Heritage List, it also becomes the focus of the State's duty to protect the site (see Sect. 3.3.1).

Like the term 'protection', the word 'conservation' lacks any clear definition under international law, and the Committee does not define it in the *Operational Guidelines*. In the context of natural heritage, the International Union for the Conservation of Nature (IUCN) has defined 'conservation' as 'the in-situ maintenance of ecosystems and natural and semi-natural habitats and of viable populations of species in their natural surroundings'.[37] In the context of cultural places, it was defined in the *1994 Nara Document on Authenticity* as 'all efforts designed to under-

---

[34] *Convention on the Law of Treaties*, opened for signature 23 May 1969, 1155 UNTS 331 (entered into force 27 January 1980) art 31.

[35] See, for example, *Convention for the Protection of Cultural Property during Armed Conflict*, opened for signature 14 May 1954, 249 UNTS 240 (entered into force 7 August 1956), which says that the protection of cultural property shall comprise the safeguarding of and respect for such property (Article 2). See also the *Intangible Cultural Heritage Convention*, which defines 'safeguarding' to mean measures aimed at ensuring the viability of the intangible cultural heritage, including the identification, documentation, research, preservation, protection, promotion, enhancement, transmission, particularly through formal and non-formal education, as well as the revitalisation of the various aspects of such heritage (Article 2(3)).

[36] *Operational Guidelines 2016*, UN Doc WHC.16/01, para 96.

[37] Nigel Dudley (ed), *Guidelines for Applying Protected Area Management Categories* (IUCN, 2008) 9.

stand cultural heritage, know its history and meaning, ensure its material safeguard and, as required, its presentation, restoration and enhancement'.[38] There is therefore some overlap between the duties of 'protection' and 'conservation', both of which aim to ensure the preservation of the property. However, 'conservation' is arguably broader, potentially encompassing management, restoration, and enhancement of the place.[39]

The duty to transmit heritage to future generations also overlaps with the duties of protection and conservation. This duty is a manifestation of the principle of intergenerational equity,[40] which underlies the concept of World Heritage (see Sect. 3.2.2). It requires State parties to protect World Heritage from damage and destruction so that it can be enjoyed by future generations.

The final duty in Article 4, the duty of 'presentation', is also not defined in the *Operational Guidelines*. It has been interpreted by the Australian High Court to mean 'conserving and arranging [the heritage sites] to bring out their potentialities to the best advantage', which could involve the provision of lighting, access, or other amenities.[41] However, the protection of the property 'is not to be sacrificed by presentation',[42] and therefore arguably the duty to protect World Heritage prevails over the obligation to present it.

The Article 4 duties therefore have no clear definitions, and they overlap. The *Operational Guidelines* create further uncertainty in that some provisions refer to World Heritage 'protection' in isolation,[43] others refer to 'protection and management',[44] and others use various combinations of the Article 4 duties.[45] This inconsistent use of terminology

---

[38] *Nara Document on Authenticity* (1994) app 2. See Sect. 3.4 for discussion of the *Nara Document*.

[39] See, for example, Ben Boer and Graeme Wiffen, *Heritage Law in Australia* (Oxford University Press, 2006) 79–80.

[40] This principle says that 'the present generation should ensure that the health, diversity and productivity of the environment is maintained or enhanced for the benefit of future generations': *Intergovernmental Agreement on the Environment* (1992) s 3.5.2.

[41] *Commonwealth v Tasmania* (1983) 46 ALR 625, 775 (Brennan J).

[42] Ibid.

[43] See, for example, *Operational Guidelines 2016*, UN Doc WHC.16/01, paras 3(e), 12, 15(c), 15(f), 49, 98, 99, 101, 103.

[44] See, for example, ibid., paras 8, 78, 96–97.

[45] See, for example, ibid., paras 1(b), 6, 40 refer to protection and conservation; para 5 refers to identification, protection, conservation, and preservation; paras 7, 15(a) refer to identification, protection, conservation, presentation, and transmission to future genera-

may simply reflect a desire for brevity, as it would be unwieldy to specify 'protection, conservation, presentation and transmission to future generations' in each instance. It does however blur any distinction between the different obligations. In practice, the umbrella term of 'protection' is commonly used by the *Convention* bodies to encompass the obligations of State parties under Article 4, and that is the approach taken in this book. The lack of any precise definition of that term means there is some scope for the World Heritage Committee and State parties to interpret it in different ways.

4.2.3.2 The Duty to Identify World Heritage

In addition to the duty of protection, Article 4 refers to a State party having an obligation to identify the World Heritage within its territory.[46] Once identified, the State party must submit an inventory of such places (known as a Tentative List) to the World Heritage Committee.[47]

The duty to identify World Heritage is closely related to the duty to protect it. A site cannot be included in the World Heritage List unless it is first identified, documented, and nominated by the State party in which it is located. States are legally required to protect *all* places falling within the definitions of cultural heritage and natural heritage in Articles 1 and 2, respectively, whether or not they have been inscribed on the World Heritage List.[48] However, as a State cannot readily protect a place that it has not identified, in practice, the duty to protect is generally considered to be limited to listed sites.[49] This means that the identification of World Heritage is a crucial precursor to protection under the *Convention*.

tions; paras 15(d), 15(g) refer to protection, conservation, and presentation; paras 28(h), 40 refer to conservation and management; para 119 refers to protection, conservation, management, and presentation; para 60(c) refers to protection, safeguarding, and management.

[46] See also *World Heritage Convention* art 3. For analysis of the duty of identification, see generally Ben Boer, 'Article 3 Identification and Delineation of World Heritage Properties' in Francesco Francioni (ed), *The 1972 World Heritage Convention: A Commentary* (Oxford University Press, 2008) 85; Kishore Rao, 'A New Paradigm for the Identification, Nomination and Inscription of Properties on the World Heritage List' (2010) 16(3) *International Journal of Heritage Studies* 161.

[47] *World Heritage Convention* art 11(1).

[48] Carducci, above n 4, 109; Federico Lenzerini, 'Article 12 Protection of Properties Not Inscribed on the World Heritage List' in Francesco Francioni (ed), *The 1972 World Heritage Convention: A Commentary* (Oxford University Press, 2008) 201, 206; *Richardson v Forestry Commission* (1988) 77 ALR 237, 245 (Mason CJ and Brennan J).

[49] Francioni and Lenzerini, above n 3, 407.

As noted in Sect. 1.3, one of the causes of the under-representation of Pacific heritage on the World Heritage List is the lack of inventories detailing heritage places in the region. While most Pacific Island States have now submitted Tentative Lists to the Committee,[50] significant gaps in knowledge concerning the region's heritage remain. Given the link between the identification and protection of heritage places, efforts to conserve the region's heritage places must be accompanied by efforts to identify and document them.

4.2.3.3 The Duty to Take Active and Effective Measures to Protect World Heritage

Article 5 of the *World Heritage Convention* requires State parties to implement 'active and effective' measures to ensure the protection of World Heritage. Among other things, this provision requires a State party to integrate World Heritage protection into planning programmes,[51] to develop services[52] and research methods[53] for its protection, and to establish centres for training in the conservation of World Heritage.[54] Importantly, Article 5 also requires State parties to 'take the appropriate legal, scientific, technical, administrative and financial measures necessary' for the protection of World Heritage.[55] This is the basis of a State party's obligation to protect World Heritage under law.

Articles 4 and 5 give a State party discretion to determine what particular steps it will take to protect World Heritage. For example, while Article 5 requires a State party to take 'legal measures' to protect heritage places, it does not specify the form of legislation that a State must enact. Indeed, it does not require the State to enact new laws if they are not 'necessary'. This feature of the *Convention* allows a State party to determine how it will comply with its duties, and is consistent with the approach taken in other treaties signed during that era.[56] It also reflects the broad scope of

---

[50] Tentative Lists have been submitted by Federated States of Micronesia, Fiji, Kiribati, Marshall Islands, Palau, Papua New Guinea, Samoa, Solomon Islands, Tonga, and Vanuatu: *Pacific World Heritage Action Plan 2016–2020* (2016) 2.

[51] *World Heritage Convention* art 5(a).

[52] Ibid., art 5(b).

[53] Ibid., art 5(c).

[54] Ibid., art 5(e).

[55] Ibid., art 5(d).

[56] See, for example, the *Convention on Wetlands of International Importance especially as Waterfowl Habitat*, opened for signature 2 February 1971, 996 UNTS 245 (entered into force 21 December 1975). This treaty imposes broad obligations on State party only (Article 4).

the concept 'World Heritage', and the need for different actions to protect different types of sites. Some more recent treaties with a narrower scope are more prescriptive in terms of the measures they require State parties to undertake.[57]

As well as not prescribing any particular steps that a State party must take to protect World Heritage, both Articles 4 and 5 are couched in qualifying terms. Article 4 refers to a State party doing 'all it can' to protect heritage, 'to the utmost of its own resources'. Similarly, Article 5 says that a State party 'shall endeavour', 'in so far as possible' and 'as appropriate for each country', to take the specified measures. While these qualifications do not give States discretion as to whether to comply with the obligations,[58] they do allow States flexibility in the manner of compliance.[59] Factors that may affect their response include economic considerations,[60] the financial and administrative capacity of the State, its geographical size, the date it signed the *Convention*,[61] the volume and significance of its cultural and natural heritage, whether the State has existing duties to identify and protect heritage under national law,[62] political and cultural considerations, and the ownership of the heritage property.[63]

As such, while a top-down State-centric model of heritage protection was prevalent in the era when the *World Heritage Convention* was drafted (see Sect. 4.2.2), State parties are not legally obliged to take that approach. This is generally a positive feature of the *Convention* for Pacific Island States, as it allows them to adopt measures appropriate to their resource capacities, their plural legal systems, and the land tenure of their heritage places. However, the corollary is that the *Convention* itself provides little guidance to State parties on how to protect World Heritage.

The *Operational Guidelines* now contain some guidance on what the Committee considers to be the appropriate approach to World Heritage management.[64] In addition, manuals prepared by the Advisory Bodies

---

[57] See, for example, *Convention on Biological Diversity* art 8; *Underwater Heritage Convention* arts 10, 12.

[58] *Commonwealth v Tasmania* (1983) 46 ALR 625, 698 (Mason J).

[59] *Richardson v Forestry Commission* (1988) 77 ALR 237, 245 (Mason CJ and Brennan J).

[60] Ibid., 242 (Mason CJ and Brennan J).

[61] Forrest, above n 3, 243.

[62] Carducci, above n 4, 113–114.

[63] Boer and Wiffen, above n 38, 72.

[64] *Operational Guidelines 2016*, UN Doc WHC.16/01, part II.F.

and others aim to assist States to develop and implement management systems for World Heritage sites (and other important heritage places) and provide some case study examples.[65] However, the manuals are high level, and detailed guidelines concerning what constitutes an 'appropriate legal measure' for the purposes of Article 5 of the *Convention* remain lacking.[66]

None of the Pacific Island States currently have specific World Heritage legislation.[67] Legislative protection of World Heritage in the Pacific is therefore often limited and piecemeal. As explored further in later chapters, Pacific Island States should be supported to develop and implement legislation that is appropriate for the nature of their heritage sites and legal systems, particularly for sites that are under customary tenure (see in particular Sect. 8.4.3).

4.2.4 The Structural Elements of the World Heritage Convention Regime

4.2.4.1 The World Heritage Committee

The World Heritage Committee, an executive decision-making body established under Article 8 of the *Convention*, effectively represents the common interest of State parties in the preservation of World Heritage.[68] It plays a central role in the *Convention* regime through its administration of the World Heritage List and the World Heritage Fund. The fact that all substantive decision-making powers are given to the Committee, as opposed to the General Assembly of State parties, is a distinguishing fea-

---

[65] See, for example, IUCN, *Management Planning for Natural World Heritage Properties: A Resource Manual for Practitioners* (IUCN, 2008); Marc Hockings et al, *Enhancing Our Heritage Toolkit: Assessing Management Effectiveness of Natural World Heritage Sites*, World Heritage Papers 23 (IUCN, 2008); UNESCO et al, above n 26; UNESCO et al, *Managing Natural World Heritage*, World Heritage Resource Manual (UNESCO, 2012); Thomas Lee and Julie Middleton, *Guidelines for Management Planning of Protected Areas* (IUCN, 2003); Grazia Borrini-Feyerabend et al, *Governance of Protected Areas: From Understanding to Action*, Best Practice Protected Area Guidelines Series 20 (IUCN, 2013).

[66] UNESCO et al, *Managing Natural World Heritage*, above n 65, 35.

[67] A *Heritage Act* has been proposed for Fiji, which would provide a framework for the identification, nomination, and management of World Heritage sites: see *Heritage Bill 2016* (no. 10 of 2016) (Fiji).

[68] Tullio Scovazzi, 'Articles 8–11 World Heritage Committee and World Heritage List' in Francesco Francioni (ed), *The 1972 World Heritage Convention: A Commentary* (Oxford University Press, 2008) 147, 149.

ture of the *Convention*.[69] It means that the composition of the Committee can significantly influence the operation of the regime.

The Committee comprises 21 State parties, elected by the General Assembly of State parties to the *Convention*.[70] Although the *Convention* requires that elections to the Committee ensure an 'equitable representation of the different regions and cultures of the world',[71] to date Pacific Islanders have not been well represented. As the main decision-making body in relation to World Heritage, many States seek membership of the Committee.[72] New Zealand was a member between 2003 and 2007,[73] and Australia has served several terms,[74] but no Pacific Island State has ever been a member.[75]

One reason for this is that the Pacific Island States only became signatories to the *Convention* relatively recently. Furthermore, it is debatable whether any such State has sufficient human and financial resources to serve effectively on the Committee.[76] The implications of the lack of Pacific representation must however be recognised. It may explain why for many years the Committee interpreted 'cultural heritage' in a manner that effectively excluded places of most significance to Pacific Islanders (see Sect. 3.4). It may also explain why the Committee traditionally favoured State-centric approaches to heritage protection, which are often inappropriate in the Pacific (discussed further in Sect. 4.3).

It is not suggested here that the Committee has deliberately sought to exclude the Pacific from the *Convention* regime, but simply that its decision-making has been influenced by the perceptions and values of the mainly industrialised States that dominated its membership. In this regard,

---

[69] Bruno S Frey and Lasse Steiner, 'World Heritage List: Does it Make Sense?' (2011) 17(5) *International Journal of Cultural Policy* 555, 557.

[70] *World Heritage Convention* art 8(1).

[71] Ibid., art 8(2).

[72] Lynn Meskell, Claudia Liuzza and Nicholas Brown, 'World Heritage Regionalism: UNESCO from Europe to Asia' (2015) 22 *International Journal of Cultural Property* 437, 451.

[73] UNESCO, *New Zealand* http://whc.unesco.org/en/statesparties/nz.

[74] UNESCO, *Australia* http://whc.unesco.org/en/statesparties/au. Australia was a member of the Committee in 1976–1983, 1983–1989, 1995–2001 and 2007–2011.

[75] Vanuatu and Palau have applied for membership, but their bids were unsuccessful.

[76] Bertacchini and Saccone have found that developed countries have greater capacity to gain membership to the World Heritage Committee than developing countries: see Enroci E Bertacchini and Donatella Saccone, 'Toward a Political Economy of World Heritage' (2012) 36(4) *Journal of Cultural Economics* 327, 334.

it is notable that the Committee adopted 'enhancing the role of communities' as one of its strategic objectives while Tumu Te Heuheu, paramount chief of Ngati Tuwharetoa (Aotearoa/New Zealand), was its chair. The recognition of this objective was significant for the Pacific, and demonstrates the impact a Pacific voice within the Committee can have on its approach to World Heritage protection.

The Committee's adoption of the *Global Strategy for a Representative, Balanced and Credible World Heritage List* brought to the fore the need for the *Convention* regime to adapt to better fit the Pacific context. In addition, developments such as the *Pacific 2009 World Heritage Programme* and the *Pacific Appeal* have helped highlight the views of Pacific Islanders to the Committee.[77] Research on cultural heritage in the Pacific in the last two decades has also contributed to the Committee's changing approach to World Heritage and its protection.[78] To ensure that the regime continues to evolve to meet the views and aspirations of Pacific Islanders, efforts to inform the Committee of the Pacific perspective must continue, even if no Pacific Island State becomes a formal member (see Sect. 8.2.1).

4.2.4.2 The World Heritage List

The World Heritage List is the most well-known component of the *Convention* regime. It is a list of sites that the World Heritage Committee has found meet the definitions of cultural heritage and natural heritage in Articles 1 and 2 of the *Convention*, respectively, and has decided to include in the List on that basis.[79] The Committee is responsible for defining the criteria by which sites may be inscribed on the World Heritage List.[80] Despite the legal scope of Article 4, in practice, only sites on the World Heritage List are generally considered to be subject to the State parties' duty to protect.[81] Thus, the Committee's decisions concerning inscriptions on the World Heritage List to a large extent delineate the scope of the regime.

---

[77] The *Global Strategy*, the *Pacific 2009 World Heritage Programme* and the *Pacific Appeal* are discussed in Sects. 1.3 and 1.5.

[78] Christian Reepmeyer et al, 'Selecting Cultural Sites for the UNESCO World Heritage List: Recent Work in the Rock Islands – Southern Lagoon Area, Republic of Palau' in Jolie Liston, Geoffrey Clark and Dwight Alexander (eds), *Pacific Island Heritage: Archaeology, Identity and Community* (ANU E Press, 2011) 85, 86.

[79] *World Heritage Convention* art 11(2).

[80] Ibid., art 11(5).

[81] Francioni and Lenzerini, above n 3, 407.

State parties also play an important role in the listing process. The Committee can only inscribe a site on the World Heritage List if it has been nominated by the State party within whose territory the site is located.[82] As such, the consent of that State party is required for the site to be brought within the scope of the *Convention* regime. This requirement is an example of the delicate balance between respect for national sovereignty and the international community's interest in World Heritage protection that the *Convention* is trying to achieve.

Importantly, no other group or individual who may have an interest in the preservation of a heritage site (including customary landowners) can nominate the site for World Heritage listing. Thus, while the conservation of Pacific heritage is often highly dependent on local action, the *Convention* regime can only be used as the framework for the protection of such sites with the consent and involvement of the State party, at least at the nomination stage. To date the rate of nomination of Pacific sites has been relatively low, and there is a continuing need to address the challenges that Pacific States face in the nomination of sites (see Sect. 1.3).

4.2.4.3 *The World Heritage Fund*

In addition to determining nominations for inclusion in the World Heritage List, the Committee assesses applications by State parties for international assistance.[83] As the primary responsibility for the protection of World Heritage rests with the State party in which the heritage is located,[84] a State is not automatically entitled to receive any assistance. Rather, it must first submit a request to the Committee, which will determine the request based on the criteria in the *Operational Guidelines*.[85] International assistance is primarily funded through the World Heritage Fund, which comprises voluntary and compulsory contributions from State parties and money from other sources.[86]

Pacific Island States fall within several of the Committee's priority areas for international assistance.[87] These include requests from Least Developed

---

[82] *World Heritage Convention* art 11(3).

[83] Ibid., art 13(3).

[84] Ibid., art 4.

[85] Ibid., art 13(3); *Operational Guidelines 2016*, UN Doc WHC.16/01, annex 9.

[86] *World Heritage Convention* chapter IV. See generally, Jehanne Phares and Cynthia Guttman, *Investing in World Heritage: Past Achievements, Future Ambitions – A Guide to International Assistance*, World Heritage Papers 2 (UNESCO, 2002).

[87] *World Heritage Convention* art 13(4); *Operational Guidelines 2016*, UN Doc WHC.16/01, paras 236–239.

Countries and Small Island Developing States[88] and requests that support the Committee's strategic objectives,[89] including the *Global Strategy.* However, the annual budget of the World Heritage Fund is very small considering there are over 1000 listed World Heritage sites.[90] Therefore, although international cooperation is a principle underpinning the *Convention* regime, the Committee's capacity to finance heritage conservation in the Pacific is limited.

Another limitation of the international assistance system is that only State parties can apply for assistance.[91] This is logical, given that the *Convention* imposes the duty to protect World Heritage on State parties. However, in practice, it means that groups such as customary landowners, non-government organisations (NGOs), and provincial governments, which may be directly involved with the conservation of a World Heritage site, are at the behest of the relevant State party to access assistance through the *Convention* regime. If the State party does not apply for assistance, the Committee cannot mobilise funds from the World Heritage Fund to help such groups conserve the site.

The case of Solomon Islands shows that State parties do not always apply for international assistance, despite the critical need for it. To date, Solomon Islands has applied for (and thus received) relatively little funding through the *Convention* regime,[92] which has been a point of frustration for some Committee members. For example, at the 2015 World Heritage Committee annual meeting in Bonn, Germany, the Turkish delegate on the Committee stated:

> Despite all the offers of money and technical help there is no response from the [Solomon Islands] State party. We are wondering why the State party is not cooperating? Some countries need assistance that they can't get. This country gets all the assistance, but do not try and receive it.

---

[88] *Operational Guidelines 2016*, UN Doc WHC.16/01, para 239(b).

[89] Ibid., paras 238, 239(e).

[90] UNESCO, *World Heritage Fund* http://whc.unesco.org/en/world-heritage-fund/.

[91] *World Heritage Convention* art 13(1); *Operational Guidelines 2016*, UN Doc WHC.16/01, para 233. An exception to this is that the Committee can provide international assistance to national or regional centres for the training of staff in heritage identification and protection: *World Heritage Convention* art 23.

[92] See UNESCO, *Solomon Islands: International Assistance* http://whc.unesco.org/en/statesparties/sb/assistance/.

The SIG may have submitted few requests for assistance because it lacks the resources and/or capacity to conduct the detailed scientific, economic, and technical studies that must precede an application,[93] or to administer the assistance once it is received. Moreover, while Article 22 of the *Convention* allows international assistance to be granted for a broad range of activities (including research, the provision of experts and labour, the training of staff, and the supply of equipment),[94] it does not necessarily extend to all initiatives that the State prioritises. For example, it does not allow for the funding of alternative livelihood projects, which are arguably necessary to ensure the long-term protection of East Rennell (see Sect. 5.3.3.3). This may be one of the reasons why Pacific Island States continue to call for the establishment of a permanent Pacific World Heritage Fund.[95]

Furthermore, as discussed further in Part III of this book, the protection of East Rennell is interrelated with a range of economic, social, and environmental issues. While one-off grants from the World Heritage Fund for specific projects may be of some benefit, addressing the full range of issues that threaten East Rennell is likely to require a larger and longer-term investment than the Committee can currently provide. As such, the SIG and others involved with the protection of East Rennell will require assistance from donors and organisations outside the *Convention* regime to safeguard the site's OUV.

4.3   THE WORLD HERITAGE COMMITTEE'S APPROACH TO THE PROTECTION OF WORLD HERITAGE

The World Heritage Committee's views on how heritage sites should be protected are significant because it has substantial decision-making powers under the *Convention*, including the power to inscribe sites on the World Heritage List and to administer the World Heritage Fund. In addition, the Committee can influence the implementation of the *Convention* by State parties, through the *Operational Guidelines* and its resolutions. Three developments in the Committee's approach to World Heritage protection of particular importance to the Pacific are focused on here. These concern

---

[93] *World Heritage Convention* arts 21(1), 24; *Operational Guidelines 2016*, UN Doc WHC.16/01, para 243, annex 8.

[94] *World Heritage Convention* art 22; *Operational Guidelines 2016*, UN Doc WHC.16/01, para 241.

[95] See, for example, *Pacific World Heritage Action Plan 2016–2020* (2016) para 24.

the relationship between heritage protection and sustainable development, the rights and roles of local communities in heritage conservation, and customary protection of World Heritage sites. As explained below, the Committee's contemporary approach is more appropriate for the Pacific than earlier top-down methods, but significant challenges remain.

4.3.1 The Relationship Between Sustainable Development and the Protection of Heritage

When the *World Heritage Convention* was adopted, wilderness areas and cultural properties in industrialised societies were most commonly protected through top-down approaches that sought to preserve the sites as 'islands' isolated from the impacts of human activities. While that approach is still used effectively in many places today,[96] since the *Convention* was signed, a new 'conservation paradigm' for heritage protection has emerged.[97] Under this new paradigm, efforts to conserve the natural environment include a wider range of actors, are approached at a broader scale, and are pursued alongside social and economic objectives.[98] Similarly, it is now widely recognised that cultural properties cannot be protected as museum pieces, separated from local communities and the broader economic and social changes occurring around them, so a more holistic, integrated approach to their preservation is required.[99]

The emergence of this new conservation paradigm was triggered in part by the growing recognition of the need for 'sustainable development', often defined as development that 'meets the needs of the present without compromising the ability of future generations to meet their own needs'.[100] This concept became widely accepted at the international level in the 1980s, through the publication of documents such as the *World*

---

[96] Lausche, above n 21, 76.

[97] UNESCO et al, above n 26, 12–15; Gonzalo Oviedo and Tatjana Puschkarsky, 'World Heritage and Rights-Based Approaches to Nature Conservation' (2012) 18(3) *International Journal of Heritage Studies* 285, 287; Phillips, above n 25, 19–20; Borrini-Feyerabend, Kothari and Oviedo, above n 22, 1.

[98] Phillips, above n 25, 19–20; Lausche, above n 21, 142; Borrini-Feyerabend, Kothari and Oviedo, above n 22, 2.

[99] UNESCO et al, above n 26, 13, 15.

[100] World Commission on Environment and Development, *Our Common Future*, UN Doc A/42/427 (1987) annex cl 27 (frequently referred to as the *Brundtland Report* after Gro Harlem Brundtland, Chairman of the Commission).

Conservation Strategy[101] and the *Brundtland Report*,[102] which explored the relationship between development and environmental protection. The signature of the *Rio Declaration*,[103] *Agenda 21*,[104] and the *Convention on Biological Diversity*[105] at the 1992 United Nations Conference on Environment and Development (UNCED) firmly embedded the concept under international law. Achieving sustainable development remains a pillar of international policy, as evidenced by the United Nations General Assembly's adoption of *Transforming Our World: the 2030 Agenda for Sustainable Development in 2015*.[106] That document arguably does not sufficiently acknowledge the contribution of heritage protection to the achievement of sustainable development. It does however note the need to 'strengthen efforts to protect and safeguard the world's cultural and natural heritage'.[107]

As the concept of sustainable development only became widely used in the 1980s, it is not referred to in the *World Heritage Convention*. The *Convention* does however reflect some of its principles. For example, Article 4 requires State parties to ensure the transmission of heritage to future generations, in accordance with the notion of intergenerational equity that lies at the heart of sustainable development. In addition, Article 5 requires State parties to adopt a general policy which aims to give World Heritage a function in the life of the community and to integrate the protection of that heritage into comprehensive planning programmes. This provision therefore supports holistic approaches to conservation, consistent with sustainable development.

Despite these linkages, it took many years for the Committee to enshrine the principles of sustainable development in its *Operational Guidelines*. A milestone in this process was the Committee's adoption of

---

[101] IUCN et al, *World Conservation Strategy* (1980).

[102] World Commission on Environment and Development, above n 100.

[103] *Rio Declaration on Environment and Development*, Report of the United Nations Conference on Environment and Development, UN Doc A/CONF.151/6/Rev.1 (1992) ('*Rio Declaration*').

[104] *Agenda 21, Report of the UNCED*, I, UN Doc. A/CONF.151/26/Rev.1 (1992) ('*Agenda 21*').

[105] *Convention on Biological Diversity*, opened for signature 5 June 1992, 1760 UNTS 79 (entered into force 29 December 1993) ('*Convention on Biological Diversity*').

[106] *Transforming Our World: The 2030 Agenda for Sustainable Development*, UNGA Res A/RES/70/L.1, UN GAOR, 70th sess, UN Doc A/RES/70/1 (21 October 2015) ('*Transforming Our World*').

[107] Ibid., 22.

the *Budapest Declaration* in 2002.[108] That document recognises the need to 'ensure an appropriate and equitable balance between conservation, sustainability and development, so that World Heritage properties can be protected through appropriate activities contributing to the social and economic development and the quality of life of our communities'.[109] More recently, the General Assembly of State parties adopted the *World Heritage Sustainable Development Policy*,[110] following the endorsement of a similar document by the Committee.[111] The adoption of that policy was a clear acknowledgement of the need for heritage conservation objectives to be promoted 'within a broader range of economic, social and environmental values and needs encompassed in the sustainable development concept'.[112] The policy contains provisions reflecting the various dimensions of sustainable development, namely environmental sustainability, inclusive social and economic development, and fostering peace and security.[113]

These developments were particularly significant for Pacific Island States. Pacific heritage often comprises landscapes and seascapes of continuing cultural significance to the areas' inhabitants and owners (see Sect. 2.2.2). For example, Chief Roi Mata's Domain in Vanuatu is a cultural landscape representing the continuing Pacific chiefly system and respect for customary authority.[114] In addition, many Pacific Islanders live subsistence lifestyles, and are highly dependent on their natural resources for their livelihoods. For example, the Rock Islands Southern Lagoon site in Palau is utilised by Palauans for subsistence harvesting of fish and fruit bats.[115] Similarly, the customary owners of East Rennell in Solomon

---

[108] *Budapest Declaration on World Heritage*, WHC Res 26 COM 9, WHC 26th sess, UN Doc WHC-02/CONF.202/25 (1 August 2002) 6 ('*Budapest Declaration*').

[109] Ibid., para 3(c).

[110] *Policy for the Integration of a Sustainable Development Perspective into the Processes of the World Heritage Convention*, WHC GA Res 20 GA 13, 20th sess, UN Doc WHC-15/20. GA/15 (20 November 2015) 7 ('*World Heritage Sustainable Development Policy*').

[111] WHC Res 39 COM 5D, WHC 39th sess, UN Doc WHC-15/39.COM/19 (8 July 2015) 7; *World Heritage and Sustainable Development*, WHC 39th sess, UN Doc WHC-15/39.COM/5D (15 May 2015) annex.

[112] *World Heritage Sustainable Development Policy*, UN Doc WHC-15/20.GA/15, para 2.

[113] Ibid., paras 13–33.

[114] Republic of Vanuatu, *Chief Roi Mata's Domain – Nomination by the Republic of Vanuatu for Inscription on the World Heritage List* (2007) 56.

[115] Republic of Palau, *The Rock Islands Southern Lagoon Nomination for Inscription on the World Heritage List* (2012) 22–23.

Islands live predominantly subsistence lifestyles, relying on resources from the forests, lake, and marine areas within the World Heritage site.[116] An approach to heritage protection that involves the exclusion of all human activity from the heritage place and/or which does not recognise the cultural values associated with the natural environment is unlikely to be appropriate in the Pacific. Consequently, the Committee's efforts to integrate World Heritage protection into the broader framework of sustainable development could make the *Convention* regime a better fit for the Pacific. As explained below however, more could be done to ensure that the Committee's change in approach has practical impact.

The *Operational Guidelines* now refer to sustainable development, but they do not fully reflect the *Budapest Declaration* or the *World Heritage Sustainable Development Policy*. For example, the *Operational Guidelines* state that the protection of World Heritage is a significant contributor to sustainable development,[117] and its principles should be integrated into heritage management systems.[118] In addition, they acknowledge that World Heritage properties may support a variety of uses that are ecologically and culturally sustainable and which may contribute to the quality of life of local communities.[119] However, they do not refer to the need for State parties to seek an equitable balance between conservation, sustainability, and development, as stated in the *Budapest Declaration* and the *Sustainable Development Policy*.[120] They also do not refer to the need to protect and promote environmental, social, economic, and cultural rights in the implementation of the *Convention*.[121] The *Operational Guidelines* should be reviewed to identify the amendments needed to fully embed the principles of sustainable development in the *Convention* regime. Indeed, following the adoption of the *Sustainable Development Policy* in 2015, the World Heritage Centre and the Advisory Bodies foreshadowed that such changes may be required.[122]

[116] Elspeth J Wingham, *Nomination of East Rennell, Solomon Islands by the Government of Solomon Islands for Inclusion in the World Heritage List Natural Sites* (1997) 27.

[117] *Operational Guidelines 2016*, UN Doc WHC.16/01, para 6.

[118] Ibid., para 132(5).

[119] Ibid., para 119.

[120] *Budapest Declaration*, UN Doc WHC-02/CONF.202/25, para 3(c); *World Heritage Sustainable Development Policy*, UN Doc WHC-15/20.GA/15, paras 1, 9.

[121] *World Heritage Sustainable Development Policy*, UN Doc WHC-15/20.GA/15, para 7.

[122] *World Heritage and Sustainable Development*, WHC 39th sess, UN Doc WHC-15/39. COM/5D (15 May 2015) para 9.

In addition, the Committee needs to ensure that its resolutions concerning World Heritage sites reflect the *Sustainable Development Policy*. For example, the Committee has repeatedly requested that Solomon Islands address the threats to East Rennell by banning logging and mining on the island, regulating the taking of species, developing a new management plan, and implementing heritage protection legislation. Until recently, there has been little acknowledgement in its resolutions of the critical role of local people in decision-making concerning World Heritage protection or their right to economic and social development.[123] This may have contributed to the SIG's failure to comply with those resolutions. Cooperation between the Committee and the SIG may improve if future Committee resolutions more closely reflect the principles of sustainable development (discussed further in Sect. 8.4.2).

4.3.2　The Rights and Roles of Local Communities in Heritage Protection

Since the *World Heritage Convention* was adopted, it has become increasingly accepted that a broad range of actors can contribute to heritage protection, including local communities.[124] The emergence of the notion of sustainable development, as well as increasing recognition of the rights of Indigenous peoples,[125] has contributed to this change.

The near-universal acceptance of the concept of 'sustainable development' has highlighted the need for more holistic approaches to heritage protection, and drawn attention to the need for effective systems of governance, involving participatory and multi-stakeholder approaches.[126] This

---

[123] Unlike earlier decisions, the Committee's 2016 and 2017 decisions concerning East Rennell acknowledge the need to support sustainable livelihood development for the East Rennellese people: see WHC Res 40 COM 7A.49, 40th sess, UN Doc WHC/16/40. COM/19 (15 November 2016) 68; WHC Res 41 COM 7A.19, 41st sess, UN Doc WHC/17/41.COM/18 (12 July 2017) 35.

[124] It is recognised that in some contexts there are critical differences between Indigenous people and local communities. However, for convenience, in this book, the term 'local communities' is used broadly to encompass Indigenous people, unless the context dictates otherwise.

[125] There is no agreed definition of Indigenous people under international law. For discussion, see, for example, Benedict Kingsbury, "Indigenous Peoples' in International Law: A Constructivist Approach to the Asian Controversy' (1998) 92 *American Journal of International Law* 414.

[126] UNESCO et al, above n 26, 13.

is based on increasing recognition that local people's 'knowledge, perceptions, and cosmologies' are important to managing heritage places,[127] as
well as ethical and moral concerns.

The important role of local people in achieving sustainable development is reflected in several documents adopted at the UNCED conference. For example, the *Rio Declaration* recognised the vital role of
Indigenous people and local communities in environmental management
because of their knowledge and traditional practices[128]; the *Convention on
Biological Diversity* called on States to maintain the knowledge, innovations, and practices of Indigenous and local communities relevant for the
conservation of biodiversity[129]; and *Agenda 21* devoted a chapter to
exploring mechanisms for strengthening the role of Indigenous people
and their communities.[130] The concept of sustainable development therefore clearly supports more decentralised approaches to heritage protection
than existed under the traditional State-centric model.

The role of Indigenous peoples in heritage protection has gained particular attention, reflecting increasing international acknowledgement of
their rights. This is demonstrated by the establishment of international
bodies such as the United Nations' Working Group on Indigenous
Populations[131] and the Permanent Forum on Indigenous Issues.[132] It is
also evident from the adoption of international instruments, including *the
International Labour Organisation's Indigenous and Tribal Peoples*

---

[127] Matthew Lauer and Shankar Aswani, 'Indigenous Ecological Knowledge as Situated
Practices: Understanding Fishers' Knowledge in the Western Solomon Islands (2009)
111(3) *American Anthropologist* 317, 317.

[128] *Rio Declaration*, UN Doc A/CONF.151/6/Rev.1, principle 22.

[129] *Convention on Biological Diversity* art 8(j).

[130] *Agenda 21*, UN Doc. A/CONF.151/26/Rev.1, ch 26.

[131] The Working Group on Indigenous Populations was established in 1982 as a subsidiary
organ to the Sub-Commission on the Promotion and Protection of Human Rights: see
United Nations Office of the High Commissioner (UN-OHC), *Mandate of the Working
Group on Indigenous Populations* http://www.ohchr.org/EN/Issues/IPeoples/Pages/
MandateWGIP.aspx. The Working Group has been discontinued and replaced by the Expert
Mechanism on the Rights of Indigenous Peoples in 1997: see United Nations Office of the
High Commissioner (UN-OHC), *Expert Mechanism on the Rights of Indigenous Peoples*
http://www.ohchr.org/EN/Issues/IPeoples/EMRIP/Pages/EMRIPIndex.aspx.

[132] The Permanent Forum on Indigenous Issues was established in 2000 and is an advisory
body to the United Nations' Economic and Social Council: see United Nations Division for
Social Policy and Development, *Permanent Forum* https://www.un.org/development/
desa/indigenouspeoples/unpfii-sessions-2.html.

Convention 1989 (ILO 169)[133] and the *United Nations Declaration on the Rights of Indigenous People (UNDRIP)*.[134]

ILO 169 is the only binding international law concerning the rights of Indigenous peoples. Among other things, it confirms their right to have their cultural values and practices protected,[135] to participate in the formulation of development plans that may affect them,[136] and to the lands traditionally occupied by them.[137] While it has limited direct application to the Pacific Island States,[138] its adoption was a significant milestone in the growing appreciation of the rights of Indigenous people at the international level. This was solidified by the United Nations General Assembly's adoption of *UNDRIP* in 2007. Although *UNDRIP* is not binding, many of its provisions reflect principles of customary international law and principles enshrined in human rights instruments.[139] Thus, it is an emerging standard for the treatment of Indigenous people.

Top-down conservation measures involving stringent restrictions on Indigenous peoples' access to and use of their lands, and measures developed without the full involvement of the affected peoples may be inconsistent with the provisions of *UNDRIP*. In contrast, more localised approaches to conservation find support in *UNDRIP* as expressions of

[133] *Convention (No. 169) Concerning Indigenous and Tribal Peoples in Independent Countries*, opened for signature 27 June 1989, 1650 UNTS 383 (entered into force 5 September 1991) ('*ILO 169*').

[134] *United Nations Declaration on the Rights of Indigenous Peoples*, GA Res 61/295, UN GAOR, 61st sess, 107th plen mtg, Supp No 49, UN Doc A/RES/61/295 (13 September 2007) ('*UNDRIP*').

[135] *ILO 169* art 5.

[136] Ibid., art 7(1).

[137] Ibid., art 16.

[138] Of the independent Pacific Island States, only Fiji has signed it. Furthermore, under Article 1, the Convention defines Indigenous peoples to include (a) tribal peoples in independent countries whose social cultural and economic conditions distinguish them from other sections of the national community, and (b) descendants of people who inhabited the area at the time of colonisation, who retain some or all of their own social, economic, cultural and political institutions (*ILO 169* art 1). In most Pacific Island States, Indigenous peoples comprise the majority of the population and government. Therefore, it is arguably not aimed at Indigenous populations in such States: see Erika Techera, 'Samoa: Law, Custom and Conservation' (2006) 10 *New Zealand Journal of Environmental Law* 361, 367.

[139] Beatriz Barreiro Carril, 'Indigenous Peoples' Participation in Decision-Making in the Context of World Heritage Sites: How International Human Rights Law Can Help?' (2016) 7(2–3) *The Historic Environment: Policy and Practice* 224, 227; Erika Techera, *Strengthening International Law to Address the Needs of Legally Pluralist Nations*, Macquarie Law Working Paper 2010–02 (Macquarie University, 2010) 16.

Indigenous peoples' self-governance, decision-making, and autonomy,[140] which are rights guaranteed by the declaration.[141] Such approaches may also be a means for Indigenous people to maintain their cultures, livelihoods, and identities.[142] As such, they may be consistent with other rights guaranteed by *UNDRIP*, including the right of Indigenous people to maintain their spiritual relationship with the land,[143] their right to practice their customs and traditions,[144] and their right to the land they traditionally owned and occupied.[145] *UNDRIP* also guarantees procedural rights, including the right of Indigenous people to participate in decision-making that affects them,[146] which supports their full involvement in efforts to conserve their lands.

In the past, World Heritage was often something that was imposed on local populations,[147] and the impacts of World Heritage listing on communities received little attention.[148] However, as instances where the rights of local communities have been abused in the implementation of the *Convention* became better known,[149] some scholars and practitioners

---

[140] Stan Stevens, 'Implementing the UN Declaration on the Rights of Indigenous Peoples and International Human Rights Law through the Recognition of ICCAs' (2010) 17 *Policy Matters* 181, 186.

[141] *UNDRIP*, UN Doc A/RES/61/295, art 4.

[142] Stevens, above n 140, 186.

[143] *UNDRIP*, UN Doc A/RES/61/295, art 25.

[144] Ibid., arts 11(2), 14, 34.

[145] Ibid., arts 26, 32.

[146] Ibid., arts 9, 10, 11(2), 18, 19, 25, 27, 32.

[147] Naomi Deegan, 'The Local-Global Nexus in the Politics of World Heritage: Space for Community Development?' in Marie-Theres Albert, Marielle Richon, Marie José Viñals and Andrea Witcomb (eds), *Community Development through World Heritage*, World Heritage Papers 31 (UNESCO, 2012) 77, 80.

[148] Gillespie, above n 33, 12.

[149] For example, the Kenya Lake System in the Great Rift Valley was listed with little effective consultation with the area's traditional owners, the Endorois people. Many of these traditional owners had been previously relocated from the area to create a wildlife reserve and tourist facilities. The African Commission on Human and Peoples Rights found that the listing violated the Endorois peoples' right to development. For discussion, see Peter Bille Larsen, *World Heritage and Evaluation Processes Related to Communities and Rights: An Independent Review* (IUCN, 2012) 19–20; Harry Jonas et al, *An Analysis of International Law, National Legislation, Judgements and Institutions as they Interrelate with Territories and Areas Conserved by Indigenous Peoples and Local Communities* (Natural Justice, 2012) 99–101. Rights violations have also been reported at other World Heritage sites, such as the Chitwan National Park World Heritage site in Nepal (see United Nations Humans Rights Council, *Report by the Special Rapporteur on the Situation of Human Rights and Fundamental*

have advocated for greater attention to be paid to such issues.[150] In addition, many international organisations have called on the Committee to amend the *Operational Guidelines* to be consistent with *UNDRIP*.[151] It is notable however that not all State parties agree with this approach.[152]

The World Heritage Committee has, to some extent, responded to calls for it to ensure compliance with *UNDRIP*. A milestone in the Committee's changing approach was its inclusion of 'enhancing the role of communities'

Freedoms of Indigenous People, James Anaya, Addendum: report on the situation of indigenous peoples in Nepal, UN Doc A/HRC/12/34/Add.3 (20 July 2009) paras 35–37), and Lhasa, Tibet (see Amund Sinding-Larsen, 'Lhasa Community, World Heritage and Human Rights' (2012) 18(3) *International Journal of Heritage Studies* 297. For other case studies concerning human rights issues at World Heritage sites, see Peter Bille Larsen (ed), *World Heritage and Human Rights: Lessons from the Asia-Pacific and Global Arena* (Routledge, 2017).

[150] See, for example, Robert James Hales et al, 'Indigenous Free Prior Informed Consent: A Case for Self Determination in World Heritage Nomination Processes' (2013) 19(3) *International Journal of Heritage Studies* 270; Stefan Disko, *Indigenous Peoples' Rights in the Context of the World Heritage Convention: The Role of IUCN* (IUCN, 2011) https://www.iucn.org/content/indigenous-peoples-rights-context-world-heritage-convention-%E2%80%93-role-iucn; Eman Assi, 'World Heritage Sites, Human Rights and Cultural Heritage in Palestine' (2012) 18(3) *International Journal of Heritage Studies* 316; Jukka Jokilehto, 'Human Rights and Cultural Heritage: Observations on the Recognition of Human Rights in the International Doctrine' (2012) 18(3) *International Journal of Heritage Studies* 226.

[151] See, for example, *Report of the Expert Mechanism on the Rights of Indigenous Peoples on its Fifth Session (Geneva, 9–13 July 2012),* Human Rights Council, 21st sess, UN Doc A/HRC/21/52 (17 August 2012) 7; *Report on the Twelfth Session (20–31 May 2013),* United Nations Permanent Forum on Indigenous Issues, UN ESCOR, 12th sess, UN Doc E/2013/43-E/C.19/2013/25 (2013) 6 [23]; IUCN, *Implementation of the United Nations Declaration on the Rights of Indigenous Peoples in the Context of the UNESCO World Heritage Convention,* WCC-2012-Res-047-EN (2012); UN Special Rapporteur on the Rights of Indigenous Peoples, *Rights of Indigenous Peoples,* UNGA 67th sess, UN Doc A/67/301 (13 August 2013) 9–12, paras 33–42.

[152] For example, the International Work Group for Indigenous Affairs (IWGIA) reported that discussions at the 2015 World Heritage Committee meeting in Bonn, Germany 'revealed strong resistance by many States Parties against adopting safeguards for the rights of indigenous peoples in the context of the World Heritage Convention'. IWGIA noted that a World Heritage Committee meeting member stated, in relation to the nomination of Kaeng Krachan Forest Complex in Thailand: '[W]e are here at a prestigious committee of culture and heritage, we are not in Geneva on the Human Rights Council': see International Work Group for Indigenous Affairs (IWGIA), *8th Session of the EMRIP: Joint Statement on Indigenous Rights and World Heritage* (22 July 2015) http://www.iwgia.org/news/search-news?news_id=1234.

as one of its five strategic objectives,[153] the other four being credibility, conservation, capacity-building, and communication.[154] The Committee decided to include the fifth strategic objective in 'recognition of the critical importance of involving indigenous, traditional and local communities in the implementation of the *Convention*'.[155] This objective was adopted when the Committee was under the chairmanship of a paramount chief of Aotearoa/New Zealand, demonstrating the impact that a Pacific voice within the Committee can have.

In 2015, the Committee formally resolved that the rights of Indigenous peoples should be respected when nominating, managing, and reporting on World Heritage sites,[156] and it made some relevant amendments to the *Operational Guidelines*. The *Operational Guidelines* now recognise that the involvement of local communities, Indigenous peoples, and other stakeholders in the World Heritage nomination process is essential for them to have a shared responsibility with the State party in the protection of the property.[157] As such, State parties are encouraged to prepare nominations with the widest possible participation of stakeholders and to 'demonstrate, as appropriate, that the free, prior and informed consent of Indigenous peoples has been obtained'.[158] The Committee also supports the involvement of a range of actors in World Heritage protection. The *Operational Guidelines* state that a partnership approach to management is preferable,[159] involving local communities, Indigenous people, NGOs, and others with an interest in the property.[160] Through these developments, the Committee has shifted towards an approach that is more likely to be appropriate in the Pacific, where Pacific Islanders have governed and

---

[153]WHC Res 31 COM 13A, WHC 31st sess, UN Doc WHC-07/31.COM/24 (31 July 2007) 193; WHC Res 31 COM 13B, WHC 31st sess, UN Doc WHC-07/31.COM/24 (31 July 2007) 193.

[154] *Operational Guidelines 2016*, UN Doc WHC.16/01, para 26.

[155] WHC Res 31 COM 13A, WHC 31st sess, UN Doc WHC-07/31.COM/24 (31 July 2007) 193, 193 para 5.

[156]WHC Res 35 COM 12E, WHC 35th sess, UN Doc WHC-11/35.COM/20 (7 July 2011) 270, 271 para 15(f).

[157] *Operational Guidelines 2016*, UN Doc WHC.16/01, para 123.

[158] Ibid.

[159] Ibid., para 39.

[160] Ibid., para 40.

managed their land and resources for millennia,[161] and where governments generally lack the capacity and resources to administer, monitor, and enforce top-down heritage protection laws.[162]

While the Committee has moved away from a purely State-centric approach to heritage protection, the provisions of the *Operational Guidelines* have their limits. Importantly, they do not require State parties to involve local communities in the nomination and protection of World Heritage sites, they merely encourage them to do so. In that sense, they fall short of what some commentators have sought.[163] Furthermore, the nomination dossier 'format and content' requirements in the *Operational Guidelines* do not require the State party to specify the extent to which local communities have been involved in the nomination process, or whether their consent has been obtained.[164] Consequently, the Committee may not have any information about these issues when assessing a nomination.

The Committee has also refused calls to establish an expert group to advise on matters concerning Indigenous people. A formal proposal to establish a 'World Heritage Indigenous Peoples Council of Experts' was

---

[161] See, for example, Ashish Kothari et al (eds), *Recognising and Supporting Territories and Areas Conserved by Indigenous Peoples and Local Communities: Global Overview and National Case Studies* (Kalpavriksh, and Natural Justice, 2012) 16; Peter Bridgewater, Salvatore Arico and John Scott, 'Biological Diversity and Cultural Diversity: The Heritage of Nature and Culture Through the Looking Glass of Multilateral Agreements' (2007) 13(4–5) *International Journal of Heritage Studies* 405, 407; K Ruddle, E Hviding, R E Johannes, 'Marine Resources Management in the Context of Customary Tenure' (1992) 7 *Marine Resource Economics* 249, 250; Marjo Vierros et al, *Traditional Marine Management Areas of the Pacific in the Context of National and International Law and Policy* (United Nations University, 2010) 7; David J Doulman, 'Community-Based Fisheries Management: Towards Restoration of Traditional Practices in the South Pacific' (March 1993) *Marine Policy* 108, 108; R E Johannes, 'Traditional Marine Conservation Methods in Oceania and their Demise' 9 (1978) *Annual Review of Ecology and Systematics* 349, 350.

[162] See generally Benjamin J Richardson, 'Environmental Law in Postcolonial Societies: Straddling the Local-Global Institutional Spectrum' (2000) 11(1) *Colorado Journal of International Environmental Law and Policy* 1. See also Sect. 2.5.1.

[163] See, for example, Stefan Disko, 'World Heritage Sites in Indigenous Peoples' Territories: Ways of Ensuring Respect for Indigenous Cultures, Values and Human Rights' in Dieter Offenhäußer, Walther Ch Zimmerli and Marie-Theres Albert (eds), *World Heritage and Cultural Diversity* (German Commission for UNESCO, 2010) 167. Disko argues that the *Operational Guidelines* should require the full and effective participation of Indigenous peoples in the identification, nomination, management, and protection of World Heritage: at 174. See also Carril, above n 139.

[164] *Operational Guidelines 2016*, UN Doc WHC.16/01, part IIIB, annex 5.

raised following a forum held in Australia in 2000. Several possible roles were discussed for the group, including ensuring consultation with local people, strengthening the management of existing sites, assisting with the development of management guidelines, and advising on the nomination and evaluation of sites.[165] However, the Committee did not support the proposal,[166] and the group is unlikely to be established in the foreseeable future.[167] As such, Indigenous peoples still do not have a formal position within the *Convention* regime, which limits their ability to influence the manner in which the treaty is implemented.

The *World Heritage Sustainable Development Policy* notes that recognising rights and fully involving Indigenous peoples and local communities, in line with international standards, is at the heart of sustainable development.[168] It refers to the need to facilitate the participation of all stakeholders and rights holders, including Indigenous peoples and local communities, in the conservation of World Heritage sites.[169] The policy's adoption may lead the Committee to make further changes to the *Operational Guidelines*, perhaps addressing the limitations referred to above.

4.3.3   Customary Protection of World Heritage Sites

4.3.3.1   Adequate Protection and Management as a Threshold Requirement for World Heritage Listing

As explained in Chap. 3, a site is only eligible for World Heritage listing if it has OUV.[170] The *Convention* does not define the term OUV, but rather empowers the Committee to determine the criteria against which a site's value will be assessed.[171] The Committee has decided that to have OUV, a

---

[165] *These Are Our Powerful Worlds*, Summary Report of the Working Group Workshop on the World Heritage Indigenous People's Council of Experts (Winnipeg, Manitoba, November 5–8 2001) http://www.whc.unesco.org/document/9474 4.

[166] *Report of the World Heritage Committee*, WHC 25th sess, UN Doc WHC-01/CONF 208/24 (8 February 2002) 57, 57 para XV.5.

[167] Lynn Meskell, 'UNESCO and the fate of the World Heritage Indigenous Peoples Council of Experts (WHIPCOE)' (2013) 20 *International Journal of Cultural Property* 155, 157.

[168] *World Heritage Sustainable Development Policy*, UN Doc WHC-15/20.GA/15, para 21.

[169] Ibid., para 9.

[170] *World Heritage Convention* arts 1, 2, 11(2).

[171] Ibid., art 11(5).

site must meet one or more of the prescribed criteria, as well as the conditions of integrity and authenticity. These requirements were analysed in Sect. 3.3. In addition, the Committee considers that a site must be adequately protected and managed to have OUV.[172] Thus, paragraph 97 of the 2016 *Operational Guidelines* states:

> All properties inscribed on the World Heritage List must have adequate long-term legislative, regulatory, institutional and/or traditional[173] protection and management to ensure their safeguarding. This protection should include adequately delineated boundaries. Similarly States Parties should demonstrate adequate protection at the national, regional, municipal, and/or traditional level for the nominated property. They should append appropriate texts to the nomination with a clear explanation of the way this protection operates to protect the property.

Paragraph 97 is supplemented by other provisions containing more detailed prescriptions about the management of sites, boundaries, and buffer zones.[174] Through these provisions, the Committee is requiring the State party to provide some assurance that it will protect its World Heritage. This is reinforced by paragraph 53 of the *Operational Guidelines*, which states that nominations for World Heritage listing must demonstrate the full commitment of the State party to preserve the heritage concerned, within its means.

The protection and management requirement in paragraph 97 is expressed as a mandatory requirement. Its mandatory nature is reinforced by UNESCO's manual on the preparation of World Heritage nominations, which states that a nomination will fail if this requirement is not met.[175] However, the extent to which the provision is strictly or consistently enforced is debatable, given that sites have been listed despite uncertainty as to how their protection regimes operate (see Sect. 5.2). Regardless, paragraph 97 of the *Operational Guidelines* and the provisions that supplement it are important because they make the protection and management of a site an issue for the Committee to consider at the listing stage. Furthermore, State parties who wish to secure a successful nomination are likely to try to ensure they meet the Committee's requirements.

---

[172] *Operational Guidelines 2016*, UN Doc WHC.16/01, para 78.
[173] In this book, the word 'customary' is used instead of 'traditional': see Sect. 1.6.3.
[174] *Operational Guidelines 2016*, UN Doc WHC.16/01, part II.F.
[175] UNESCO/ICCROM/ICOMOS/IUCN, *Preparing World Heritage Nominations* (UNESCO, 2nd ed, 2011) 87.

As such, these provisions provide the Committee with an avenue to influence how State parties manage and protect their sites.

4.3.3.2 The Committee's Recognition of Customary Protection and Management of World Heritage Sites

Like the criteria for World Heritage listing, the protection and management requirements for the inscription of sites on the World Heritage List have changed over time. In the Pacific context, the most significant change occurred when the Committee recognised that a site protected and managed through 'traditional' (referred to in this book as 'customary'—see Sect. 1.6.3) systems could satisfy these requirements. This amendment to the *Operational Guidelines* was a manifestation of changing attitudes towards the notion of cultural heritage (see Sect. 3.4) as well as the growing recognition of the need for sustainable development and the rights and roles of local people in heritage protection (see Sects. 4.3.1 and 4.3.2, respectively).

Under the 1978 version of the *Operational Guidelines*, all nomination dossiers had to outline the 'means of preservation' of the nominated site.[176] At that time, the *Operational Guidelines* stated that the Committee must consider the 'state of preservation' of cultural sites nominated for World Heritage listing,[177] but there was no requirement for such places to be protected to any particular standard in order to be listed.

Adequate protection and management became a mandatory requirement for World Heritage listing under the 1988 *Operational Guidelines*.[178] This change was made to align the *Operational Guidelines* with the Committee's practice in implementing the *Convention*.[179] On several previous occasions the Committee had deferred nominations on the grounds that the sites were inadequately protected, on the recommendation of the

---

[176] UNESCO, *Operational Guidelines for the Implementation of the World Heritage Convention*, UN Doc WHC/2 (1978) para 13(iv).

[177] Ibid., para 8.

[178] WHC Res CONF 001 VIII.20–27, WHC 12th sess, UN Doc SC-88/CONF.001.13 (23 December 1988) 5–6; UNESCO, *Operational Guidelines for the Implementation of the World Heritage Convention*, UN Doc WHC/2/Revised (December 1988) paras 24(b) (ii), 36(b) (vi).

[179] *Report of the World Heritage Committee*, WHC 12th sess, UN Doc SC-88/CONF.001.13 (23 December 1988) 5.

relevant Advisory Body.[180] The *Operational Guidelines* were therefore amended to state that protection legislation was essential for nominated cultural sites,[181] and natural sites required long-term legislative, regulatory, or institutional protection.[182]

In 1994, the *Operational Guidelines* were further amended so that a cultural heritage site under customary protection and management could qualify for World Heritage listing.[183] This change occurred around the time the *Global Strategy* was adopted, and can be seen as part of the Committee's efforts to make the List more responsive to the diversity of the world's heritage. Importantly, the change coincided with the Committee's introduction of 'cultural landscapes' as a category of World Heritage site (discussed in Sect. 3.4). As Smith and Jones have stated, 'Many landscapes of the Pacific Islands are managed according to customary practices and these practices will be the key to sustaining their values.'[184] It was therefore logical that the Committee's recognition of cultural landscapes was accompanied by recognition of the customary systems that shape and protect such places.

The amendment of the *Operational Guidelines* to allow for the listing of sites protected through customary systems was initially restricted to cultural sites. However, during this era, there was also increasing recognition of the role of customary systems in protecting natural areas. This is particularly evident in the work of the IUCN. Its 1994 guidelines on protected areas defined a 'protected area' as '[a]n area of land and/or sea especially dedicated to the protection and maintenance of biological diversity, and of natural and associated cultural resources, and managed through legal or other effective means'.[185] By including the words 'other effective means' in the definition, IUCN was acknowledging that protected areas could be managed through mechanisms other than legislation, including

---

[180] *Revision of the Operational Guidelines,* WHC 12th sess, UN Doc SC-88/CONF.007/12 (9 May 1988) 3.

[181] UNESCO, *Operational Guidelines for the Implementation of the World Heritage Convention,* UN Doc WHC/2/Revised (December 1988) para 24(b) (ii).

[182] Ibid., para 36(b) (vi).

[183] WHC Res CONF 003 XIV.3, WHC 18th sess, UN Doc WHC-94/CONF.003.16 (31 January 1995) 64–66; UNESCO, *Operational Guidelines for the Implementation of the World Heritage Convention,* UN Doc WHC/2/Revised (February 1994) para 24(b) (ii).

[184] Anita Smith and Kevin L Jones (eds), *Cultural Landscapes of the Pacific Islands* (ICOMOS, 2007) 120.

[185] IUCN, *Guidelines for Protected Area Management Categories* (IUCN, 1994).

customary systems.[186] This was reiterated by IUCN's inclusion of 'Indigenous Community Conservation Areas' (ICCAs) in its list of protected area governance types (alongside governance by states, private governance, and shared governance).[187] ICCAs are ecosystems 'voluntarily conserved by Indigenous peoples and local communities, both sedentary and mobile, through customary laws or other effective means'.[188] Given that IUCN is an Advisory Body under the *Convention*, these developments no doubt influenced the Committee's approach to World Heritage protection. In 1998, the Committee further amended the *Operational Guidelines* so that natural sites protected under customary mechanisms could also qualify for World Heritage listing.[189]

The Committee's recognition of customary protection of World Heritage sites enabled the listing of East Rennell, which at the time had little State legislative protection (see Sect. 5.2.1). Customary protection was also recognised in the listing of other Pacific Island sites. The Rock Islands Southern Lagoon site in Palau enjoys some protection under traditional cultural controls, such as *bul* (which are temporary restrictions imposed by village chiefs on certain activities).[190] The heritage of Chief Roi Mata's Domain in Vanuatu continues to be protected through *tapu* restrictions determined by the area's chiefs, which seek to prevent the over-exploitation of natural resources.[191] The Kuk Early Agricultural Site in Papua New Guinea (PNG) was found to have sufficient protection to warrant World Heritage listing in part because of the protection provided through customary farming practices.[192] The Nan Madol site in the Federated States of Micronesia is also subject to some customary protection.[193] This development therefore substantially increased the potential for the *Convention* to be utilised effectively in the Pacific.

---

[186] Dudley (ed), above n 37, 8.

[187] Ibid., 26.

[188] Grazia Borrini-Feyerabend et al, above n 65, 40.

[189] WHC Res CONF 203 XIV.3, WHC 22nd sess, UN Doc WHC-98/CONF.203/18 (29 January 1999) 56.

[190] Republic of Palau, above n 115, 109.

[191] Republic of Vanuatu, above n 114, 96.

[192] ICOMOS, *Evaluations of Nominations of Cultural and Mixed Properties to the World Heritage List*, WHC 32nd sess, UN Doc WHC-08/32.COM/INF.8B1 (2008) 84 (Kuk Early Agricultural Site, Papua New Guinea, Advisory Body Evaluation 887) 89.

[193] Federated States of Micronesia, *Nan Madol: Ceremonial Center of Eastern Micronesia – As Nominated by the Federated States of Micronesia for inscription on the World Heritage List* (2015) 10.

The listing of sites under customary protection does however present challenges not experienced at sites under private or State ownership. For example, some provisions of the *Operational Guidelines* concerning site management, boundaries, and buffer zones may prove problematic for such sites (see Sects. 6.4 and 6.5). In addition, the role of the State in heritage conservation must be carefully considered when a site under customary protection is nominated for World Heritage listing. It cannot be presumed that the State will be willing and able to take the steps necessary to conserve the OUV of such a site (see Sect. 5.3.3.2). These issues are explored further in Part III of this book.

4.4 CONCLUSION

The *World Heritage Convention* regime reflects the era in which the treaty was drafted. It reveals an attempt to balance respect for State sovereignty with the international community's interest in the protection of World Heritage. It also focuses on delineating the roles of State parties and the international community in achieving heritage protection, while making no mention of the role of non-State actors operating at the local level. This reflects the traditional centralised approach to heritage protection, which was widely accepted when the *Convention* was adopted, but which is often inappropriate in the Pacific.

Over time however, the *Convention* regime has evolved to become a better fit for the Pacific context. Since the *Convention* was adopted, support has grown for a holistic approach to heritage protection, under which the heritage place is considered in its economic, social, and environmental context, and the rights and roles of local people are respected. In response, the World Heritage Committee has revised the *Operational Guidelines* to encourage States to approach heritage protection through the framework of sustainable development, and to involve local communities in the nomination and protection of sites. The Committee's decision that sites under customary protection are eligible for World Heritage listing was also significant for Pacific Island States.

This evolution has been possible because the *Convention* text just establishes a framework for that regime, giving the Committee and State parties significant powers and discretions to implement its provisions in accordance with contemporary views. Some challenges associated with the *Convention* text remain, including the inherent tension between national sovereignty and the international community's interest in World Heritage protection, and the limitations of the international assistance system. Furthermore, the provisions of the 2016 *Operational Guidelines* have their limitations and further amendments are warranted. Notwithstanding this, the dynamic nature of the *Convention* regime has allowed it to become a more useful tool for the preservation of Pacific heritage.

Ultimately, however, it is the Pacific Island States, not the Committee, who dictate how World Heritage sites in the region will be protected. They must strive to develop measures that achieve an appropriate balance between heritage conservation and economic and social development. They must respect the rights of local communities whilst also ensuring the preservation of OUV, and they must identify approaches that are appropriate given their resource constraints, the nature of their heritage, their plural legal systems, and the land tenure of their heritage sites. The analysis of the implementation of the *Convention* by Solomon Islands in Part III of this book demonstrates that these are not easy tasks.

REFERENCES

Articles, Books and Reports

Affolder, Natasha, 'Democratising or Demonising the World Heritage Convention?' (2007) 39 *University of Wellington Law Review* (2007) 341

Assi, Eman, 'World Heritage Sites, Human Rights and Cultural Heritage in Palestine' (2012) 18(3) *International Journal of Heritage Studies* 316

Barreiro Carril, Beatriz, "Indigenous Peoples' Participation in Decision-Making in the Context of World Heritage Sites: How International Human Rights Law Can Help?' (2016) 7(2–3) *The Historic Environment: Policy and Practice* 224

Battini, Stefano, 'The Procedural Side of Legal Globalisation: The Case of the World Heritage Convention' (2011) 9(2) *International Journal of Constitutional Law* 340

Bertacchini, Enroci E and Donatella Saccone, 'Toward a Political Economy of World Heritage' (2012) 36(4) *Journal of Cultural Economics* 327

Boer, Ben and Graeme Wiffen, *Heritage Law in Australia* (Oxford University Press, 2006)

Boer, Ben, 'Article 3 Identification and Delineation of World Heritage Properties' in Francesco Francioni (ed), *The 1972 World Heritage Convention: A Commentary* (Oxford University Press, 2008) 85

Borrini-Feyerabend, Grazia, Ashish Kothari and Gonzalo Oviedo, *Indigenous and Local Communities and Protected Areas: Towards Equity and Enhanced*

Conservation, Best Practice Protected Area Guidelines 11 (World Conservation Union, 2004)

Bourdeau, Phillipe, 'The Man-Nature Relationship and Environmental Ethics' (2004) 72 *Journal of Environmental Radioactivity* 9

Bridgewater, Peter, Salvatore Arico and John Scott, 'Biological Diversity and Cultural Diversity: The Heritage of Nature and Culture Through the Looking Glass of Multilateral Agreements' (2007) 13(4–5) *International Journal of Heritage Studies* 405

Buzzini, Gionata P and Luigi Condorelli, 'Article 11 List of World Heritage in Danger and Deletion of a Property from the World Heritage List' in Francesco Francioni (ed), *The 1972 World Heritage Convention: A Commentary* (Oxford University Press, 2008) 175

Cameron, Christina, 'The Strengths and Weaknesses of the World Heritage Convention' (1992) 28(3) *Nature and Resources* 18

Carducci, Guido, 'Articles 4–7 National and International Protection of the Cultural and Natural Heritage' in Francesco Francioni (ed), *The 1972 World Heritage Convention: A Commentary* (Oxford University Press, 2008) 103

Carew-Reid, Jeremy, 'Conservation and Protected Areas on South-Pacific Islands: The Importance of Tradition' (1990) 17(1) *Environmental Conservation* 29

Deegan, Naomi, 'The Local-Global Nexus in the Politics of World Heritage: Space for Community Development?' in Marie-Theres Albert, Marielle Richon, Marie José Viñals and Andrea Witcomb (eds), *Community Development through World Heritage*, World Heritage Papers 31 (UNESCO, 2012) 77

Disko, Stefan, 'World Heritage Sites in Indigenous Peoples' Territories: Ways of Ensuring Respect for Indigenous Cultures, Values and Human Rights' in Dieter Offenhäußer, Walther Ch Zimmerli and Marie-Theres Albert (eds), *World Heritage and Cultural Diversity* (German Commission for UNESCO, 2010) 167

Disko, Stefan, *Indigenous Peoples' Rights in the Context of the World Heritage Convention: The Role of IUCN* (IUCN, 2011) https://www.iucn.org/content/indigenous-peoples-rights-context-world-heritage-convention-%E2%80%93-role-iucn

Doulman, David J, 'Community-Based Fisheries Management: Towards Restoration of Traditional Practices in the South Pacific' (March 1993) *Marine Policy* 108

Dudley, Nigel (ed), *Guidelines for Applying Protected Area Management Categories* (IUCN, 2008)

Federated States of Micronesia, *Nan Madol: Ceremonial Center of Eastern Micronesia – As Nominated by the Federated States of Micronesia for inscription on the World Heritage List* (2015)

Forrest, Craig, *International Law and the Protection of Cultural Heritage* (Routledge, 2011)

Francioni, Francesco, 'The 1972 World Heritage Convention: An Introduction' in Francesco Francioni (ed), *The 1972 World Heritage Convention: A Commentary* (Oxford University Press, 2008) 3

Francioni, Francesco, 'The Preamble' in Francesco Francioni (ed), *The 1972 World Heritage Convention: A Commentary* (Oxford University Press, 2008) 11

Francioni, Francesco and Federico Lenzerini, 'The Future of the World Heritage Convention: Problems and Prospects' in Francesco Francioni (ed), *The 1972 World Heritage Convention: A Commentary* (Oxford University Press, 2008) 401

Frey, Bruno S and Lasse Steiner, 'World Heritage List: Does it Make Sense?' (2011) 17(5) *International Journal of Cultural Policy* 555

Galis, Allan, 'UNESCO Documents and Procedure: The Need to Account for Political Conflict When Designating World Heritage Sites' (2009–2010) 38 *Georgia Journal of International and Comparative Law* 205

Gillespie, Josephine, 'Legal Pluralism and World Heritage Management at Angkor, Cambodia' (2012) 14(1&2) *Asia Pacific Journal of Environmental Law* 1

Goodwin, Edward J, 'The World Heritage Convention, the Environment and Compliance' (2008–2009) 20 *Colorado Journal of International Environmental Law and Policy* 157

Borrini-Feyerabend, Grazia, Nigel Dudley, Tilman Jaeger, Barbara Lassen, Neema Pathak Broome, Adrian Phillips and Trevor Sandwith, *Governance of Protected Areas: From Understanding to Action*, Best Practice Protected Area Guidelines Series 20 (IUCN, 2013)

Hales, Robert James, John Rynne, Cathy Howlett, Jay Devine and Vivian Hauser, 'Indigenous Free Prior Informed Consent: A Case for Self Determination in World Heritage Nomination Processes' (2013) 19(3) *International Journal of Heritage Studies* 270

Hockings, Marc, Robyn James, Sue Stolton, Nigel Dudley, Vinod Mathur, John Makombo, Jose Courrau and Jeffrey Parish, *Enhancing Our Heritage Toolkit: Assessing Management Effectiveness of Natural World Heritage Sites*, World Heritage Papers 23 (IUCN, 2008)

IUCN, *Guidelines for Protected Area Management Categories* (IUCN, 1994)

IUCN, *Management Planning for Natural World Heritage Properties: A Resource Manual for Practitioners* (IUCN, 2008)

Johannes, R E, 'Traditional Marine Conservation Methods in Oceania and their Demise' 9 (1978) *Annual Review of Ecology and Systematics* 349

Jokilehto, Jukka, 'Human Rights and Cultural Heritage: Observations on the Recognition of Human Rights in the International Doctrine' (2012) 18(3) *International Journal of Heritage Studies* 226

Jonas, Harry, Eli J Makagon, Stephanie Booker and Holly Shrumm, *An Analysis of International Law, National Legislation, Judgements and Institutions as they Interrelate with Territories and Areas Conserved by Indigenous Peoples and Local Communities* (Natural Justice, 2012)

Kingsbury, Benedict, "Indigenous Peoples' in International Law: A Constructivist Approach to the Asian Controversy' (1998) 92 *American Journal of International Law* 414

Kothari, Ashish, Colleen Corrigan, Harry Jonas, Aurélie Neumann and Holly Shrumm (eds), *Recognising and Supporting Territories and Areas Conserved by Indigenous Peoples and Local Communities: Global Overview and National Case Studies* (Kalpavriksh, and Natural Justice, 2012)

Larsen, Peter Bille, *World Heritage and Evaluation Processes Related to Communities and Rights: An Independent Review* (IUCN, 2012)

Larsen, Peter Bille (ed), *World Heritage and Human Rights: Lessons from the Asia-Pacific and Global Arena* (Routledge, 2017)

Lauer, Matthew and Shankar Aswani, 'Indigenous Ecological Knowledge as Situated Practices: Understanding Fishers' Knowledge in the Western Solomon Islands (2009) 111(3) *American Anthropologist* 317

Lausche, Barbara, *Guidelines for Protected Area Legislation* (IUCN, 2011)

Lee, Thomas and Julie Middleton, *Guidelines for Management Planning of Protected Areas* (IUCN, 2003)

Lenzerini, Federico, 'Article 12 Protection of Properties Not Inscribed on the World Heritage List' in Francesco Francioni (ed), *The 1972 World Heritage Convention: A Commentary* (Oxford University Press, 2008) 201

Meskell, Lynn, 'UNESCO and the fate of the World Heritage Indigenous Peoples Council of Experts (WHIPCOE)' (2013) 20 *International Journal of Cultural Property* 155

Meskell, Lynn, Claudia Liuzza and Nicholas Brown, 'World Heritage Regionalism: UNESCO from Europe to Asia' (2015) 22 *International Journal of Cultural Property* 437

Oviedo, Gonzalo and Tatjana Puschkarsky, 'World Heritage and Rights-Based Approaches to Nature Conservation' (2012) 18(3) *International Journal of Heritage Studies* 285

Phares, Jehanne and Cynthia Guttman, *Investing in World Heritage: Past Achievements, Future Ambitions – A Guide to International Assistance*, World Heritage Papers 2 (UNESCO, 2002)

Phillips, Adrian, 'Cultural Landscapes: IUCN's Changing Vision of Protected Areas' in *Cultural Landscapes: The Challenges of Conservation*, World Heritage Papers 7 (UNESCO, 2003)

Phillips, Adrian, 'Turning Ideas on their Head: The New Paradigm for Protected Areas' (2003) 20(2) *The George Wright Forum* 8

Rao, Kishore, 'A New Paradigm for the Identification, Nomination and Inscription of Properties on the World Heritage List' (2010) 16(3) *International Journal of Heritage Studies* 161

Reepmeyer, Christian, Geoffrey Clark, Dwight Alexander, Ilebrang U Olkeriil, Jolie Liston and Ann Hillmann Kitalong, 'Selecting Cultural Sites for the

UNESCO World Heritage List: Recent Work in the Rock Islands – Southern Lagoon Area, Republic of Palau' in Jolie Liston, Geoffrey Clark and Dwight Alexander (eds), *Pacific Island Heritage: Archaeology, Identity and Community* (ANU E Press, 2011) 85

Republic of Palau, *The Rock Islands Southern Lagoon Nomination for Inscription on the World Heritage List* (2012)

Republic of Vanuatu, *Chief Roi Mata's Domain – Nomination by the Republic of Vanuatu for Inscription on the World Heritage List* (2007)

Richardson, Benjamin J, 'Environmental Law in Postcolonial Societies: Straddling the Local-Global Institutional Spectrum' (2000) 11(1) *Colorado Journal of International Environmental Law and Policy* 1

Ruddle, K, E Hviding, R E Johannes, 'Marine Resources Management in the Context of Customary Tenure' (1992) 7 *Marine Resource Economics* 249

Scovazzi, Tullio, 'Articles 8–11 World Heritage Committee and World Heritage List' in Francesco Francioni (ed), *The 1972 World Heritage Convention: A Commentary* (Oxford University Press, 2008) 147

Shearing, Susan, 'Here Today, Gone Tomorrow? Climate Change and World Heritage' (2008) 12(2) *The Australasian Journal of Natural Resources Law and Policy* 161

Sinding-Larsen, Amund, 'Lhasa Community, World Heritage and Human Rights' (2012) 18(3) *International Journal of Heritage Studies* 297

Smith, Anita and Kevin L Jones (eds), *Cultural Landscapes of the Pacific Islands* (ICOMOS, 2007)

Stevens, Stan, 'Implementing the UN Declaration on the Rights of Indigenous Peoples and International Human Rights Law through the Recognition of ICCAs' (2010) 17 *Policy Matters* 181

Strasser, Ian, 'Putting Reform into Action: Thirty Years of the World Heritage Convention: How to Reform a Convention without Changing its Regulations' (2002) 11(2) *International Journal of Cultural Property* 215

Techera, Erika, 'Samoa: Law, Custom and Conservation' (2006) 10 *New Zealand Journal of Environmental Law* 361

Techera, Erika, *Strengthening International Law to Address the Needs of Legally Pluralist Nations*, Macquarie Law Working Paper 2010–02 (Macquarie University, 2010)

These Are Our Powerful Worlds, Summary Report of the Working Group Workshop on the World Heritage Indigenous People's Council of Experts (Winnipeg, Manitoba, November 5–8 2001) http://www.whc.unesco.org/document/9474

UNESCO/ICCROM/ICOMOS/IUCN, *Preparing World Heritage Nominations* (UNESCO, 2nd ed, 2011)

UNESCO/ICCROM/ICOMOS/IUCN, *Managing Natural World Heritage*, World Heritage Resource Manual (UNESCO, 2012)

UNESCO/ICCROM/ICOMOS/IUCN, *Managing Cultural World Heritage*, World Heritage Resource Manual (UNESCO, 2013)

Vierros, Marjo, Alifereti Tawake, Francis Hickey, Ana Tiraa and Rahera Noa, *Traditional Marine Management Areas of the Pacific in the Context of National and International Law and Policy* (United Nations University, 2010)

Vrdoljak, Ana Filipa, 'Article 14 The Secretariat and Support of the World Heritage Committee' in Francesco Francioni (ed), *The 1972 World Heritage Convention: A Commentary* (Oxford University Press, 2008) 243

Wingham, Elspeth J, *Nomination of East Rennell, Solomon Islands by the Government of Solomon Islands for Inclusion in the World Heritage List Natural Sites* (1997)

CASES

Commonwealth v Tasmania (1983) 46 ALR 625
Richardson v Forestry Commission (1988) 77 ALR 237

LEGISLATION AND BILLS: FIJI

Heritage Bill 2016 (no. 10 of 2016)

CONVENTIONS

Convention Concerning the Protection of the World Cultural and Natural Heritage, opened for signature 16 November 1972, 1037 UNTS 151 (entered into force 17 December 1975)

Convention for the Protection of Cultural Property during Armed Conflict, opened for signature 14 May 1954, 249 UNTS 240 (entered into force 7 August 1956)

Convention (No. 169) Concerning Indigenous and Tribal Peoples in Independent Countries, opened for signature 27 June 1989, 1650 UNTS 383 (entered into force 5 September 1991)

Convention on Biological Diversity, opened for signature 5 June 1992, 1760 UNTS 79 (entered into force 29 December 1993)

Convention on the Law of Treaties, opened for signature 23 May 1969, 1155 UNTS 331 (entered into force 27 January 1980)

Convention on the Protection of the Underwater Cultural Heritage, opened for signature 2 November 2001, 2562 UNTS 48 (entered into force 2 January 2009)

Convention for the Safeguarding of the Intangible Cultural Heritage, opened for signature 17 October 2003, 2368 UNTS 3 (entered into force 20 April 2006)

Convention on Wetlands of International Importance especially as Waterfowl Habitat, opened for signature 2 February 1971, 996 UNTS 245 (entered into force 21 December 1975)

UNITED NATIONS DOCUMENTS

Agenda 21, Report of the UNCED, I, UN Doc. A/CONF.151/26/Rev.1 (1992)

Budapest Declaration on World Heritage, WHC Res 26 COM 9, WHC 26th sess, UN Doc WHC-02/CONF.202/25 (1 August 2002) 6

ICOMOS, *Evaluations of Nominations of Cultural and Mixed Properties to the World Heritage List*, WHC 32nd sess, UN Doc WHC-08/32.COM/INF.8B1 (2008) 84 (Kuk Early Agricultural Site, Papua New Guinea, Advisory Body Evaluation 887)

IUCN, *Implementation of the United Nations Declaration on the Rights of Indigenous Peoples in the Context of the UNESCO World Heritage Convention*, WCC-2012-Res-047-EN (2012)

Policy for the Integration of a Sustainable Development Perspective into the Processes of the World Heritage Convention, WHC GA Res 20 GA 13, 20th sess, UN Doc WHC-15/20.GA/15 (20 November 2015) 7

Report of the Expert Mechanism on the Rights of Indigenous Peoples on its Fifth Session (Geneva, 9–13 July 2012), Human Rights Council, 21st sess, UN Doc A/HRC/21/52 (17 August 2012) 7

Report of the World Heritage Committee, WHC 12th sess, UN Doc SC-88/CONF.001.13 (23 December 1988)

Report of the World Heritage Committee, WHC 25th sess, UN Doc WHC-01/CONF.208/24 (8 February 2002)

Report on the Twelfth Session (20–31 May 2013), United Nations Permanent Forum on Indigenous Issues, UN ESCOR, 12th sess, UN Doc E/2013/43-E/C.19/2013/25 (2013)

Revision of the Operational Guidelines, WHC 12th sess, UN Doc SC-88/CONF.007/12 (9 May 1988)

Rio Declaration on Environment and Development, Report of the United Nations Conference on Environment and Development, UN Doc A/CONF.151/6/Rev.1 (1992)

Transforming Our World: The 2030 Agenda for Sustainable Development, UNGA Res A/RES/70/L.1, UN GAOR, 70th sess, UN Doc A/RES/70/1 (21 October 2015)

UNESCO, *Operational Guidelines for the Implementation of the World Heritage Convention*, UN Doc WHC/2 (1978)

UNESCO, *Operational Guidelines for the Implementation of the World Heritage Convention*, UN Doc WHC/2/Revised (December 1988)

UNESCO, *Operational Guidelines for the Implementation of the World Heritage Convention*, UN Doc WHC/2/Revised (February 1994)
UNESCO, *Operational Guidelines for the Implementation of the World Heritage Convention*, UN Doc WHC.16/01 (26 October 2016)
United Nations Declaration on the Rights of Indigenous Peoples, GA Res 61/295, UN GAOR, 61st sess, 107th plen mtg, Supp No 49, UN Doc A/RES/61/295 (13 September 2007)
United Nations Humans Rights Council, *Report by the Special Rapporteur on the Situation of Human Rights and Fundamental Freedoms of Indigenous People, James Anaya*, Addendum: report on the situation of indigenous peoples in Nepal, UN Doc A/HRC/12/34/Add.3 (20 July 2009)
UN Special Rapporteur on the Rights of Indigenous Peoples, *Rights of Indigenous Peoples*, UNGA 67th sess, UN Doc A/67/301 (13 August 2013)
WHC Res CONF 001 VIII.20–27, WHC 12th sess, UN Doc SC-88/CONF.001.13 (23 December 1988) 5
WHC Res CONF 003 XIV.3, WHC 18th sess, UN Doc WHC-94/CONF.003.16 (31 January 1995) 64
WHC Res CONF 203 XIV.3, WHC 22nd sess, UN Doc WHC-98/CONF.203/18 (29 January 1999) 56
WHC Res 31 COM 13A, WHC 31st sess, UN Doc WHC-07/31.COM/24 (31 July 2007a) 193
WHC Res 31 COM 13B, WHC 31st sess, UN Doc WHC-07/31.COM/24 (31 July 2007b) 193
WHC Res 35 COM 12E, WHC 35th sess, UN Doc WHC-11/35.COM/20 (7 July 2011) 270
WHC Res 39 COM 5D, WHC 39th sess, UN Doc WHC-15/39.COM/19 (8 July 2015) 7
WHC Res 40 COM 7A.49, 40th sess, UN Doc WHC/16/40.COM/19 (15 November 2016) 68
WHC Res 41 COM 7A.19, 41st sess, UN Doc WHC/17/41.COM/18 (12 July 2017) 35
World Commission on Environment and Development, *Our Common Future*, UN Doc A/42/427 (1987)
World Heritage and Sustainable Development, WHC 39th sess, UN Doc WHC-15/39.COM/5D (15 May 2015)

INTERNET MATERIALS

UNESCO, *Australia* http://whc.unesco.org/en/statesparties/au
UNESCO, *New Zealand* http://whc.unesco.org/en/statesparties/nz
UNESCO, *Solomon Islands: International Assistance* http://whc.unesco.org/en/statesparties/sb/assistance/

UNESCO, *World Heritage Fund* http://whc.unesco.org/en/world-heritage-fund/

United Nations Division for Social Policy and Development, *Permanent Forum* https://www.un.org/development/desa/indigenouspeoples/unpfii-sessions-2.html

United Nations Office of the High Commissioner (UN-OHC), *Expert Mechanism on the Rights of Indigenous Peoples* http://www.ohchr.org/EN/Issues/IPeoples/EMRIP/Pages/EMRIPIndex.aspx

United Nations Office of the High Commissioner (UN-OHC), *Mandate of the Working Group on Indigenous Populations* http://www.ohchr.org/EN/Issues/IPeoples/Pages/MandateWGIP.aspx

OTHER

Intergovernmental Agreement on the Environment (1992)

International Work Group for Indigenous Affairs (IWGIA), *8th Session of the EMRIP: Joint Statement on Indigenous Rights and World Heritage* (22 July 2015) http://www.iwgia.org/news/search-news?news_id=1234

IUCN et al, *World Conservation Strategy* (1980)

Nara Document on Authenticity (1994)

Pacific World Heritage Action Plan 2016–2020 (2016)

PART III

The Case of Solomon Islands

CHAPTER 5

The Listing and Protection of the East Rennell World Heritage Site

5.1 Introduction

The International Union for the Conservation of Nature (IUCN) has contended that lessons learned from East Rennell should be identified and communicated, to assist with the implementation of the *World Heritage Convention*[1] at similar sites.[2] This is particularly pertinent now that the site is on the List of World Heritage in Danger.[3] This Part of the book therefore explores the listing and protection of East Rennell (Chaps. 5, 6, and 7), with a view to identifying lessons that can be learned from Solomon Islands' experience (Chap. 8).

In this chapter, East Rennell's listing is analysed with reference to the 1997 version of the *Operational Guidelines for the Implementation of the World Heritage Convention*[4] (which applied when the site was nominated) (see Sect. 5.2). The chapter explains how the values for which the site was

---

[1] *Convention Concerning the Protection of the World Cultural and Natural Heritage*, opened for signature 16 November 1972, 1037 UNTS 151 (entered into force 17 December 1975) ('*World Heritage Convention*').

[2] T Badman et al, *Outstanding Universal Value: Standards for Natural World Heritage* (IUCN, 2008) 25.

[3] WHC Res 37 COM 7B.14, WHC 37th sess, UN Doc WHC-13/37.COM/20 (5 July 2013) 68.

[4] UNESCO, *Operational Guidelines for the Implementation of the World Heritage Convention*, UN Doc WHC 97/2 (February 1997) ('*Operational Guidelines 1997*').

© The Author(s) 2018   165
S. C. Price, *World Heritage Conservation in the Pacific*,
Palgrave Series in Asia and Pacific Studies,
https://doi.org/10.1007/978-981-13-0602-0_5

listed and the site's boundaries influence contemporary conservation efforts. It suggests that there was some uncertainty surrounding the site's protection regime when it was nominated. In the future, when sites under customary tenure are nominated, their protection regimes should be closely scrutinised, particularly to understand the relationship between customary protection, management plans, and legislation.

The protection of East Rennell is then explored, laying a foundation for more detailed analysis in later chapters. The chapter discusses East Rennell's state of conservation, explaining how logging, mining, the over-harvesting of certain species, invasive species, and climate change threaten the site's outstanding universal value (OUV) (Sect. 5.3.1). It outlines the Solomon Islands government's (SIG's) plan for addressing these threats, which was endorsed by the World Heritage Committee in 2017[5] (Sect. 5.3.2). Key social, cultural, and economic issues influencing the site's protection are also discussed (Sect. 5.3.3). The chapter argues that conservation initiatives are unlikely to succeed unless they are appropriate for Solomon Islands' context. In particular, the economic constraints of the SIG, the development aspirations of the East Rennellese people, and Solomon Islanders' reverence for the rights of customary landowners cannot be ignored.

5.2   The Listing of the East Rennell World Heritage Site

5.2.1   East Rennell's Eligibility for World Heritage Listing

Under the 1997 version of the *Operational Guidelines for the Implementation of the World Heritage Convention*, to qualify for listing as a natural World Heritage site, a place had to

1. meet one or more of the specified criteria[6];
2. meet the conditions of integrity, including having suitable boundaries[7]; and
3. have an adequate protection and management regime.[8]

---

[5] WHC Res 41 COM 7A.19, WHC 41st sess, UN Doc WHC/17/41.COM/18 (12 July 2017) 35.
[6] *Operational Guidelines 1997*, UN Doc WHC 97/2, para 44(a).
[7] Ibid., para 44(b) (i)–(iv).
[8] Ibid., para 44(b) (v)–(vi).

An analysis of the listing of East Rennell with reference to these require-ments sheds light on some of the challenges currently associated with the site's protection.

5.2.1.1 The Criteria for the Assessment of Outstanding Universal Value

East Rennell was listed on the basis that it met the criterion in paragraph 44(a) (ii) of the 1997 *Operational Guidelines*, which referred to sites demonstrating

> significant on-going ecological and biological processes in the evolution and development of terrestrial, freshwater, coastal and marine ecosystems and communities of plants and animals.[9]

East Rennell was found to meet this provision because it is a 'stepping-stone in the migration and evolution of species in the western Pacific'[10] and thus significant speciation processes occur there.

The island of Rennell is an illustration of the theory of island biogeog-raphy.[11] In simple terms, this widely accepted theory posits that the num-ber of species on an island is linked to its size and its distance from the mainland (the source of species).[12] Evidence of the theory can be seen in the western Pacific, where as one moves eastwards, the islands become smaller and more isolated, and biodiversity decreases.[13] With a length of 87 km and an average width of 10 km, Rennell is the largest outlying island in the Solomon Islands group. The island's isolation made inhabita-

---

[9] This is now para 77(ix) in the 2016 version of the *Operational Guidelines*. See UNESCO, *Operational Guidelines for the Implementation of the World Heritage Convention*, UN Doc WHC.16/01 (26 October 2016) para 77(ix).

[10] IUCN, *Evaluations of Nominations of Natural and Mixed Properties to the World Heritage List*, WHC 22nd sess (1998) 79, 82; WHC Res CONF 203 VIII.A.1, WHC 22nd sess, UN Doc WHC-98/CONF/203/18 (29 January 1999) 25, 25.

[11] See, for example, Elspeth J Wingham, *Nomination of East Rennell, Solomon Islands by the Government of Solomon Islands for Inclusion in the World Heritage List Natural Sites* (1997) 35.

[12] Robert J MacArthur and Edward O Wilson, 'An Equilibrium Theory of Insular Zoogeography' (1963) 17 *Evolution* 373. Other factors also affect the biological diversity of islands, such as the island's age, its isolation, and its environmental heterogeneity. See, for example, Kostas A Triantis et al, 'Measurements of Area and the (Island) Species-Area Relationship: New Directions for an Old Pattern' (2008) 117 *Oikos* 1555.

[13] See, for example, Barry Cox and Peter Moore, *Biogeography: An Ecological and Evolutionary Approach*. Blackwell (Oxford, 1980) 109–11.

tion by new species rare, but when species did arrive, they often adapted to their environment by evolving to form new species.[14] Consequently, and because there are few natural predators on Rennell,[15] many endemic species can now be found there, including plants, birds, bats, land snails, and a sea snake.[16] The island is particularly renowned for its unique avifauna (bird life).[17]

It is not clear from the nomination dossier for East Rennell or the records of the World Heritage Committee's deliberations why the site was not nominated on the basis of any of the cultural criteria in the *Operational Guidelines*. It may simply be that those who prepared the nomination dossier considered that the site did not meet any such criteria. The existence of a substantial body of scientific research concerning the environment of Rennell may have also contributed to the decision. Smith has noted that the first sites nominated by Marshall Islands, Papua New Guinea (PNG), and Fiji were well-researched before they were considered for World Heritage listing, which enabled nomination dossiers to be developed with few resources and within a relatively short timeframe.[18] From the 1920s, Rennell was visited by several scientific missions,[19] and was the subject of

---

[14] See, for example, Wingham, above n 11, 35.

[15] Steve Turton, *East Rennell World Heritage Area: Assessment of the State of Conservation of World Heritage Values. Final Field Report* (James Cook University, 2014) 7.

[16] See, for example, Wingham, above n 11, 14–22.

[17] See, for example, Christopher E Filardi et al, 'New Behavioral, Ecological, and Biogeographic Data on the Avifauna of Rennell, Solomon Islands' (1999) 53(4) *Pacific Science* 319; J M Diamond, 'The Avifauna of Rennell Island' in Torben Wolff (ed), *The Natural History of Rennell Island, British Solomon Islands* (Danish Science Press, vol 8, 1984).

[18] Anita Smith, 'World Heritage and Outstanding Universal Value in the Pacific Islands' (2015) 21(2) *International Journal of Heritage Studies* 177, 183.

[19] These include the American Whitney Expeditions in 1928 and 1930, the American Templeton-Crocker Expedition in 1933, the Danish Rennell Expedition in 1951, and the British Museum (Natural History) Expedition in 1953. For a discussion of early expeditions, see Torben Monberg, 'Research on Rennell and Bellona: A Preliminary Report' (1960) 2 *Folk* 71; T Wolff, 'The Fauna of Rennell and Bellona, Solomon Islands' (1969) 255(800) *Philosophical Transactions of the Royal Society of London, Series B, Biological Sciences* 321; Torben Wolff (ed), *The Natural History of Rennell Islands, British Solomon Islands. Scientific Results of the Danish Rennell Expedition, 1951 and the British Museum (Natural History) Expedition 1953* (Danish Science Press, volumes 1–4, 1958–1962); Torben Wolff (ed), *The Natural History of Rennell Island, British Solomon Islands. Scientific Results of the Noona Dan Expedition (Rennell Section, 1962) and The Danish Rennell Expedition 1965* (Danish Science Press, volumes 5–8, 1968).

subsequent research exploring its flora and fauna.[20] This work would have helped those preparing East Rennell's nomination to demonstrate that the site met the natural criteria for World Heritage listing. The 1997 *Operational Guidelines* also presented a barrier to East Rennell's nomination as a cultural site. They stated that cultural sites could only be nominated if they were first included in the State party's Tentative List.[21] As Solomon Islands did not have a Tentative List at that time, the nomination of East Rennell as a cultural site would have been inconsistent with that requirement.[22] The *Operational Guidelines* did not however prevent the site's nomination as a mixed site, and indeed when the Bureau of the Committee[23] reviewed the nomination dossier, it recommended that the SIG assess whether this was feasible.[24] The government indicated it would consider this,[25] but ultimately East Rennell was nominated as a natural site.

As a result, there is significant variation between the international and local heritage value of East Rennell. This is clearly evident from the site's Statement of OUV, which was adopted by the World Heritage Committee in 2012.[26] It describes Rennell island as a 'true natural laboratory for scientific study', and notes that many endemic species can be found there.[27] While it acknowledges that the East Rennellese own the site, it does not

[20] For a comprehensive bibliography of literature on the natural environment of Rennell, see Rolf Kuschel, Torben Monberg, and Torben Wolff, *Bibliography of Rennell and Bellona Islands* (University of Copenhagen, 2nd ed, 2001) http://www.bellona.dk/pdf/publications//bibliography_2nd.pdf.

[21] *Operational Guidelines 1997*, UN Doc WHC 97/2, para 7.

[22] Anita Smith, 'East Rennell World Heritage Site: Misunderstandings, Inconsistencies and Opportunities in the Implementation of the World Heritage Convention in the Pacific Islands' (2011) 17(6) *International Journal of Heritage Studies* 592, 599.

[23] The Bureau of the World Heritage Committee coordinates the Committee's work. It comprises 7 of the 21 State parties that are members of the Committee. See UNESCO, *The World Heritage Committee* http://whc.unesco.org/en/committee/.

[24] *Information on Tentative Lists and Examination of Nominations of Cultural and Natural Properties to the List of World Heritage in Danger and the World Heritage List*, WHC 22nd sess, UN Doc WHC-98/CONF.203/10Rev (29 November 1998) 3.

[25] Letter from Moses K Mose, Permanent Secretary of Solomon Islands Ministry of Commerce, Employment and Tourism, to Bernd von Droste, Director of the UNESCO World Heritage Centre (1 September 1998) attached as supplementary information to Elspeth J Wingham, *Nomination of East Rennell, Solomon Islands by the Government of Solomon Islands for Inclusion in the World Heritage List Natural Sites* (1997) 1.

[26] *Adoption of Retrospective Statements of Outstanding Universal Value*, WHC 36th sess, UN Doc WHC-12/36.COM/8E (15 June 2012) 55–56 (East Rennell, Solomon Islands).

[27] Ibid.

refer to their cultural heritage or the cultural significance of the place. A Statement of OUV is intended to form the basis for a World Heritage site's protection and management.[28] It therefore follows that the World Heritage Committee's key priority regarding East Rennell is ensuring that the threats to the site's terrestrial and marine ecosystems are addressed.

In contrast, the East Rennellese are more concerned about the preservation of their cultural identity, as expressed through their land tenure system, environmental knowledge, traditional resource use, crafts, songs, and dance.[29] They are confused about how their land could be inscribed on the World Heritage List 'without them',[30] which has fuelled their misunderstanding of and disenchantment with the *World Heritage Convention*.

Some East Rennellese would like to see the World Heritage listing expanded to encompass their cultural heritage values.[31] There is precedent for this. Both Uluru-Kata Tjuta National Park in Australia and the Tongariro National Park in New Zealand were initially listed as natural World Heritage sites and subsequently re-listed as cultural landscapes. However, East Rennell is unlikely to be re-nominated in the short term, in part because substantial resources would be required to prepare a new nomination dossier. Furthermore, no study has assessed whether East Rennell meets any of the cultural criteria, so it is unclear whether it would qualify for listing as a cultural or mixed site.

As such, for the foreseeable future, the disparity between the global and local significance of East Rennell is likely to remain. This is contributing to World Heritage not being highly valued at the local level, which in turn limits the SIG's willingness to implement conservation measures. Efforts to protect the site must recognise this, and try to accommodate both global and local values and objectives (discussed further in Sect. 5.3.3).

5.2.1.2 The Conditions of Integrity, Site Boundaries, and Buffer Zones
Under the 1997 *Operational Guidelines*, a site nominated for World Heritage listing had to meet the 'conditions of integrity', which varied depending on the criterion upon which the site was nominated. As East Rennell was nominated based on the criterion in paragraph 44(a) (ii), to

---

[28] UNESCO, *Operational Guidelines for the Implementation of the World Heritage Convention*, UN Doc WHC.16/01 (26 October 2016) paras 154–155.

[29] Smith, above n 22, 605.

[30] Ibid., 597.

[31] Laurie Wein, *East Rennell World Heritage Site Management Plan* (Solomon Islands National Commission for UNESCO, 2007) 14.

meet this requirement, it had to be of 'sufficient size to demonstrate the key aspects of processes that are essential for the long-term conservation of the ecosystems and the biological diversity they contain'.[32] The *Operational Guidelines* also stated that a site's boundaries should reflect the 'spatial requirements of habitats, species, processes or phenomena' that provide the basis for its nomination.[33] In addition, if necessary for the proper conservation of the property, a buffer zone around the property should be established.[34]

The western boundary of the East Rennell World Heritage site is the border between provincial wards 2 and 3 on Rennell island.[35] No buffer zone around the site exists. Notwithstanding this, the nomination dossier contended that the boundaries were sufficient because the site contained the habitats required to maintain its flora and fauna, and there were no large-scale development plans for the island.[36]

IUCN recommended that the site be listed, despite several of its reviewers noting that the area was too small to ensure the long-term survival of endemic birds.[37] In support of its recommendation, IUCN stated that the major feature of the site (Lake Tegano) is in East Rennell, and in any event, the nomination of the entire island was not feasible (because the listing of West Rennell was not supported by the West Rennellese people).[38] The record of the Committee's decision to list the site does not detail any discussion about boundaries or buffer zones, so it is unclear whether the Committee considered that the requirements were met or should be waived.

Several recent reports confirm that East Rennell is too small to ensure the long-term conservation of its OUV.[39] Of particular concern is the

---

[32] *Operational Guidelines 1997*, UN Doc WHC 97/2, para 44(b) (ii).

[33] Ibid., para 44(b) (vi).

[34] Ibid., para 17.

[35] Wingham, above n 11, 38.

[36] Ibid.

[37] IUCN, *Evaluations of Nominations of Natural and Mixed Properties to the World Heritage List*, WHC 22nd sess (1998) 79, 81.

[38] Ibid.

[39] See, for example Turton, above n 15, 7–8, 10–11; Paul Dingwall, *Report on the Reactive Monitoring Mission to East Rennell, Solomon Islands, 21–29 October 2012* (IUCN, 2013) 16; *Adoption of Retrospective Statements of Outstanding Universal Value*, WHC 36th sess, UN Doc WHC-12/36.COM/8E (15 June 2012) 55–56 (East Rennell, Solomon Islands); IUCN, *Natural World Heritage Sites: The Pacific's Challenges* (13 June 2014) https://www. iucn.org/content/natural-world-heritage-sites-pacific%E2%80%99s-challenges;

potential for logging and mining in West Rennell to degrade the site's OUV (see Sect. 5.3.1). Given this, it is questionable whether the site would be found to meet the conditions of integrity if it was nominated today.

The provisions of the *Operational Guidelines* concerning boundaries and buffer zones can be difficult to comply with, particularly for a site under customary tenure. Consequently, in some circumstances, it may be appropriate for them to be applied flexibly, to accommodate the listing of such sites (discussed further in Sect. 6.4). However, the implications of any deviation from the requirements cannot be ignored. The fact that the conservation of East Rennell requires the regulation of activities in West Rennell presents a significant challenge for the site's conservation. The East Rennellese people have no control over that land. Furthermore, to date the SIG has done little to protect the World Heritage site from activities in West Rennell, in part because of its reverence for the rights of the West Rennellese customary owners (see Sect. 5.3.3.2). Protecting East Rennell against the impacts of development occurring outside the site's boundaries will be an ongoing challenge, and will likely only be achieved with the involvement and agreement of the West Rennellese people. Thus, the West Rennellese need to be included in conservation and development initiatives aimed at safeguarding the site.

5.2.1.3 *Management and Protection*

The World Heritage Committee has amended the management and protection requirements in the *Operational Guidelines* several times (see Sect. 4.3.3). In 1997, the *Operational Guidelines* stated that sites nominated for World Heritage listing should have a management plan, but if they did not, the State party should indicate when a plan would be prepared and how it would be resourced.[40] They also said that a nominated natural site should have legislative, regulatory, or institutional protection.[41]

In 1998, at the same meeting at which East Rennell was inscribed on the World Heritage List, the Committee amended the *Operational Guidelines* to provide that natural sites under traditional protection

---

International Centre on Space Technologies for Natural and Cultural Heritage, *Report of the Technical Consultation Meeting on East Rennell World Heritage Site in Danger, Sanya, Hainan Province, China, 1–2 February 2016* (2016) 21.

[40] *Operational Guidelines 1997*, UN Doc WHC 97/2, para 44(b) (v).
[41] Ibid., para 44(b) (vi).

(referred to in this book as customary protection) could qualify for World Heritage listing.[42] This change facilitated the inscription of East Rennell, which at the time had no management plan and little protection under legislation. It was anticipated that a management plan and legislation would be developed to strengthen the site's protection.[43] However, as explained below, it appears that when East Rennell was listed, there was uncertainty about how the site's protection regime would operate.

Customary Protection

East Rennell's nomination dossier contains little information concerning the site's customary protection. It states that the use and management of flora and fauna is regulated through the customary land tenure system and land use practices of the East Rennellese. These practices include seasonal bans on hunting and fishing, *tambus* (prohibitions) on the killing and eating of particular species, and the exclusion of outsiders from communal territory. The dossier contends that these practices were developed to ensure 'sustainable and continued use of natural resources into the future'.[44] It also notes that all major land use decisions are made by the area's chiefs, who make up the Council of Chiefs, which is headed by a Paramount Chief.[45]

However, the dossier contains little information upon which the IUCN and the Committee could assess the scope and strength of the site's customary protection. For example, it does not document the land tenure system or provide details of traditional practices, such as which species they relate to or the extent to which they are complied with. It also provides no basis for the assertion that customary practices are conducive to the conservation of natural resources, nor does it comment on the strength of customary governance.

IUCN expressed concern about the dossier's lack of detail. It noted that customary ownership can provide effective protection, but that presumes that customary practices are favourable to conservation and that 'traditional ownership powers and community support are not being eroded'.[46] This is not an assumption that should be made in the Pacific. As

[42] WHC Res CONF 203 XIV.3, WHC 22nd sess, UN Doc WHC-98/CONF.203/18 (29 January 1999) 56.
[43] Wingham, above n 11, 45.
[44] Ibid.
[45] Ibid., 5.
[46] IUCN, *Evaluations of Nominations of Natural and Mixed Properties to the World Heritage List*, WHC 22nd sess (1998) 79, 81.

explained in Sect. 2.5.2, customary practices in some places were developed to ensure the sustainable use of resources, but the motivation behind other practices included the allocation of resources, and customary and religious beliefs.[47] Furthermore, most customary systems have been significantly influenced by outside contact with the islanders,[48] often limiting their ability to contribute to World Heritage protection.

The record of the World Heritage Committee's decision to inscribe East Rennell notes that Committee members viewed the nomination as 'breaking new ground', and after a 'considerable debate' on customary protection, they agreed to support it.[49] The document does not specify the substance of this debate, but given the dossier's lack of detail, it is unlikely that the Committee had sufficient information to discuss the specifics of East Rennell's customary protection. It therefore appears that the site was listed without a clear understanding of how the customary legal system of the East Rennellese would protect the site against current and foreseeable threats.

Management Plan

While East Rennell had no management plan when it was nominated, the dossier stated that a plan based on customary practices would be prepared.[50] It contended that the plan would have the status of customary law when approved by the Council of Chiefs, so it would strengthen customary protection.[51]

In its review of the dossier, IUCN commented that in the absence of any document detailing objectives and management prescriptions for the

[47] See, for example, K Ruddle, E Hviding and R E Johannes, 'Marine Resources Management in the Context of Customary Tenure' (1992) 7 *Marine Resource Economics* 249, 262.

[48] See, for example, Christophe Sand, 'Melanesian Tribes vs Polynesian Chiefdoms: Recent Archaeological Assessment of a Classic Model of Socio-Political Types in Oceania' (2002) 41(2) *Asian Perspectives* 284, 291; Jennifer Corrin Care and Jean G Zorn, 'Legislating pluralism: Statutory "Developments" in Melanesian Customary Law' (2001) 46 *Journal of Legal Pluralism* 49, 51.

[49] WHC Res CONF 203 VIII.A.1, WHC 22nd sess, UN Doc WHC-98/CONF/203/18 (29 January 1999) 25, 26. The only recorded dissent to the Committee's decision came from the delegate from Thailand, who noted that customary tenure does not guarantee effective protection. The Thai delegate also opposed the listing on the basis that it did not comply with the requirements in the 1997 *Operational Guidelines*. This dissent was reasonable because the Committee's decision to amend the *Operational Guidelines* to allow for the listing of natural sites under customary protection was made after its decision to list East Rennell, albeit at the same meeting.

[50] Wingham, above n 11, 38.

[51] Ibid.

site, it was impossible to confirm how customary practices would provide any protection.[52] Presumably in response to that comment, a document entitled 'Resource Management Objectives and Guidelines' was attached as supplementary information to the dossier.[53] While the document set out broad resource management guidelines, it did not contain any new information about customary protection, instead highlighting that further research on traditional practices was required.[54] IUCN commented that while the document was a good beginning for a management regime, it was unclear whether the East Rennellese would support it.[55]

It therefore appears that when East Rennell was listed, there was uncertainty surrounding when a management plan would be prepared, how it would be resourced, and importantly how it would relate to and strengthen customary protection. As explained in Sect. 6.5, while a management plan was prepared for the site in 2007,[56] it has been relatively ineffective, in part because it has no basis under customary or State law.

State Legislation

When East Rennell was nominated, Solomon Islands had no World Heritage protection legislation. The site's nomination dossier stated that the SIG would enact a *World Heritage Cultural and Natural Sites Act*,[57] but did not specify what form this legislation would take or how it would interact with customary law. A *World Heritage Properties Conservation Bill* was prepared.[58] However, by the time IUCN finalised its review of the dossier, it had received advice that the SIG was not pursuing this

---

[52] IUCN, *Evaluations of Nominations of Natural and Mixed Properties to the World Heritage List*, WHC 22nd sess (1998) 79, 82.

[53] Elspeth J Wingham, *Resource Management Objectives and Guidelines for East Rennell, Solomon Islands (May 1998)*, attached as attachment 1 to Elspeth J Wingham, *Nomination of East Rennell, Solomon Islands by the Government of Solomon Islands for Inclusion in the World Heritage List Natural Sites* (1997).

[54] Ibid., 1, 19.

[55] IUCN, *Evaluations of Nominations of Natural and Mixed Properties to the World Heritage List*, WHC 22nd sess (1998) 79, 83.

[56] Wein, above n 31.

[57] Wingham, above n 11, 38.

[58] Ben Boer, 'Solomon Islands' in Ben Boer (ed), *Environmental Law in the South Pacific: Consolidated Report of the Reviews of Environmental Law in the Cook Islands, Federated States of Micronesia, Kingdom of Tonga, Republic of the Marshall Islands and Solomon Islands* (South Pacific Regional Environment Programme and IUCN Environmental Law Centre, 1996) 189, 193; Ben Boer, *Solomon Islands: Review of Environmental Law* (SPREP, 1993) 11.

legislation.[59] IUCN expressed concern about this, noting that appropriate legislation would reinforce customary rights and ensure some legal commitment to World Heritage at the national level. It also recognised that implementing such a law would be challenging, stating that the land tenure of Rennell 'makes it difficult (but not impossible) for national government legislation to be effective in terms of management'.[60] The Committee ultimately accepted that East Rennell could be listed notwithstanding the lack of World Heritage protection legislation, but it recommended that such a law be developed.[61]

As explained in Chap. 7, the SIG never passed the *World Heritage Properties Conservation Bill* and the site remains only weakly protected under State law. Importantly, complex issues concerning the relationship between heritage protection legislation and customary law remain to be addressed.

5.2.2   The Listing of East Rennell in Context

The analysis above shows that East Rennell was listed despite there being a lack of clarity concerning the site's protection regime. Other sites in the Pacific have also been listed notwithstanding the relevant Advisory Body recommending to the Committee that the nominations be deferred to allow the State party to strength the protection of the property. For example, in its review of the nomination dossier for the Chief Roi Mata's Domain site in Vanuatu, the International Council on Monuments and Sites (ICOMOS) commented that 'the lack of legal protection for the core and buffer zone is a cause for concern'.[62] Regarding the Rock Islands Southern Lagoon site in Palau, ICOMOS stated that 'legal protection in place is not yet adequate and thus overall the protective measures for the property are not adequate'.[63] Similarly, IUCN commented in relation to

---

[59] IUCN, *Evaluations of Nominations of Natural and Mixed Properties to the World Heritage List*, WHC 22nd sess (1998) 79, 83.

[60] Ibid.

[61] WHC Res CONF 203 VIII.A.1, 22nd sess, UN Doc WHC-98/CONF/203/18 (29 January 1999) 25.

[62] ICOMOS, *Evaluations of Nominations of Cultural and Mixed Properties to the World Heritage List*, WHC 32nd sess, UN Doc WHC-08/32.COM/INF/8B1 (2008) 92 (Chief Roi Mata's Domain, Vanuatu, Advisory Body Evaluation 1280) 98.

[63] ICOMOS, *Evaluations of Nominations of Cultural and Mixed Properties to the World Heritage List*, WHC 36th sess, UN Doc WHC-12/36.COM/INF.8B1 (2012) 21 (Rock Islands Southern Lagoon, Republic of Palau, Advisory Body Evaluation 1386) 28.

the Phoenix Islands Protected Area in Kiribati that the property did not fully meet the requirements of the *Operational Guidelines* in relation to protection and management.[64] Despite these concerns, these sites were all inscribed on the World Heritage List.

It may be that the Committee's desire to list sites in the Pacific to help address the imbalances in the World Heritage List (in accordance with the *Global Strategy for a Representative, Balanced and Credible World Heritage List*) influenced its decision-making. Indeed, when the Committee resolved to list East Rennell, several State party delegates noted the contribution the listing would make to the implementation of that strategy.[65] Additionally, before East Rennell was nominated, the only sites inscribed on the World Heritage List based on customary protection were 'cultural landscapes',[66] which had different heritage values and management requirements.[67] Consequently, there were no analogous precedents against which the nomination dossier could be compared.

Whatever the reason, when future sites are nominated, it may be beneficial for their protection regimes to be more comprehensively investigated. As explored further in Chap. 6, this should involve considering the scope of customary laws, the strength of customary governance, and if and how customary protection can be supplemented by a management plan and/or legislation. Customary tenure presents unique challenges for World Heritage protection not experienced at sites under State and private ownership and control. Enthusiasm to support customary protection should not translate into an assumption that customary landowners and the relevant State parties are willing and able to protect World Heritage to the same standard as other sites. A thorough assessment of the site's protection regime at the nomination stage may assist all stakeholders to agree upon feasible and appropriate conservation objectives, and to anticipate and address challenges concerning the site's protection.

It is also evident that East Rennell was listed before the SIG had established the administrative structures and legal instruments required to fully

---

[64] IUCN, *Evaluations of Nominations of Natural and Mixed Properties to the World Heritage List*, WHC 34th sess, UN Doc WHC/10/34.COM/INF.8B2 (2010) 19 (Phoenix Islands Protected Area, Kiribati, Advisory Body Evaluation 1325) 22.

[65] WHC Res CONF 203 VIII.A.1, WHC 22nd sess, UN Doc WHC-98/CONF/203/18 (29 January 1999) 25, 26.

[66] Tongariro National Park (New Zealand), Uluru-Kata Tjuta National Park (Australia) and the Rice Terraces of the Philippine Cordilleras (The Philippines).

[67] Smith, above n 22, 600.

implement the *Convention*. The SIG must now try to 'catch up', by developing measures that ideally should have been in place when the site was nominated. This is difficult, in part because the SIG has limited resources to dedicate to the task, and other pressing priorities. Civil conflict, political instability, and governance issues have also hindered its progress in complying with its *Convention* obligations (discussed further in Sect. 5.3.3). If the *Convention* is to be implemented effectively in the region, Pacific Island States need support to not only prepare nominations, but also develop and implement the measures needed to manage and protect their listed sites (see Sect. 8.4.3).

5.3   The Protection of the East Rennell World Heritage Site

For several years after East Rennell was listed, IUCN and the World Heritage Committee considered that East Rennell's heritage values were relatively intact.[68] However, following a reactive monitoring mission to the area in 2012,[69] IUCN contended that the threats to the site were sufficiently serious as to warrant the site's inclusion in the List of World Heritage in Danger.[70] The World Heritage Committee inscribed the site on that List in 2013,[71] where it remains in 2018.[72] As explained here, East Rennell is unlikely to be removed from that List unless its protection under customary and State law is strengthened. Initiatives to achieve this must be designed to fit the Solomon Islands context.

---

[68] Salamat Ali Tabbasum and Paul Dingwall, *Report on the Mission to East Rennell World Heritage Property and Marovo Lagoon, Solomon Islands*, 30 March–10 April 2005 (IUCN and World Heritage Centre, 2005) 5, 11.

[69] Reactive monitoring missions are conducted by the Advisory Bodies and other groups, at the request of the Committee to ascertain a site's state of conservation: see UNESCO, *Operational Guidelines for the Implementation of the World Heritage Convention*, UN Doc WHC.16/01 (26 October 2016) Part IV.A.

[70] For the report of this mission, see Dingwall, above n 39.

[71] WHC Res 37 COM 7B.14, WHC 37th sess, UN Doc WHC-13/37.COM/20 (5 July 2013) 68.

[72] WHC Res 41 COM 7A.19, WHC 41st sess, UN Doc WHC/17/41.COM/18 (12 July 2017) 35.

5.3.1   The State of Conservation of East Rennell

In 2014, IUCN ranked East Rennell's status as 'critical', and contended that the conditions required for the site to qualify for listing may no longer be in place.[73] Recent studies have found that the heritage values of East Rennell remain relatively intact, perhaps calling into question IUCN's dire assessment of the site's current state of conservation.[74] These studies do however confirm that East Rennell's OUV is under threat from logging and mining, the over-harvesting of certain species, invasive species, and climate change.

5.3.1.1   Logging and Mining

Rennell is increasingly attracting logging companies, which find the island's 'pencil cedar' (*Palaquium* sp.) most lucrative.[75] As the forests across the island are intrinsically linked, logging and mining in either East or West Rennell could impact the World Heritage site.[76] Logging only commenced relatively recently in West Rennell,[77] and has involved selective rather than clear felling, and thus forest cover in that part of the island remains above 90%.[78] Logging in West Rennell could however affect the World Heritage site by reducing the forest cover required to maintain bird populations, changing groundwater hydrology, decreasing the island's resilience to cyclones, and facilitating the introduction of invasive spe-

[73] IUCN, *World Heritage Outlook – East Rennell* http://www.worldheritageoutlook.iucn. org/explore-sites/wdpaid/168242.

[74] Simon Albert et al, *Survey of the Condition of the Marine Ecosystem within the East Rennell World Heritage Area, Solomon Islands* (University of Queensland, Solomon Islands Marine Ecology Laboratory, Griffith University and WWF-Solomon Islands, 2013) 36; Turton, above n 15, 10.

[75] Nils Finn Munch-Petersen, 'An Island Saved, At Least for Some Time? The Advent of Tourism to Rennell, Solomon Islands' in Godfrey Baldacchino and Daniel Niles (eds), *Island Futures: Conservation and Development Across the Asia-Pacific Region* (Springer, 2011) 169, 173.

[76] Dingwall, above n 39, 13–18; Turton, above n 15, 7, 11.

[77] The first logging licence for West Rennell was granted in 2008: Dingwall, above n 39, 13. This is relatively recent compared to other parts of the Solomon Islands. See, for example, Ian Frazer, 'The Struggle for Control of Solomon Island Forests' (1997) 9(1) *Contemporary Pacific* 39. Frazer notes that large-scale logging in Solomon Islands began in the 1960s, and accelerated in the 1980s when companies started to operate on customary land: at 46.

[78] International Centre on Space Technologies for Natural and Cultural Heritage, above n 39, 1, 18, 20.

cies.[79] Logging has not yet occurred in East Rennell, but applications for licences to log within the site have been made.

Interest in mining on Rennell stretches back to the protectorate era.[80] Prospecting conducted on Rennell between 1969 and 1977 revealed substantial reserves of bauxite, but mining did not proceed at that time.[81] Prospecting commenced again in West Rennell in 2014, and since then two companies have commenced mining there.[82] These operations involved targeting bauxite pocket soil deposits, which according to the SIG minimises environmental impact.[83] However, this is unproven, and mining could have similar impacts on East Rennell's OUV as logging.[84] Companies have expressed interest in mining within the World Heritage site, but it appears that no operations have been approved yet.[85]

5.3.1.2 Over-harvesting

East Rennell's OUV is also threatened by the over-harvesting of certain species. A key species of concern is coconut crab (*Birgus latro*, locally known as kasusu), which is caught by the Rennellese people for consumption and sale.[86] These crabs are susceptible to over-exploitation because they mature very slowly.[87] A recent report suggests that at East Rennell they are harvested all year around, including when females are carrying

---

[79] Dingwall, above n 39, 4; Turton, above n 15, 7–8, 10–11, 14.

[80] For discussion of history of mining on Rennell, see generally Peter Larmour, 'Sharing the Benefits: Customary Landowners and Natural Resource Projects in Melanesia' (1989) 36 *Pacific Viewpoint* 56; David Ruthven, 'Rennell Bauxite' in Peter Larmour (ed), *Land in Solomon Islands* (Institution of Pacific Studies and Ministry of Agriculture and Lands, 1979) 94; Colin Filer, 'Between a Rock and a Hard Place: Mining, "Indigenous People" and the Development of States' in Benedict Y Imbun and Paul A McGavin (eds), *Mining in Papua New Guinea: Analysis and Policy Implications* (University of Papua New Guinea Press, 2001) 7.

[81] For discussion of the reasons for this, see Larmour, above n 80; John McKinnon, *Solomon Islands World Heritage Site Proposal: Report on a Fact Finding Mission (4–22 February 1990)* (Victoria University of Wellington, 1990) 17; John Smith, *An Island in the Autumn* (Librario Publishing Ltd, 2012) 59.

[82] Solomon Islands Government, *State Party Report on the State of Conservation of East Rennell (Solomon Islands)* (SIG, 2017) 5.

[83] Ibid., 6.

[84] Turton, above n 15, 14.

[85] Aatai John, 'No Mining at Lake Tegano', *The Solomon Star* (online), 20 February 2017 http://www.solomonstarnews.com/news/national/12281-no-mining-at-lake-tegano.

[86] Turton, above n 15, 10; Dingwall, above n 39, 21–22, 32.

[87] Dingwall, above n 39, 22.

eggs.[88] Coconut crabs are no longer found in West Rennell, and there is a risk they will be harvested to extinction in East Rennell as well.[89]

The over-exploitation of marine resources is also a concern.[90] Albert et al. found that subsistence fishing is unlikely to significantly affect marine ecosystems in the short to medium term, because the island has a low population and accessing the ocean from most villages is relatively difficult.[91] Commercial and artisanal fishing pressures are also low, but could substantially increase if access to markets improves.[92] The most significant current concern is the over-harvesting of commercially valuable invertebrate species, including beche de mer (which is processed from *holothurians*, commonly known as sea cucumbers) and trochus (*Trochus niloticus*).[93]

5.3.1.3 Invasive Species

Invasive species, particularly the black ship rat (*Rattus rattus*) and the giant African snail (*Achatina* spp.), are a significant threat to the OUV of East Rennell. Ship rats have recently been observed within the World Heritage site.[94] Some reports say that the animal was probably introduced into West Rennell from logging vessels,[95] but a recent study contends its introduction predates the commencement of logging.[96] Regardless, logging and mining create habitats favoured by the rats and thus increase their spread.[97] They could potentially affect the site's OUV by reducing endemic bird and snail populations.[98]

The giant snail is another species of concern. They are now prevalent in Honiara, and could be introduced to Rennell on ships and aircraft.[99] If

---

[88] Ibid.

[89] Ibid., 5; Turton, above n 15, 10.

[90] Dingwall, above n 39, 19–21; WHC Res 37 COM 7B.14, WHC 37th sess, UN Doc WHC-13/37.COM/20 (5 July 2013) 68, 68.

[91] Albert et al, above n 74, 36.

[92] Ibid.

[93] Ibid. Albert et al. do however note that the low abundance they encountered could be a result of the sampling method used: at 28.

[94] International Centre on Space Technologies for Natural and Cultural Heritage, above n 39, 22.

[95] Dingwall, above n 39, 4; Turton, above n 15, 12.

[96] International Centre on Space Technologies for Natural and Cultural Heritage, above n 39, 22.

[97] Ibid.

[98] Dingwall, above n 39, 4. Turton, above n 15, 13.

[99] Dingwall, above n 39, 4. Turton, above n 15, 13–14.

that occurred, the snails could compete with native fauna of the island.[100] Both ship rats and giant snails could also destroy crops, affecting the food security of the East Rennellese people.[101]

5.3.1.4 Climate Change

Climate change is becoming one of the most significant threats facing World Heritage sites,[102] and East Rennell has been identified as 1 of the 19 such places at most risk.[103] Predicted impacts include an increase in the level and salinity of Lake Tegano, which could affect aquatic and lakeside ecology.[104]

5.3.2 Achieving the Desired State of Conservation for the Removal of East Rennell from the List of World Heritage in Danger

The World Heritage Committee has repeatedly called upon Solomon Islands to do more to address the threats to East Rennell. For example, it has urged the SIG to strengthen the regulation of logging and mining on the island,[105] ensure that the harvesting of species is sustainable,[106] implement biosecurity controls,[107] and incorporate climate change

---

[100] Dingwall, above n 39, 4. Turton, above n 15, 13.

[101] Ibid.

[102] See, for example, A Markham et al, *World Heritage and Tourism in a Changing Climate* (UNEP and UNESCO, 2016) 9.

[103] Jim Perry, 'World Heritage Hot Spots: A Global Model Identifies the 16 Natural Heritage Properties on the World Heritage List Most at Risk From Climate Change' (2011) 17(5) *International Journal of Heritage Studies* 426, 426.

[104] Dingwall, above n 39, 22–24. Turton, above n 15, 7.

[105] WHC Res 34 COM 7B.17, WHC 34th sess, UN Doc WHC-10/34.COM/20 (3 September 2010) 71, 71; WHC Res 36 COM 7B.15, WHC 36th sess, UN Doc WHC-12/36.COM/19 (June–July 2012) 63, 63; WHC Res 37 COM 7B.14, WHC 37th sess, UN Doc WHC 13/37.COM/20 (5 July 2013) 68, 68; WHC Res 38 COM 7A.29, WHC 38th sess, UN Doc WHC-14/38.COM/16 (7 July 2014) 39, 40; WHC Res 39 COM 7A.16, WHC 39th sess, UN Doc WHC-15/39.COM/19 (8 July 2015) 30, 30; WHC Res 40 COM 7A.49, WHC 40th sess, UN Doc WHC-16/40.COM/19 (15 November 2016) 68, 69.

[106] WHC Res 37 COM 7B.14, WHC 37th sess, UN Doc WHC-13/37.COM/20 (5 July 2013) 68, 68.

[107] WHC Res 41 COM 7A.19, WHC 41st sess, UN Doc WHC/17/41.COM/18 (12 July 2017) 35, 35.

adaptation and mitigation measures into the site's management plan.[108] These requests remain largely unfulfilled.

Recently, however, the SIG took a significant step by developing a strategic framework for safeguarding the site. That strategy (referred to as the *Desired State of Conservation for the Removal of East Rennell from the List of World Heritage in Danger* or *DSOCR*) was endorsed by the Committee in 2017.[109] It sets out indicators for the removal of the property from the List of World Heritage in Danger, concerning the maintenance of forest cover, resource development, invasive species, harvesting of certain species, and site management. It also specifies measures for verifying whether those indicators have been achieved. They cover a range of techniques, including the use of satellite imagery and scientific assessments, and importantly steps to improve the site's protection. For example, the *DSOCR* calls for the implementation of sustainable harvesting limits based on customary resource use regimes. It states that a new management plan should be developed, supported by an action plan to help the East Rennellese people undertake income generating projects. It also calls for stronger regulation of approval processes for logging and mining in West Rennell, and the implementation of the *Protected Areas Act 2010* at the site. These measures are explored further in later chapters.

While the adoption of the *DSOCR* was a significant step, the document is a framework only. Much work remains to be done to develop and implement the specific measures required to facilitate East Rennell's removal from the List of World Heritage in Danger. SIG has been granted funding through the World Heritage Fund to assist with the process of developing an action plan for achieving the *DSOCR*.[110] Like any conservation initiative, this action plan is unlikely to be effective unless it is appropriate for the Solomon Islands context.

---

[108] WHC Res 37 COM 7B.14, WHC 37th sess, UN Doc WHC-13/37.COM/20 (5 July 2013) 68, 68.

[109] *State of Conservation of Properties Inscribed on the World Heritage List*, WHC 41st sess, UN Doc WHC/17/41.COM/7A.Add (2 June 2017) 26 (East Rennell, Solomon Islands); WHC Res 41 COM 7A.19, WHC 41st sess, UN Doc WHC/17/41.COM/18 (12 July 2017) 35.

[110] *State of Conservation of Properties Inscribed on the World Heritage List*, UN Doc WHC/17/41.COM/7A.Add, 28.

5.3.3   The Protection of East Rennell in Context

Chapter 2 outlined some the key challenges associated with implementing the *World Heritage Convention* in the Pacific. As was explained, many countries in the region have experienced low or even negative economic growth, and thus economic and social development is generally a higher priority than heritage conservation. Governance issues, such as political instability and corruption, have plagued some Pacific Island States, impeding the development and implementation of policies on national issues such as World Heritage. The lack of relevance of the national government to many Pacific Islanders and the difficulties associated with implementing and enforcing heritage protection legislation also present challenges.

Building on Chap. 2, three further issues of particular relevance to Solomon Islands are explored here. Firstly, the impact of civil conflict and State-building challenges on SIG's involvement in World Heritage initiatives. Secondly, Solomon Islanders' perceptions of the role of the State in the protection of sites under customary tenure. Finally, the priorities of the East Rennellese people. These issues must be taken into account in the design of any future conservation initiatives for East Rennell.

5.3.3.1   Civil Conflict and State-Building Challenges in Solomon Islands

Soon after East Rennell was listed, civil conflict (commonly referred to as 'the tensions') broke out in Solomon Islands.[111] Beginning in late 1998, regular skirmishes between armed militia from Guadalcanal and Malaita occurred in and around Honiara.[112] The fighting escalated in 2000, when militants from Malaita seized control of Honiara and the Prime Minister

---

[111] For analysis of the tensions, see generally Judith Bennett, *Roots of Conflict in Solomon Islands – Though Much is Taken, Much Abides: Legacies of Tradition and Colonialism*, State, Society and Governance in Melanesia Discussion Paper (Australian National University, 2002); Sinclair Dinnen, 'State-Building in a Post-Colonial Society: The Case of Solomon Islands' (2008) 9 *Chicago Journal of International Law* 51; Sinclair Dinnen, 'The Solomon Islands Intervention and the Instabilities of the Post-Colonial State' (2008) 20(3) *Global Change, Peace and Security* (formerly *Pacific Review: Peace, Security and Global Change*) 338; Clive Moore, 'Pacific View: The Meaning of Governance and Politics in the Solomon Islands' (2008) 62(3) *Australian Journal of International Affairs* 386; John Braithwaite et al, *Pillars and Shadows: Statebuilding as Peacebuilding in Solomon Islands* (ANU E Press, 2010); Matthew G Allen, 'Land, Identity and Conflict on Guadalcanal, Solomon Islands' (2012) 43(2) *Australian Geographer* 153.

[112] See, for example, Dinnen, 'State-Building in a Post-Colonial Society', above n 111, 61.

was forced to resign.[113] Despite attempts by Australia and New Zealand to broker peace talks, the conflict continued, and Solomon Islands' central and provincial governments effectively ceased to function. The tensions caused the country's gross domestic product to fall by 24%, and by 2002 the government was insolvent.[114]

The violence caused by the tensions mainly occurred on Guadalcanal and Malaita, allowing people on other islands (including Rennell) to continue to live subsistence lifestyles,[115] pursuant to their customary legal systems.[116] However, as the SIG was dysfunctional during this period, it was not involved with any World Heritage activities and it had little communication with the World Heritage Committee or the East Rennellese people.[117] The outbreak of the tensions also led to the cancellation of New Zealand's World Heritage programme in Solomon Islands, and a Japanese funded project to assess Rennell's cultural values.[118] There was also little activity concerning East Rennell at the international level, with the Committee making no resolutions relating to the site until 2003.

By mid-2003, Australia saw the situation in Solomon Islands as a threat to Australian and regional security,[119] and the Regional Assistance Mission to the Solomon Islands (RAMSI) was formed. RAMSI quelled the fighting relatively quickly, enabling the SIG to recommence its engagement with the *World Heritage Convention* regime. However, its involvement with World Heritage remains limited.

In 2003, the SIG established the Solomon Islands National Commission, to manage its programmes associated with the United Nations Educational, Scientific and Cultural Organisation (UNESCO), including World

---

[113] See, for example, Moore, above n 111, 387.

[114] Daniel Gay (ed), *Solomon Islands Diagnostic Trade Integration Study 2009 Report* (Solomon Islands Government, 2009) 19.

[115] See, for example, Moore, above n 111, 387.

[116] Graham Baines, *Beneath the State: Chiefs of Santa Isabel, Solomon Islands, Coping and Adapting*, State, Society and Governance Working Paper 2014/2 (Australian National University, 2014) 3.

[117] Tabbasum and Dingwall, above n 68, 5; *State of Conservation of Properties Inscribed on the World Heritage List*, WHC 22nd sess, UN Doc WHC-03/27.COM/7B (12 June 2003) 11 (East Rennell, Solomon Islands) 11; Wein, above n 31, 7.

[118] *State of Conservation of Properties Inscribed on the World Heritage List*, WHC 22nd sess, UN Doc WHC-03/27.COM/7B (12 June 2003) 11 (East Rennell, Solomon Islands) 11.

[119] See, for example, Dinnen, 'State-Building in a Post-Colonial Society', above n 111, 63.

Heritage.[120] The National Commission later set up a sub-commission to coordinate the SIG's World Heritage activities,[121] but the sub-commission was inactive for many years. Confusion as to which government Ministry was responsible for World Heritage also contributed to the government's lack of engagement. This has also been a problem in other Pacific Island States, including Fiji.[122] The situation in Solomon Islands improved in 2011 when SIG confirmed that responsibility was shared between the Ministries for Environment and Culture.[123] More recently, the SIG established an inter-ministerial 'core team' to administer the *DSOCR*.[124] It remains to be seen however whether this will lead to greater government involvement in the management and protection of East Rennell.

Like most other Pacific Island States, the Solomon Islands government has very limited financial and human resources to dedicate to heritage conservation. The nature of the political system in Solomon Islands, in which elected members often feel pressured to provide direct benefits to the constituents who voted them in,[125] also impedes the development and implementation of policies concerning national issues. Additionally, governance problems, including extreme political instability and corruption, have contributed to the government's lack of engagement with World Heritage. These issues exist at both the national and the provincial level. Indeed, the Rennell and Bellona provincial government was suspended in 2014 following allegations of financial and administrative mismanagement.[126] As such, conservation measures that are highly dependent on substantial long-term government input are unlikely to succeed.

---

[120] Interview by the author with an officer in the Ministry of Education, who was formerly the focal point for World Heritage within the Solomon Islands National Commission for UNESCO (Honiara, 28 July 2013).

[121] Ibid.; Solomon Islands Government, *State Party Report on the State of Conservation of the East Rennell World Heritage Area (Solomon Islands)* (SIG, 2012) 3.

[122] Erika J Techera, 'Safeguarding cultural heritage: Law and policy in Fiji' (2011) 12 *Journal of Cultural Heritage* 329, 331.

[123] Letter from Aseri Yalangono, Deputy Secretary General of National Commission for UNESCO Solomon Islands to the Director of the UNESCO World Heritage Centre (31 August 2011) 1.

[124] Solomon Islands Government, above n 82, 2.

[125] See, for example, Dinnen, 'State-Building in a Post-Colonial Society', above n 111.

[126] Minister for Provincial Government and Institutional Strengthening, 'Rennell and Bellona Provincial Government (Suspension of Executive Powers) Order 2014' in Solomon Islands, *Extraordinary Gazette*, No 81, 5 September 2014, 184.

5.3.3.2 The Role of the State in Protecting Sites Uunder Customary Tenure

The *World Heritage Convention* imposes an obligation on State parties to implement the legal measures required to protect World Heritage[127] (see Sect. 4.2.3). However, it gives State parties discretion to tailor the measures to fit their legal, economic, and political context. This discretion is not diminished by the *Operational Guidelines*. In relation to the protection of World Heritage under law, they merely state that legislative and regulatory measures should ensure that the property is protected from pressures or changes that might negatively impact its OUV.[128]

From a legal point of view, the fact that a site is listed based on its customary protection does not derogate from the State party's duty to protect the site. Having ratified the *Convention*, a State party must implement the treaty in good faith,[129] and it cannot justify any failure to do so on the basis of its domestic law.[130] Consequently, a State party is still required to implement the legal measures necessary to protect its World Heritage even if the site qualified for World Heritage listing because of its customary protection. Furthermore, from a practical point of view, successful heritage protection will often require a combination of both customary and State approaches. The State party is therefore faced with the task of developing laws and other measures to comply with its *Convention* obligations, whilst also respecting and supporting the customary system that enabled the site to be listed in the first place.

As the East Rennell case study demonstrates, achieving this in practice can be challenging. The State party may consider itself unable and/or unwilling to implement the measures that the Committee and the Advisory Bodies consider are necessary to protect OUV, because of the site's customary ownership. For example, the SIG has repeatedly noted the central role of the East Rennellese in ensuring the conservation of the World Heritage site. In a letter to the World Heritage Centre attached to the East Rennell nomination dossier, a representation of the SIG wrote:

---

[127] *World Heritage Convention* arts 4–5.

[128] UNESCO, *Operational Guidelines for the Implementation of the World Heritage Convention*, UN Doc WHC.16/01 (26 October 2016) para 98.

[129] *Convention on the Law of Treaties*, opened for signature 23 May 1969, 1155 UNTS 331 (entered into force 27 January 1980) art 26.

[130] Ibid., art 27.

It should be emphasized that the proposed East Rennell World Heritage site is in customary land ownership and the long term wise management of the site will depend on the commitment made by the local people.[131]

This was reiterated in a letter from the Solomon Islands' National Commissioner for Culture to the World Heritage Centre in 2004, in which the Commissioner indicated that

[it is] not appropriate for the national government to prepare national legislation to regulate a property governed by customary ownership where land is protected by traditional laws recognized by the National Constitution.[132]

Current and former SIG employees have made similar comments, when interviewed by the author. For example, an employee within the Ministry of Culture stated:

We [the government] cannot throw up a management plan from here or pick it from anywhere and go to East Rennell and say this is how we do it. What they [the East Rennellese people] say about their land is just as strong as us.[133]

Joe Horokou (Director of the Environment and Conservation Division of the Ministry of Environment) commented that the Ministry of Environment has 'no direct authority over the site', and the East Rennellese people have the right to make the final decision about their resources.[134] A conservation officer within the Ministry of Environment stated that it was difficult for the State to require good resource management at the site, because the government does not own the natural resources.[135] She added

---

[131] Letter from Moses K Mose, Permanent Secretary of Solomon Islands Ministry of Commerce, Employment and Tourism, to Bernd von Droste, Director of the UNESCO World Heritage Centre (1 September 1998) attached as supplementary information to Elspeth J Wingham, *Nomination of East Rennell, Solomon Islands by the Government of Solomon Islands for Inclusion in the World Heritage List Natural Sites* (1997) 2.

[132] *State of Conservation of Properties Inscribed on the World Heritage List*, WHC 28th sess, UN Doc WHC-04/28.COM/15B (15 June 2004) 15 (East Rennell, Solomon Islands) 15.

[133] Interview by the author with an officer in the Ministry of Culture (Honiara, 26 July 2013).

[134] Interview by the author with Joe Horokou, Director of the Environment and Conservation Division of the Ministry of Environment (Honiara, 15 August 2013).

[135] Interview by the author with a conservation officer in the Ministry of Environment (Honiara, 2 August 2013).

that in her view Solomon Islands' governance system is not a conducive environment for implementation of the *World Heritage Convention*.[136] SIG officials have also stressed that because the East Rennellese people rely on their natural resources for their livelihoods, the State would have to compensate them if it were to restrict resource use or development on the island. For example, Malchoir Mataki (Permanent Secretary of the Ministry of Environment) commented on the lack of employment opportunities on Rennell, noting that people in East Rennell 'cannot go and work in a factory'.[137] He said that if the government constrained the people from using their resources, it would need to provide them with opportunities elsewhere.[138] An employee within the Ministry of Culture contended that World Heritage protection would be easier if the East Rennellese people had access to alternative livelihood options.[139] Another SIG employee contrasted East Rennell and Tetepare island, in the Western province of Solomon Islands (see Fig. 1.3). He noted that because Tetepare is uninhabited, its customary owners can access resources to support their livelihoods elsewhere, whereas East Rennell is all that the East Rennellese have.[140] Other similar comments from SIG officials have included:

Protecting the site is difficult because they [the East Rennellese people] use the resources we [the government] want to conserve.[141]

The government has an obligation to allow people to grow and development.[142]

It is not practical to deny people from harvesting some of the things they require from the environment. It's their livelihood.[143]

---

[136] Ibid.

[137] Interview by the author with Malchoir Mataki, Permanent Secretary of the Ministry of Environment (Honiara, 1 October 2013).

[138] Ibid.

[139] Interview by the author with an officer in the Ministry of Culture (Honiara, 26 July 2013).

[140] Interview by the author with an officer in the Ministry of Education, who was formerly the focal point for World Heritage within the Solomon Islands National Commission for UNESCO (Honiara, 28 July 2013). Tetepare is discussed further in Sect. 8.2.

[141] Ibid.

[142] Interview by the author with Bradley Tovosia, Minister for Environment (Honiara, 24 September 2013).

[143] Interview by the author with Joe Horokou, Director of the Environment and Conservation Division of the Ministry of Environment (Honiara, 15 August 2013).

> If logging were banned, it would require UNESCO to go there and provide alternative livelihood options for the East Rennellese people.[144]

> A request to ban logging must come with a responsibility from the international community to assist with that process.[145]

These comments suggest that the SIG is unlikely to implement any legislation that substantially restrains local peoples' use and development of their land, unless the measures have widespread local support. They help explain why (for example) the World Heritage Committee's calls for SIG to unilaterally ban logging and mining on Rennell have for many years fallen on deaf ears (see Sect. 7.3.1).

The SIG's approach does however place the government in a difficult position. Under the *World Heritage Convention*, it has an obligation to implement the legal measures required to protect East Rennell's OUV, yet it perceives that it lacks the authority (legal or otherwise) to dictate how customary landowners use their land and resources. As a conservation officer within the Ministry of Environment stated:

> Communities will always say they have a need for subsistence and income. Government will always say that it has international obligations. Getting the two to match up is difficult.[146]

It is unhelpful to advocate for the SIG to implement conservation measures that fundamentally diverge from the views of Solomon Islanders concerning the rights of customary landowners, or that will significantly impinge on the livelihoods of the East Rennellese people. Conservation efforts must seek to resolve the tension between SIG's international obligations and its reverence for customary rights, and must be accompanied by initiatives to support the livelihood needs of the local communities (see Sects. 8.2.2 and 8.4.4.4).

---

[144] Interview by the author with an officer in the Ministry of Culture (Honiara, 26 July 2013).

[145] Interview by the author with a conservation officer in the Ministry of Environment (Honiara, 2 August 2013).

[146] Ibid.

5.3.3.3 Local Priorities and Aspirations

The SIG supported the nomination of East Rennell in part because of the economic benefits that it anticipated would flow from a World Heritage site. A tourism proposal prepared for Rennell in the late 1980s included a recommendation that the island be nominated for World Heritage listing.[147] Around this time, the SIG was increasing its efforts to establish a tourism industry in Solomon Islands, and World Heritage was viewed as a means of achieving this.[148] The fact that the SIG chose its newly established Ministry of Tourism and Aviation to manage the nomination of East Rennell demonstrates a strong economic rationale behind its decision to support the listing.

The SIG's perception of World Heritage as a mechanism for enhancing economic development was shared by the East Rennellese people. In conjunction with preparation of the nomination dossier, the New Zealand government supported the development of ecotourism in East Rennell. It funded the construction of guesthouses (see Fig. 5.1), the purchase of canoes, and the establishment of small businesses such as bee keeping, a bakery, and a poultry farm.[149] These initiatives contributed to the high level of local support for the nomination (which was estimated at 80% of the adult population[150]). As the nomination dossier states:

> [T]he small business component of the [World Heritage] project is the area that is of the most interest to local people. Some people are interested in looking after the environment but they all require a means to make money.[151]

Despite their early enthusiasm, many East Rennellese people are now disenchanted with World Heritage. This is partly because their cultural heritage was not recognised in the site's listing (discussed in Sect. 5.2.1). In addition, they are disappointed that the listing of their land brought them few tangible benefits.[152] None of the small-scale ecotourism projects funded

---

[147] Nils Finn Munch-Petersen, above n 75, 173.

[148] McKinnon, above n 81, 35–36.

[149] Dingwall, above n 39, 8.

[150] Wingham, above n 11, 39.

[151] Ibid., 17.

[152] Smith, above n 22, 592, 597; Tabbasum and Dingwall, above n 68, 13; Scott Alexander Stanley, *REDD Feasibility Study for East Rennell World Heritage Site, Solomon Islands* (Secretariat of the Pacific Community and Deutsche Gesellschaft für Internationale Zusammenarbeit, 2013) 12; Kasia Gabrys and Mike Heywood, 'Community and Governance

Fig. 5.1 Tourist accommodation, East Rennell World Heritage site (Stephanie Price, 2013)

by the New Zealand government in conjunction with the nomination were successful,[153] and today only a handful of tourists travel there each year.[154] The failure of the World Heritage programme to deliver economic development has reduced support for conservation at the local level. This has also been the experience at the Kuk Early Agricultural site in PNG.[155]

in the World Heritage Property of East Rennell' in Anita Smith (ed), *World Heritage in a Sea of Islands: Pacific 2009 Programme*, World Heritage Papers 34 (UNESCO, 2012) 60, 62; Jacob Zikuli and Hazel Clothier, *Community Attitudes and Perceptions Towards the East Rennell World Heritage Programme* (Live and Learn Environmental Education, 2008) 12; Maria Ana Borges et al, *Sustainable Tourism and Natural World Heritage* (IUCN, 2011) 10.

  [153] Dingwall, above n 39, 8.
  [154] Smith, above n 22, 598. See also Stanley, above n 152. Stanley states that less than ten tourists visited the site in 2012; at 12.
  [155] John Denham, Tim Muke and Vagi Genorupa, 'Nominating and Managing a World Heritage Site in the Highlands of Papua New Guinea' (2007) 39(3) *World Archaeology* 324, 330.

The current level of support for World Heritage among the East Rennellese people is not known. In 2018, the Tuhunui Tribe of East Rennell (which claims that it owns a significant portion of the World Heritage site) wrote to the World Heritage Centre stating that the tribe has decided to "withdraw all its customary land from the World Heritage Program Site in East Rennell".[156] While it is not clear whether other tribal leaders and landowners share this view, the letter does suggest there is a significant level of opposition to World Heritage among the East Rennellese communities.

As for many rural Solomon Islanders, livelihood issues are the primary concern for the people of East Rennell.[157] Most live predominantly subsistence lifestyles, relying on tilapia fish from the lake, food from their gardens, coconut crabs, marine resources, and occasionally birds and bats.[158] Growing crops is challenging because the island is extremely rocky and has limited fertile soil and few water courses. Climate change and invasive species pose further threats to food security.

The people of East Rennell are therefore increasingly looking to participate in the cash economy, to bolster their food security, and to meet other expenses such as education and health care. Opportunities to earn cash income on Rennell are however very limited because of the island's isolation and geography. The island is 180 km south of Guadalcanal, and is difficult to access because it has no port and in most places limestone cliffs at the coast drop straight down to the sea (see Fig. 5.2). This impedes the development of any industry requiring the import or export of products.[159]

While West Rennell hosts an airstrip connecting the island with Honiara, flights only operate a few times a week and can be irregular. Furthermore, the 90 km trip from the airstrip to East Rennell can take many hours due to the poor condition of the road and vehicles (see Fig. 5.3). West Rennell has some phone and internet coverage, but there is none within the World

---

[156] *State of Conservation of the Properties Inscribed on the List of World Heritage in Danger*, WHC 42nd sess, UN Doc WHC/18/42.COM/7A.Add.2 (15 June 2018) 17 (East Rennell, Solomon Islands) 18.

[157] Gabrys and Heywood, above n 152, 62; Zikuli and Clothier, above n 152, 13. See also Smith, above n 22, who notes that livelihood issues dominated the meetings with community members that she was involved with to discuss the World Heritage programme: at 598. Similarly, livelihood issues dominated many of the meetings the author attended in East Rennell (in her capacity as legal advisor for a non-government organisation) concerning the protection of the World Heritage site.

[158] See, for example, Wingham, above n 11, 27.

[159] Ibid.

Fig. 5.2 Aerial view of the south-eastern end of Rennell, showing limestone cliffs dropping to the sea along much of the coast (Stephanie Price, 2013)

Heritage site. Due to these constraints, there is essentially no private sector on Rennell, and most local people rely on cash sent by relatives in Honiara and small-scale commercial activities to support their livelihoods.[160]

Many conservationists still adhere to 'romantic notions' of rural Solomon Islanders as people who are satisfied with their subsistence lifestyle and who have limited material and financial aspirations.[161] In reality, many Solomon Islanders want to participate in the cash economy,[162] and view developments such as logging and mining as a means of achiev-

[160] Smith, above n 22, 598; John Foimua, 'Renbel (Rennell-Bellona) Province, Provincial Profile' in David Lawrence and Matthew Allen (eds), *Hem Nao, Solomon Islands, Tis Taem – Community Sector Program – Volume 1, Provincial Profiles* (2006) 131, 144–145. See also Solomon Islands Government, *Volume I Report on 2009 Population and Housing Census: Basic Tables and Census Description*, Statistical Bulletin 6/2012 (Solomon Islands Government, 2012).

[161] Simon Foale, 'Where's Our Development? Landowner Aspirations and Environmentalist Agendas in Western Solomon Islands' (2001) 2(2) *Asia Pacific Journal of Anthropology* 44, 49.

[162] Ibid.

Fig. 5.3 The road linking the capital of Rennell (Tigoa) and the East Rennell World Heritage site (Stephanie Price, 2013)

ing that. The royalties and other fees that landowners receive from resource companies make an important contribution to some local economies. Landowners may also be persuaded to sign logging and mining contracts by a company's promise to fund or construct local infrastructure, which the government is unwilling or unable to provide.[163]

The protection of World Heritage must be considered in the context of the pressing livelihood issues that the East Rennellese people face and the limited development options available to them. These issues are likely to influence their adherence with customary laws concerning natural resource

---

[163] For example, Rennell's airstrip, located in the capital Tigoa, was built by a company that conducted prospecting for bauxite there in the 1970s (Phillip Iro Tagini, *The Search for King Solomon's Gold: An Examination of the Policy and Regulatory Framework for Mining in Solomon Islands* (PhD Thesis, The Australian National University, 2007) 61). When prospecting began, a road was constructed between Tigoa and Lavagu (in central Rennell). It was later extended to reach Lake Tegano with funding from the European Union (Wingham, above n 11, 29). More recently, a mining company built a road linking the four East Rennell villages.

use. They may also reduce the impetus for local people to take steps to protect the site's OUV, such as complying with legislation imposing harvesting restrictions, opposing logging and mining developments, and agreeing to the establishment of a protected area over their land. In addition, as noted in (B) above, they also influence the steps the SIG is willing to take to ensure the site's conservation.

Until recently, the World Heritage Committee's resolutions made no reference to the East Rennellese people's right to and desire for development. However, the Committee is now encouraging the SIG to develop an action plan for implementing alternative incomegenerating mechanisms to ensure that local communities derive benefits from the conservation of East Rennell's OUV.[164] This reflects the Committee's increasing appreciation of the need to approach World Heritage conservation through the lens of sustainable development (discussed in Sect. 4.3.1). In accordance with the Committee's request, the *DSOCR* calls for the preparation of a new management plan integrating OUV conservation with the development needs of local communities.[165] This is critical, as any management plan that does not align with the priorities of the East Rennellese people is unlikely to be successful (discussed further in Sect. 6.5).

5.4   Conclusion

This chapter identified some key lessons that can be learned from East Rennell.

Firstly, the values for which a site is nominated can significantly impact the site's protection. Within the *Convention* regime, East Rennell is recognised primarily because of the rich biodiversity and endemic species it hosts. However, the East Rennellese are confused as to how their land could be listed 'without them', and are more concerned about the preservation of their customs.[166] This fuels their disenchantment with the World Heritage system, which hinders conservation efforts.

Secondly, in some circumstances, the boundary and buffer zone requirements for World Heritage listing in the *Operational Guidelines* should be applied flexibly, particularly for sites under customary tenure (discussed

---

[164] WHC Res 40 COM 7A.49, 40th sess, UN Doc WHC-16/40.COM/19 (15 November 2016) 68. See also WHC Res 41 COM 7A.19, 41st sess, UN Doc WHC/17/41.COM/18 (12 July 2017) 35.

[165] *State of Conservation of Properties Inscribed on the World Heritage List*, UN Doc WHC/17/41.COM/7A.Add, 31–32.

[166] Smith, above n 22, 597.

further in Sect. 6.4). However, the implications of any deviation from those requirements must be carefully considered when a site is nominated. The fact that safeguarding East Rennell's OUV requires measures to be implemented in West Rennell (such as biosecurity measures, and restrictions on logging and mining) presents a significant challenge for the site's protection. World Heritage conservation initiatives should not only focus on the East Rennellese communities. West Rennellese people should also be encouraged and supported to participate in such projects.

Thirdly, the protection regime of a site should be closely scrutinised at the nomination stage, particularly if the site is under customary tenure. This includes understanding the relationship between customary protection and the site's management plan and legislation (discussed further in Chap. 6).

Fourthly, Pacific Island States need to be supported to not only identify and nominate sites, but supplement and strengthen customary protection of sites. East Rennell was listed before SIG had established the administrative and legal measures required to fully implement the *Convention*. Developing such measures has proved extremely challenging, in part because of the pressing economic and social issues that the government faces. Thus, 20 years after East Rennell was listed, Solomon Islands still does not have a strong legal framework for the site's conservation.

Finally, given the strong reverence of Solomon Islanders for the rights of customary owners, the SIG is unlikely to implement any conservation measures that do not enjoy broad local support. East Rennell's OUV therefore cannot be safeguarded in the long term without the support and involvement of the local people. Their current disenchantment with the World Heritage system and their (understandable) economic aspirations present challenges for conservation, both under customary law and under State legislation. Efforts to address the threats to the site's OUV must therefore be pursued in conjunction with initiatives that accord with the priorities and aspirations of the local communities, in both East and West Rennell.

REFERENCES

ARTICLES, BOOKS AND REPORTS

Albert, Simon, Peter Ramohia, Andrew Olds, Tingo Leve, Charlotte Kvennefors, *Survey of the Condition of the Marine Ecosystem within the East Rennell World Heritage Area, Solomon Islands* (University of Queensland, Solomon Islands Marine Ecology Laboratory, Griffith University and WWF-Solomon Islands, 2013)

Allen, Matthew G, 'Land, Identity and Conflict on Guadalcanal, Solomon Islands' (2012) 43(2) *Australian Geographer* 153

Badman, T, B Bomhard, A Fincke, J Langley, P Rosabal and D Sheppard, *Outstanding Universal Value: Standards for Natural World Heritage* (IUCN, 2008)

Baines, Graham, *Beneath the State: Chiefs of Santa Isabel, Solomon Islands, Coping and Adapting*, State, Society and Governance Working Paper 2014/2 (Australian National University, 2014)

Bennett, Judith, *Roots of Conflict in Solomon Islands – Though Much is Taken, Much Abides: Legacies of Tradition and Colonialism*, State, Society and Governance in Melanesia Discussion Paper (Australian National University, 2002)

Boer, Ben, *Solomon Islands: Review of Environmental Law* (SPREP, 1993)

Boer, Ben, 'Solomon Islands' in Ben Boer (ed), *Environmental Law in the South Pacific: Consolidated Report of the Reviews of Environmental Law in the Cook Islands, Federated States of Micronesia, Kingdom of Tonga, Republic of the Marshall Islands and Solomon Islands* (South Pacific Regional Environment Programme and IUCN Environmental Law Centre, 1996) 189

Borges, Maria Ana, Giulia Carbone, Robyn Bushell and Tilman Jaeger, *Sustainable Tourism and Natural World Heritage* (IUCN, 2011)

Braithwaite, John, Sinclair Dinnen, Matthew Allen, Valerie Braithwaite and Hilary Charlesworth, *Pillars and Shadows: Statebuilding as Peacebuilding in Solomon Islands* (ANU E Press, 2010)

Corrin Care, Jennifer and Jean G Zorn, 'Legislating pluralism: Statutory 'Developments' in Melanesian Customary Law' (2001) 46 *Journal of Legal Pluralism* 49

Cox, Barry and Peter Moore, *Biogeography: An Ecological and Evolutionary Approach.* Blackwell (Oxford, 1980)

Denham, John, Tim Muke and Vagi Genorupa, 'Nominating and Managing a World Heritage Site in the Highlands of Papua New Guinea' (2007) 39(3) *World Archaeology* 324

Diamond, J M, 'The Avifauna of Rennell Island' in Torben Wolff (ed), *The Natural History of Rennell Island, British Solomon Islands* (Danish Science Press, vol 8, 1984)

Dingwall, Paul, *Report on the Reactive Monitoring Mission to East Rennell, Solomon Islands, 21–29 October 2012* (IUCN, 2013)

Dinnen, Sinclair, 'State-Building in a Post-Colonial Society: The Case of Solomon Islands' (2008) 9 *Chicago Journal of International Law* 51

Dinnen, Sinclair, 'The Solomon Islands Intervention and the Instabilities of the Post-Colonial State' (2008) 20(3) *Global Change, Peace and Security* (formerly *Pacific Review: Peace, Security and Global Change*) 338

Filardi, Christopher E, Catherine E Smith, Andrew W Kratter, David W Steadman and H Price Webb, 'New Behavioral, Ecological, and Biogeographic Data on the Avifauna of Rennell, Solomon Islands' (1999) 53(4) *Pacific Science* 319

Filer, Colin, 'Between a Rock and a Hard Place: Mining, 'Indigenous People' and the Development of States' in Benedict Y Imbun and Paul A McGavin (eds), *Mining in Papua New Guinea: Analysis and Policy Implications* (University of Papua New Guinea Press, 2001) 7

Foale, Simon, 'Where's Our Development? Landowner Aspirations and Environmentalist Agendas in Western Solomon Islands' (2001) 2(2) *Asia Pacific Journal of Anthropology* 44

Foimua, John, 'Renbel (Rennell-Bellona) Province, Provincial Profile' in David Lawrence and Matthew Allen (eds), *Hem Nao, Solomon Islands, Tis Taem – Community Sector Program – Volume 1, Provincial Profiles* (2006) 131

Frazer, Ian, 'The Struggle for Control of Solomon Island Forests' (1997) 9(1) *Contemporary Pacific* 39

Gabrys, Kasia and Mike Heywood, 'Community and Governance in the World Heritage Property of East Rennell' in Anita Smith (ed), *World Heritage in a Sea of Islands: Pacific 2009 Programme*, World Heritage Papers 34 (UNESCO, 2012) 60

Gay, Daniel (ed), *Solomon Islands Diagnostic Trade Integration Study 2009 Report* (Solomon Islands Government, 2009)

International Centre on Space Technologies for Natural and Cultural Heritage, *Report of the Technical Consultation Meeting on East Rennell World Heritage Site in Danger, Sanya, Hainan Province, China, 1–2 February 2016* (2016) 21

John, Aatai, 'No Mining at Lake Tegano', *The Solomon Star* (online), 20 February 2017 http://www.solomonstarnews.com/news/national/12281-no-mining-at-lake-tegano

Kuschel, Rolf, Torben Monberg, and Torben Wolff, *Bibliography of Rennell and Bellona Islands* (University of Copenhagen, 2nd ed, 2001) http://www.bellona.dk/pdf/publications//bibliography_2nd.pdf

Larmour, Peter, 'Sharing the Benefits: Customary Landowners and Natural Resource Projects in Melanesia' (1989) 36 *Pacific Viewpoint* 56

MacArthur, Robert J and Edward O Wilson, 'An Equilibrium Theory of Insular Zoogeography' (1963) 17 *Evolution* 373

Markham, A, E Osipova, K Lafrenz Samuels and A Caldas, *World Heritage and Tourism in a Changing Climate* (UNEP and UNESCO, 2016)

McKinnon, John, *Solomon Islands World Heritage Site Proposal: Report on a Fact Finding Mission (4–22 February 1990)* (Victoria University of Wellington, 1990)

Monberg, Torben, 'Research on Rennell and Bellona: A Preliminary Report' (1960) 2 *Folk* 71

Moore, Clive, 'Pacific View: The Meaning of Governance and Politics in the Solomon Islands' (2008) 62(3) *Australian Journal of International Affairs* 386

Munch-Petersen, Nils Finn, 'An Island Saved, At Least for Some Time? The Advent of Tourism to Rennell, Solomon Islands' in Godfrey Baldacchino and Daniel Niles (eds), *Island Futures: Conservation and Development Across the Asia-Pacific Region* (Springer, 2011) 169

Perry, Jim, 'World Heritage Hot Spots: A Global Model Identifies the 16 Natural Heritage Properties on the World Heritage List Most at Risk From Climate Change' (2011) 17(5) *International Journal of Heritage Studies* 426

Ruddle, K, E Hviding and R E Johannes, 'Marine Resources Management in the Context of Customary Tenure' (1992) 7 *Marine Resource Economics* 249

Ruthven, David, 'Rennell Bauxite' in Peter Larmour (ed), *Land in Solomon Islands* (Institution of Pacific Studies and Ministry of Agriculture and Lands, 1979) 94

Sand, Christophe, 'Melanesian Tribes vs Polynesian Chiefdoms: Recent Archaeological Assessment of a Classic Model of Socio-Political Types in Oceania' (2002) 41(2) *Asian Perspectives* 284

Smith, Anita, 'East Rennell World Heritage Site: Misunderstandings, Inconsistencies and Opportunities in the Implementation of the World Heritage Convention in the Pacific Islands' (2011) 17(6) *International Journal of Heritage Studies* 592

Smith, Anita, 'World Heritage and Outstanding Universal Value in the Pacific Islands' (2015) 21(2) *International Journal of Heritage Studies* 177

Smith, John, *An Island in the Autumn* (Librario Publishing Ltd, 2012)

Solomon Islands Government, *State Party Report on the State of Conservation of the East Rennell World Heritage Area (Solomon Islands)* (SIG, 2012)

Solomon Islands Government, *Volume I Report on 2009 Population and Housing Census: Basic Tables and Census Description*, Statistical Bulletin 6/2012 (Solomon Islands Government, 2012)

Solomon Islands Government, *State Party Report on the State of Conservation of East Rennell (Solomon Islands)* (SIG, 2017)

Stanley, Scott Alexander, *REDD Feasibility Study for East Rennell World Heritage Site, Solomon Islands* (Secretariat of the Pacific Community and Deutsche Gesellschaft für Internationale Zusammenarbeit, 2013)

Tabbasum, Salamat Ali and Paul Dingwall, *Report on the Mission to East Rennell World Heritage Property and Marovo Lagoon, Solomon Islands*, 30 March–10 April 2005 (IUCN and World Heritage Centre, 2005) 5

Techera, Erika J, 'Safeguarding cultural heritage: Law and policy in Fiji' (2011) 12 *Journal of Cultural Heritage* 329

Triantis, Kostas A, David Nogués-Bravo, Joaquín Hortal, Paulo A V Borges, Henning Adsersen, José Maria Fernandez-Palacios, Miguel B. Araujo and

Robert J Whittaker, 'Measurements of Area and the (Island) Species-Area Relationship: New Directions for an Old Pattern' (2008) 117 *Oikos* 1555

Turton, Steve, *East Rennell World Heritage Area: Assessment of the State of Conservation of World Heritage Values. Final Field Report* (James Cook University, 2014)

Wingham, Elspeth J, *Nomination of East Rennell, Solomon Islands by the Government of Solomon Islands for Inclusion in the World Heritage List Natural Sites* (1997)

Wingham, Elspeth J, *Resource Management Objectives and Guidelines for East Rennell, Solomon Islands (May 1998)*, attached as attachment 1 to Elspeth J Wingham, *Nomination of East Rennell, Solomon Islands by the Government of Solomon Islands for Inclusion in the World Heritage List Natural Sites* (1997)

Wolff, Torben (ed), *The Natural History of Rennell Islands, British Solomon Islands. Scientific Results of the Danish Rennell Expedition, 1951 and the British Museum (Natural History) Expedition 1953* (Danish Science Press, volumes 1–4, 1958–1962)

Wolff, Torben (ed), *The Natural History of Rennell Island, British Solomon Islands. Scientific Results of the Noona Dan Expedition (Rennell Section, 1962) and The Danish Rennell Expedition 1965* (Danish Science Press, volumes 5–8, 1968)

Wolff, T, 'The Fauna of Rennell and Bellona, Solomon Islands' (1969) 255(800) *Philosophical Transactions of the Royal Society of London, Series B, Biological Sciences* 321

Zikuli, Jacob and Hazel Clothier, *Community Attitudes and Perceptions Towards the East Rennell World Heritage Programme* (Live and Learn Environmental Education, 2008)

CONVENTIONS

Convention Concerning the Protection of the World Cultural and Natural Heritage, opened for signature 16 November 1972, 1037 UNTS 151 (entered into force 17 December 1975)

Convention on the Law of Treaties, opened for signature 23 May 1969, 1155 UNTS 331 (entered into force 27 January 1980)

UNITED NATIONS DOCUMENTS

Adoption of Retrospective Statements of Outstanding Universal Value, WHC 36th sess, UN Doc WHC-12/36.COM/8E (15 June 2012a) 55 (East Rennell, Solomon Islands)

ICOMOS, *Evaluations of Nominations of Cultural and Mixed Properties to the World Heritage List*, WHC 32nd sess, UN Doc WHC-08/32.COM/INF/8B1

(2008) 92 (Chief Roi Mata's Domain, Vanuatu, Advisory Body Evaluation 1280)

ICOMOS, *Evaluations of Nominations of Cultural and Mixed Properties to the World Heritage List*, WHC 36th sess, UN Doc WHC-12/36.COM/INF.8B1 (2012) 21 (Rock Islands Southern Lagoon, Republic of Palau, Advisory Body Evaluation 1386)

Information on Tentative Lists and Examination of Nominations of Cultural and Natural Properties to the List of World Heritage in Danger and the World Heritage List, WHC 22nd sess, UN Doc WHC-98/CONF.203/10Rev (29 November 1998)

IUCN, *Evaluations of Nominations of Natural and Mixed Properties to the World Heritage List*, WHC 22nd sess (1998) 79 (East Rennell, Solomon Islands)

IUCN, *Evaluations of Nominations of Natural and Mixed Properties to the World Heritage List*, WHC 34th sess, UN Doc WHC/10/34.COM/INF.8B2 (2010) 19 (Phoenix Islands Protected Area, Kiribati, Advisory Body Evaluation 1325)

State of Conservation of Properties Inscribed on the World Heritage List, WHC 22nd sess, UN Doc WHC-03/27.COM/7B (12 June 2003) 11 (East Rennell, Solomon Islands)

State of Conservation of Properties Inscribed on the World Heritage List, WHC 28th sess, UN Doc WHC-04/28.COM/15B (15 June 2004) 15 (East Rennell, Solomon Islands)

State of Conservation of Properties Inscribed on the World Heritage List, WHC 41st sess, UN Doc WHC/17/41.COM/7A.Add (2 June 2017) 26 (East Rennell, Solomon Islands)

State of Conservation of the Properties Inscribed on the List of World Heritage in Danger, WHC 42nd sess, UN Doc WHC/18/42.COM/7A.Add.2 (15 June 2018) 17 (East Rennell, Solomon Islands)

UNESCO, *Operational Guidelines for the Implementation of the World Heritage Convention*, UN Doc WHC 97/2 (February 1997)

UNESCO, *Operational Guidelines for the Implementation of the World Heritage Convention*, UN Doc WHC 16/01 (26 October 2016)

WHC Res CONF 203 VIII.A.1, WHC 22nd sess, UN Doc WHC-98/CONF/203/18 (29 January 1999a) 25

WHC Res CONF 203 XIV.3, WHC 22nd sess, UN Doc WHC-98/CONF.203/18 (29 January 1999b) 56

WHC Res 34 COM 7B.17, WHC 34th sess, UN Doc WHC-10/34.COM/20 (3 September 2010) 71

WHC Res 36 COM 7B.15, WHC 36th sess, UN Doc WHC-12/36.COM/19 (June–July 2012b) 63

WHC Res 37 COM 7B.14, WHC 37th sess, UN Doc WHC-13/37.COM/20 (5 July 2013) 68

WHC Res 38 COM 7A.29, WHC 38th sess, UN Doc WHC-14/38.COM/16 (7 July 2014) 39

WHC Res 39 COM 7A.16, WHC 39th sess, UN Doc WHC-15/39.COM/19 (8 July 2015) 30

WHC Res 40 COM 7A.49, WHC 40th sess, UN Doc WHC-16/40.COM/19 (15 November 2016) 68

WHC Res 41 COM 7A.19, WHC 41st sess, UN Doc WHC-17/41.COM/18 (12 July 2017) 35

LEGISLATION AND BILLS: SOLOMON ISLANDS

Protected Areas Act (2010)

World Heritage Properties Conservation Bill

THESES

Tagini, Phillip Iro, *The Search for King Solomon's Gold: An Examination of the Policy and Regulatory Framework for Mining in Solomon Islands* (PhD Thesis, The Australian National University, 2007)

INTERNET MATERIALS

IUCN, *Natural World Heritage Sites: The Pacific's Challenges* (13 June 2014) https://www.iucn.org/content/natural-world-heritage-sites-pacific%E2%80%99s-challenges

IUCN, *World Heritage Outlook – East Rennell* http://www.worldheritageoutlook.iucn.org/explore-sites/wdpaid/168242

UNESCO, *The World Heritage Committee* http://whc.unesco.org/en/committee/

INTERVIEWS

Interview by the author with an officer in the Ministry of Culture (Honiara, 26 July 2013)

Interview by the author with an officer in the Ministry of Education, who was formerly the focal point for World Heritage within the Solomon Islands National Commission for UNESCO (Honiara, 28 July 2013)

Interview by the author with a conservation officer in the Ministry of Environment (Honiara, 2 August 2013)

Interview by the author with Joe Horokou, Director of the Environment and Conservation Division of the Ministry of Environment (Honiara, 15 August 2013)

Interview by the author with Bradley Tovosia, Minister for Environment (Honiara, 24 September 2013)

Interview by the author with Malchoir Mataki, Permanent Secretary of the Ministry of Environment (Honiara, 1 October 2013)

OTHER

Letter from Aseri Yalangono, Deputy Secretary General of National Commission for UNESCO Solomon Islands to the Director of the UNESCO World Heritage Centre (31 August 2011)

Letter from Moses K Mose, Permanent Secretary of Solomon Islands Ministry of Commerce, Employment and Tourism, to Bernd von Droste, Director of the UNESCO World Heritage Centre (1 September 1998) attached as supplementary information to Elspeth J Wingham, *Nomination of East Rennell, Solomon Islands by the Government of Solomon Islands for Inclusion in the World Heritage List Natural Sites* (1997)

Minister for Provincial Government and Institutional Strengthening, 'Rennell and Bellona Provincial Government (Suspension of Executive Powers) Order 2014' in Solomon Islands, *Extraordinary Gazette*, No 81, 5 September 2014, 184

Wein, Laurie, *East Rennell World Heritage Site Management Plan* (Solomon Islands National Commission for UNESCO, 2007)

Protecting the East Rennell World Heritage Site: Customary Protection and Management Planning

6.1 INTRODUCTION

As explained in Chap. 4, the decision to allow sites under customary protection to qualify for World Heritage listing substantially increased the scope for the *World Heritage Convention*[1] to be utilised by Pacific Island States. However, custom alone can rarely protect a site against all activities that threaten its outstanding universal value (OUV).[2] Furthermore, customary tenure presents some challenges for World Heritage conservation not experienced at sites under State or private ownership or control. Thus, while customary protection of World Heritage should be supported, we need to understand its limitations, and how it can be strengthened and

---

[1] *Convention Concerning the Protection of the World Cultural and Natural Heritage*, opened for signature 16 November 1972, 1037 UNTS 151 (entered into force 17 December 1975) ('*World Heritage Convention*').

[2] See, for example, Chris Ballard and Meredith Wilson, 'Unseen Monuments: Managing Melanesian Cultural Landscapes' in Ken Taylor and Jane L Lennon (eds), *Managing Cultural Landscapes* (Routledge, Oxon 2012) 130, 132; Anita Smith, 'The World Heritage Pacific 2009 Programme' in Anita Smith (ed), *World Heritage in a Sea of Islands: Pacific 2009 Programme*, World Heritage Papers 34 (UNESCO, Paris 2012) 2, 5; Pepe Clarke and Charles Taylor Gillespie, *Legal Mechanisms for the Establishment and Management of Terrestrial Protected Areas in Fiji* (IUCN, Suva 2009) 2; Jonathan M Lindsay, *Creating Legal Space for Community-Based Management: Principles and Dilemmas* (Food and Agriculture Organization of the United Nations, Rome 1998) 3.

© The Author(s) 2018
S. C. Price, *World Heritage Conservation in the Pacific*,
Palgrave Series in Asia and Pacific Studies,
https://doi.org/10.1007/978-981-13-0602-0_6

supplemented. Through analysis of the protection of East Rennell under custom and the site's management plan, this chapter provides insights into the opportunities and challenges presented by customary protection of World Heritage.

The customary protection of East Rennell has not been comprehensively researched despite several reports recommending that this be done.[3] This is a significant gap in knowledge, given that the site was inscribed on the World Heritage List on the basis of its protection under customary law. Questions about the possible conservation functions served by customary systems can only be answered fully through 'intensive, localized, and multidisciplinary field research'.[4] There remains a critical need for this type of research at East Rennell, to document relevant customs and to understand how the East Rennellese people can be assisted to conserve their land and resources. In the absence of such work, the analysis of customary laws and governance in this chapter is necessarily based on the relatively limited available literature.

This chapter begins by exploring customary law (Sect. 6.2) and customary governance (Sect. 6.3) at East Rennell. It explains that some of the key threats to the site's OUV cannot be addressed through the customary system alone. Furthermore, weak customary governance presents a challenge for the protection of the site.

The chapter then considers the boundaries of the East Rennell World Heritage site and how they impact conservation efforts (Sect. 6.4). It demonstrates that while the provisions of the *Operational Guidelines for the Implementation of the World Heritage Convention* concerning site boundaries and buffer zones need to be applied flexibly for sites under customary protection, the implications of any non-compliance should be carefully considered when such places are nominated.

---

[3] For example, in 2004, the World Heritage Committee requested that IUCN document and assess the effectiveness of the site's customary protection: WHC Res 28 COM 15B.12, WHC 28th sess, UN Doc WHC-04.28.COM/26 (29 October 2004) 84. The 2007 East Rennell management plan identifies documenting the traditional knowledge and customary practices of the East Rennellese communities as a future management action: Laurie Wein, *East Rennell World Heritage Site Management Plan* (Solomon Islands National Commission for UNESCO, 2007) 20; Dingwall refers to the need for the 'systematic cataloguing and documentation of cultural values and traditional resource use and conservation practices': Paul Dingwall, *Report on the Reactive Monitoring Mission to East Rennell, Solomon Islands, 21–29 October 2012* (IUCN, 2013) 28.

[4] K Ruddle, E Hviding and R E Johannes, 'Marine Resources Management in the Context of Customary Tenure' (1992) 7 *Marine Resource Economics* 249, 267.

Management planning for sites under customary protection is then explored, through analysis of East Rennell's 2007 management plan[5] (Sect. 6.5). The chapter suggests that future management plans may be more effective if they better reflect the customs, cultural values, and development aspirations of the East Rennellese people, and if they have status under State law.

6.2 Customary Laws and World Heritage Protection at East Rennell

Land and resource use at East Rennell is regulated through a system of customary land tenure and other customs and practices.[6] The island's land tenure system differs from most other parts of Solomon Islands. In the predominantly Melanesian Solomon Islands, customary land is commonly owned by a group such as a family, line, or clan.[7] However, on Rennell (where people are of Polynesian decent), land was traditionally held individually by male members of the lineage, and passed down from father to first-born son, or if the man had no sons, to his brother's sons.[8] Landowners (referred to as *matu'a*) had certain powers over their land, including deciding whether to cultivate the land, what to plant, and whether to grant rights to others over their land.[9]

Early anthropological literature evidences some customs relevant to the protection of the island's natural heritage values. For example, traditionally people allowed garden areas to lay fallow for four to six years, to ensure soil integrity was maintained.[10] Wild ducks, snakes, geckos, and skinks were not commonly eaten.[11] In addition, East Rennellese who became Seventh Day Adventists following their conversion to Christianity did not partake in activities such as shark fishing, eel netting, flying-fox snaring, gathering

---

[5] Wein, above n 3.

[6] See, for example, K A J Birket-Smith, *An Ethnological Sketch of Rennell Island: A Polynesian Outlier in Melanesia* (Munksgaard, 2nd ed, 1969).

[7] Jennifer Corrin and Don Paterson, *Introduction to South Pacific Law* (Palgrave Macmillan, 3rd ed, 2011) 274–275.

[8] Samuel H Elbert and Torben Monberg, *From the Two Canoes: Oral Traditions of Rennell and Bellona Islands* (Danish National Museum and University of Hawaii Press, 1965) 10.

[9] Ibid., 11.

[10] Ibid., 16.

[11] Birket-Smith, above n 6, 75.

shell fish and longicorns, and catching coconut crabs.[12] However, the extent to which these customs are practised today is yet to be verified. Like elsewhere in the Pacific, the customs of the Rennellese changed substantially following contact with outsiders. However, on Rennell, this did not occur for many years after the island was discovered by Europeans in the 1790s.[13] Following its discovery, Rennell was visited by whalers and recruiters seeking workers for plantations in Queensland, but the establishment of European settlements was impeded by the island's isolation, poor soils, and the lack of freshwater and safe anchorage sites.[14] In addition, the Rennellese were not considered to be good workers, so few were taken to work on plantations.[15] Consequently, at a time when many parts of the Pacific were undergoing significant change at the hands of colonisers, Rennell remained relatively unaffected.

Rennell's isolation ended in 1938 when events transpired leading most Rennellese to rapidly convert to Christianity. Missionaries first began visiting the island in 1856,[16] but their attempts to establish Christianity were unsuccessful for many years. In the early 1930s, some young Rennellese men were recruited by missionaries to undertake religious studies, and were taken away from the island.[17] They returned in 1936, and for a few years, the people worshipped Christianity alongside their ancient gods.[18] In 1938, after a short period of chaos and hysteria, most Rennellese came to believe that the Christian God was more powerful than their old deities, and they swiftly accepted Christianity.[19]

Following their conversion, the islanders moved from their scattered settlements to larger villages (see Figs. 6.1 and 6.2), impacting traditional

---

[12] Elbert and Monberg, above n 8, 19.

[13] For discussion of the discovery of Rennell by Europeans, see Rolf Kuschel, 'Early Contacts Between Bellona and Rennell Islands and the Outside World' (1988) 23(2) *Journal of Pacific History* 191.

[14] T Wolff, 'The Fauna of Rennell and Bellona, Solomon Islands' (1969) 255(800) *Philosophical Transactions of the Royal Society of London, Series B, Biological Sciences* 321, 321.

[15] Kuschel, above n 13, 196; Judith Bennett, *Wealth of the Solomons: A History of a Pacific Archipelago, 1800–1978* (University of Hawaii Press, 1988) 272.

[16] Wolff, above n 14, 321.

[17] Kuschel, above n 13, 199.

[18] Elspeth J Wingham, *Nomination of East Rennell, Solomon Islands by the Government of Solomon Islands for Inclusion in the World Heritage List Natural Sites* (1997) 24.

[19] This period is described in detail in Torben Monberg, 'Crisis and Mass Conversion on Rennell Island in 1938' (1962) 71(2) *Journal of Polynesian Society* 145.

Fig. 6.1   Hutuna village, East Rennell World Heritage site (Stephanie Price, 2012)

land tenure systems.[20] Traditional culture broke down quickly, as people abandoned old rituals and social structures changed.[21] Writing in 1960, Monberg (a Danish anthropologist who conducted extensive research on the island) wrote that the social structure of the Rennellese had almost completely changed from that which existed 20 years earlier.[22] Changes to the customs of the Rennellese further accelerated after World War II, when improved shipping services made it easier for the islanders to travel to other places, exposing them to new ideas.[23] As the population grew, the land areas owned by individuals decreased in size, and disagreements over the rules concerning land tenure increased.[24] The customs of the East Rennellese were therefore significantly impacted by contact with outsiders and it cannot be assumed that traditional practices are still adhered to.

[20] Wingham, above n 18, 26.
[21] Monberg, above n 19, 149.
[22] Torben Monberg, 'Research on Rennell and Bellona: A Preliminary Report' (1960) 2 *Folk* 71, 71.
[23] Kuschel, above n 13, 199.
[24] Torben Monberg, 'Bellona and Rennell Islanders' in Melvin Ember, Carol R Ember and Jan Skoggard (eds), *Encyclopedia of World Cultures Supplement* (Macmillan, 2002) 46, 48.

Fig. 6.2   Nuipani village, East Rennell World Heritage site (Stephanie Price, 2013)

Little information concerning the customary laws of the East Rennellese can be gleaned from recent reports and documents regarding the World Heritage site. The site's World Heritage nomination dossier states that the use of flora and fauna is regulated by the customary legal tenure system and traditional practices,[25] but contains little further details. The site's 2007 management plan says that the East Rennellese people employ many traditional practices, including customary fishing methods.[26] However, the plan does not document them and indeed notes that it is unknown whether methods like seasonal closures and other restrictions are implemented.[27] A draft management plan for East Rennell prepared in 2014 contains a series of rules regulating resource use, for example, bans on the hunting of birds on breeding islands, the use of gillnets, and the taking of animals carrying eggs.[28] However, it is unclear whether these rules reflect custom or whether they are merely proposed management measures.

---

[25] Wingham, above n 18, 45.
[26] Wein, above n 3, 16.
[27] Ibid.
[28] Anna Price, *(Draft) Management Plan – East Rennell, Solomon Islands* (2014).

Gabrys and Heywood, Australian advisors who lived at East Rennell for 18 months between 2008 and 2009,[29] suggest that few customary practices supporting natural heritage protection are widely implemented at the site today.[30] Their consultations with local communities led them to state that there is 'little evidence of sustainable utilisation practices or customary conservation management, especially in relation to wild food harvesting'.[31] Similarly, in 2012, Dingwall wrote that there are no community-based controls on the harvesting of coconut crabs in the World Heritage site.[32] The absence of such practices may be because the island's population has always been too small to foster a strong conservation ethic. As Gabrys and Heywood noted:

> Several Rennellese talked about how abundant their resources were in the past, which meant that they did not have to worry about managing certain species for their long-term survival.[33]

The *Desired State of Conservation for the Removal of East Rennell from the List of World Heritage in Danger* (*DSOCR*) (discussed in Sect. 5.3.2) calls for the Solomon Islands government (SIG) to implement measures to ensure that 'species are harvested in a sustainable manner based on traditional resource use regimes'.[34] This reflects best practice: it is often said that the integration of customary and modern systems can lead to the best conservation outcomes.[35] However, the first step when designing such a

---

[29] The advisors were volunteers through the Australian Volunteers International programme: see International Heritage Section, Department of Sustainability, Environment, Water, Population and Communities, Australian Government, 'Australian Capacity Building Support for East Rennell World Heritage Area 2007–2013' in Anita Smith (ed), *World Heritage in a Sea of Islands: Pacific 2009 Programme*, World Heritage Papers 34 (UNESCO, 2012) 66.

[30] Kasia Gabrys and Mike Heywood, 'Community and Governance in the World Heritage Property of East Rennell' in Anita Smith (ed), *World Heritage in a Sea of Islands: Pacific 2009 Programme*, World Heritage Papers 34 (UNESCO, 2012) 60, 62.

[31] Ibid.

[32] Dingwall, above n 3, 5.

[33] Gabrys and Heywood, above n 30, 62.

[34] *State of Conservation of Properties Inscribed on the World Heritage List*, WHC 41st sess, UN Doc WHC/17/41.COM/7A.Add (2 June 2017) 26 (East Rennell, Solomon Islands) 31.

[35] See, for example, Grazia Borrini-Feyerabend, Ashish Kothari and Gonzalo Oviedo, *Indigenous and Local Communities and Protected Areas: Towards Equity and Enhanced Conservation* (IUCN, Gland, 2004) 46.

regime is to study and document 'community conservation values, knowledge, skills, resources and institutions'.[36] Before harvesting restrictions based on traditional resource use regimes can be implemented at East Rennell (as called for in the *DSOCR*), such regimes need to be researched and documented to understand whether they can support sustainable harvesting. The existing literature referred to above raises doubts as to whether this is the case. Further research is also needed to understand who is harvesting the species under threat and, if they are from outside East Rennell, the extent to which customary laws can be enforced against them.

The role of custom in addressing the other threats to East Rennell's OUV also needs to be considered. The World Heritage Committee has repeatedly called upon the SIG to ensure that no logging or mining impacts the World Heritage site.[37] Under Solomon Islands' legislation, unless an exception applies, a person cannot obtain government approval to conduct logging or mining without the approval of the relevant customary landowners.[38] Therefore, the customary land tenure system and decision-making processes of the East Rennellese people influence whether the World Heritage site will be affected by these activities.

The ability and willingness of the East Rennellese people to protect the World Heritage site against the impacts of logging and mining is explored in Chap. 7. As explained in that chapter, the implementation of landowner consent provisions in logging and mining legislation is often problematic due to their inconsistency with customary decision-making processes. Thus, in practice, there is a risk that these activities may occur within the World Heritage site even if many East Rennellese people object. Furthermore, the East Rennellese have limited capacity to influence whether these activities occur in West Rennell, despite their potential impacts on the site's OUV (see Sect. 7.3.1.2).

Whether any customs could be utilised to address the threats posed by invasive species such as ship rats and African snails also requires further investigation. It is however evident that while the East Rennellese people could implement biosecurity measures on their land, they cannot regulate activities elsewhere. Halting the introduction and spread of invasive species will require actions such as establishing monitoring systems across the island, baiting log storage areas in West Rennell, and implementing biose-

---

[36] Ibid.

[37] See, for example, WHC Res 41 COM 7A.19, 41st sess, UN Doc WHC/17/41. COM/18 (12 July 2017) 35.

[38] *Forest Resources and Timber Utilisation Act (Cap. 40)* s 5; *Mines and Minerals Act (Cap. 42)* ss 21, 36(a). These provisions are analysed in Sect. 7.3.1.

curity measures at places of disembarkation.[39] The East Rennellese peoples' capacity to address this threat through custom is therefore limited, making the implementation of relevant State legislation crucial (see Sect. 7.3.3). A range of adaptation and mitigation measures have been suggested to deal with the impacts of climate change at East Rennell, particularly to ensure that food security is maintained. These include monitoring the salinity of Lake Tegano, monitoring tilapia populations in the lake, and replanting lakeside areas with varieties of plants that are tolerant to changing conditions.[40] The customary legal system could also potentially help facilitate adaptation and mitigation. For example, the inundation of lakeside areas may disrupt the implementation of customary laws regulating access to land and rights to resources. The customary system may need to evolve to ensure that these laws remain workable, and everyone retains access to viable land to support their livelihoods. Mitigating the impacts of climate change also requires retention of a high degree of forest cover across the island, as this will make the island's ecosystems more resilient to extreme weather events.[41] Addressing this threat is therefore intimately related to the issue of logging and mining on the island, discussed above.

The questions that exist over the role of custom in addressing the threats to East Rennell's OUV highlight the critical need for interdisciplinary research exploring customary protection at the site.

6.3 CUSTOMARY GOVERNANCE AND WORLD HERITAGE PROTECTION AT EAST RENNELL

Customary laws derive their force from uniform practice and the peoples' subjective belief that they must be complied with.[42] Therefore, issues such as the extent of social cohesion within the community[43] and the strength

---

[39] Dingwall, above n 3, 33; Steve Turton, *East Rennell World Heritage Area: Assessment of the State of Conservation of World Heritage Values. Final Field Report* (James Cook University, 2014) 19.

[40] Dingwall above n 3, 34; Turton, above n 39, 15, 20.

[41] Turton, above n 39, 8.

[42] Michael A Ntumy, 'The Dream of a Melanesian Jurisprudence: The Purpose and Limits of Law Reform' in Jonathan Aleck and Jackson Rannells (eds), *Custom at the Crossroads* (University of Papua New Guinea, 1995) cited in Jennifer Corrin Care and Jean G Zorn, 'Legislating for the Application of Customary Law in Solomon Islands' (2005) 34 *Common Law World Review* 144, 149.

[43] See, for example, Matthew Allen et al, *Justice Delivered Locally: Systems, Challenges and Innovations in Solomon Islands* (World Bank, 2013) 69.

of local governance bodies[44] will influence compliance. In Solomon Islands, customary governance bodies do not have a formal position under State legislation,[45] so their strength is determined by their legitimacy within the local community.

In Rennell, authority within a line traditionally lay with all *matu'a* (landowners).[46] However, one *matu'a* could assume a higher position because of his seniority within his generation, the seniority of his father's generation, or his possession of special skills.[47] Such a person was called a *hakahua* (now commonly referred to as a chief).[48] In the pre-contact period, the *matu'a* of the lineage were not compelled to obey the *hakahua*.[49] However, as a *hakahua* was often a more senior member of the lineage and had more land at his disposal than other *matu'a*, he generally had a higher status.[50] In the pre-contact period, there was also no supreme chief nor any collective body of chiefs, with the chiefs considering themselves to be autonomous.[51] Today however, customary authority is exercised by the chiefs and a Council of Chiefs, headed by a paramount chief.[52]

Literature suggests that customary governance at East Rennell is currently weak. In 2008/2009, Gabrys and Heywood observed that the Council of Chiefs was losing its authority, in part because of 'increasing pressures to engage with the cash economy, internal disputes over land

---

[44] See, for example, Joeli Veitayaki et al, 'On Cultural Factors and Marine Managed Areas in Fiji' in Jolie Liston, Geoffrey Clark and Dwight Alexander (eds), *Pacific Island Heritage: Archaeology, Identity and Community* (ANU E Press, 2011) 37, 38; Shankar Aswani, 'Customary Sea Tenure in Oceania as a Case of Rights-Based Fishery Management: Does it Work?' (2005) 15 *Reviews in Fish Biology and Fisheries* 285, 289; Pepe Clarke and Stacy D Jupiter, 'Law, Custom and Community-Based Natural Resource Management in Kubulau District (Fiji)' (2010) 37(1) *Environmental Conservation* 98, 104.

[45] The only role for chiefs recognised under State legislation is in the resolution of disputes over rights to customary land: *Local Courts Act (Cap. 19)* s 12(1). This situation can be contrasted with Samoa, for example, where Indigenous governance structures are recognised and empowered under the *Village Fono Act 1990*. For discussion, see, for example, Erika J Techera, 'Samoa: Law, Custom and Conservation' (2006) 10 *New Zealand Journal of Environmental Law* 361.

[46] Elbert and Monberg, above n 8, 11.

[47] Ibid.

[48] Ibid.; Monberg, above n 22, 77.

[49] Elbert and Monberg, above n 8, 12.

[50] Ibid.

[51] Monberg, above n 22, 77; Allen et al, above n 43, 38; Elbert and Monberg, above n 8, 14.

[52] Wingham, above n 18, 5.

ownership and increasing church authority'.[53] Consequently, many local people viewed it as 'ineffective or dysfunctional'.[54] More recent field work by Allen et al. found that there is 'an almost complete collapse of community governance mechanisms' in Rennell,[55] and many community members do not trust the chiefs.[56] These governance issues are likely to be reducing adherence with customary laws, and weakening the site's customary protection.

In an attempt to strengthen local governance, several community organisations have been established at East Rennell, including the Tegano Management and Conservation Committee and the East Rennell World Heritage Trust Board.[57] The organisation established most recently, and the only one that remains functional, is the Lake Tegano World Heritage Site Association (LTWHSA), which was registered under the *Charitable Trusts Act (Cap. 55)* in 1999. It aims to safeguard the OUV of East Rennell and ensure local people benefit from the World Heritage programme.[58]

The power to make decisions on behalf of the LTWHSA is vested in a committee of ten members, comprising two representatives from each of the four villages in East Rennell (Tebaitahe, Nuipani, Tegano, and Hutuna) and two representatives of the Rennell Bellona provincial government.[59] Chiefs have no formal role in the association (unless they are elected to the committee), but can attend committee meetings in a non-voting capacity.[60] Church leaders also have no formal role, despite their status within the communities.

---

[53] Gabrys and Heywood, above n 30, 61.

[54] Ibid.

[55] Allen et al, above n 43, 24.

[56] Ibid., 38.

[57] The Tegano Management and Conservation Committee was established with the assistance of the New Zealand government: Elspeth J Wingham and Ben Devi, 'The Involvement of Local People in the Management of a Proposed World Heritage Site at East Rennell, Solomon Islands' in Hans D Thulstrup (ed), *World Natural Heritage and the Local Community: Case Studies from Asia Pacific, Australia and New Zealand* (UNESCO, 1999) 79, 80. In 2001, the East Rennell Environment and Conservation Trust Board was established: Salamat Ali Tabbasum and Paul Dingwall, *Report on the Mission to East Rennell World Heritage Property and Marovo Lagoon, Solomon Islands, 30 March–10 April 2005* (IUCN and World Heritage Centre, 2005) 9. It was renamed to the East Rennell World Heritage Trust Board; Wein, above n 3, 10. The Board is no longer functional.

[58] *Lake Tegano World Heritage Site Association Constitution and Rules* (2009) cl 2.1.

[59] Ibid., cl 5.1(a).

[60] Ibid., cl 5.1(g).

The LTWHSA is operational, but its involvement in conservation activities has been limited. Its work has been impeded by ambiguities and gaps in its Constitution,[61] a lack of funds to convene meetings,[62] and allegations of financial mismanagement.[63] In addition, because the LTWHSA's mandate includes some matters traditionally dealt with by the chiefs, the relationship between the committee and community leaders has at times been tense. While the association's decision-making processes are democratic in a Western sense, they are at odds with customary governance.[64] To help address this, some East Rennellese have suggested that the association's Constitution be amended to give chiefs and church leaders a formal role on the committee.

Other local organisations established at East Rennell and elsewhere have suffered from similar issues. For example, the establishment of the East Rennell World Heritage Trust Board 'brought new factions of power and authority into the community that many were not happy with'.[65] Similar concerns have been raised about the Gold Ridge Landowners' Association, which was set up to manage the royalties from the Gold Ridge mining project on Guadalcanal.[66] Because members of that association were chosen on the basis of educational level, elders lost their leadership role and customary authority was undermined.[67]

---

[61] For example, the LTWHSA's Constitution tries to ensure equality between the four East Rennell villages by guaranteeing equal representation on the committee for each village: *Lake Tegano World Heritage Site Association Constitution and Rules* cl 5.1(a). The chairperson of the committee is elected by the (eight) committee members: cl 5.1(d). The Constitution does not prescribe how the chairperson is to be elected if the vote of committee members results in a stalemate, which has led to problems with decision-making.

[62] The committee is dependent on receiving funds from the SIG or donors to convene meetings. The government has now allocated a fixed annual amount for the LTWHSA (*Pacific World Heritage Action Plan 2016–2020* (2016) 16), which may improve the situation.

[63] At meetings of the members and committee of the LTWHSA in 2013, the author observed several accusations of mismanagement regarding funds provided to the committee by the SIG.

[64] Anita Smith, 'East Rennell World Heritage Site: Misunderstandings, Inconsistencies and Opportunities in the Implementation of the World Heritage Convention in the Pacific Islands' (2011) 17(6) *International Journal of Heritage Studies* 592, 601.

[65] Jacob Zikuli and Hazel Clothier, *Community Attitudes and Perceptions Towards the East Rennell World Heritage Programme* (Live and Learn Environmental Education, 2008) 13.

[66] John Naitoro, 'Mineral Resource Policy in Solomon Islands: The "Six Feet" Problem' (2000) 15(1) *Pacific Economic Bulletin* 132.

[67] Ibid., 136.

As explained in Sect. 2.4.4, most customary legal systems have undergone profound changes since the pre-contact period, and they continue to evolve. In some circumstances, the establishment of a new local governance association can be an appropriate part of this process, assisting local people to meet the contemporary challenges they face. However, any new body that is established must be coherent with customary governance structures.[68] Importantly, the respective mandates of the new body and customary structures must be clear.[69] A failure to address these issues can weaken local governance and cause conflict.

At present, a lack of strong local governance at East Rennell presents a significant challenge for World Heritage protection. Strengthening customary protection requires exploring if and how the legitimacy of the chiefs within the communities can be improved, and clarifying the relationship between customary structures and the LTWHSA. It is imperative that all governance bodies have clear mandates so they can operate cooperatively together. This is particularly important if the LTWHSA is recognised as the site's management committee under the *Protected Areas Act 2010*, as that would give the association the power to make some legally binding decisions that prevail over custom (see Sect. 7.2).

6.4   Boundaries and Buffer Zones at the East Rennell World Heritage Site

The *Operational Guidelines* state that the boundaries of a World Heritage site should be drawn to ensure all the attributes that convey the site's OUV are within the property.[70] In addition, if necessary, a buffer zone subject to legal or customary protection should be established around the site.[71] As explained below, these provisions may prove problematic for some sites under customary protection.

---

[68] Anita Smith and Cate Turk, 'Customary Systems of Management and World Heritage in the Pacific Islands' in Sue O'Connor, Denis Byrne and Sally Brockwell (eds), *Transcending the Culture-Nature Divide in Cultural Heritage: Views from the Asia-Pacific Region* (ANU E Press, 2012) 22, 29.

[69] Albert Mumma, 'Legal Aspects of Cultural Landscape Protection in Africa' in *Cultural Landscapes: The Challenges of Conservation*, World Heritage Papers 7 (UNESCO, 2003) 156, 156.

[70] UNESCO, *Operational Guidelines for the Implementation of the World Heritage Convention*, UN Doc WHC.16/01 (26 October 2016) para 99. See also paras 100–102.

[71] Ibid., para 103. See also paras 104–107. For a discussion of the history of the buffer zone requirement in the *Operational Guidelines*, see Josephine Gillespie, 'Buffering for

Customary land tenure boundaries may not correspond with the heritage attributes in the area. Consequently, compliance with the *Operational Guidelines* may result in the site encompassing the land of several landowner groups governed under different customary legal systems. Coordinating the management of such an area could be difficult. For example, it might require the establishment of a new organisation to manage the relationship between the various customary governance bodies. This organisation would not necessarily have customary authority, so ensuring compliance with its decisions could be challenging.

Customary management systems generally work better where the landowning group is relatively small.[72] As such, in some circumstances, the boundary requirements in the *Operational Guidelines* may need to be relaxed, to allow the delineation of a World Heritage site subject to one customary legal system. This might lead to better conservation outcomes than if a large site under fragmented ownership was created.

Implementing the buffer zone provisions in the *Operational Guidelines* may also be challenging for a site under customary protection. At any site, the creation of a buffer zone can be contentious because it potentially intrudes on property rights.[73] The fact that the buffer zone requirements in the *Operational Guidelines* are often not enforced perhaps demonstrates a lack of consensus among States about this requirement.[74] Compliance with the provisions can be particularly challenging if the land within the buffer zone is owned by a different customary group from the World Heritage site. The buffer zone owners may not accept restrictions on the use and development of their land, especially if they receive no tangible benefits from the World Heritage listing. In such circumstances, the creation of a buffer zone could create conflict and/or be ineffective. As Trau, Ballard, and Wilson have noted in their analysis of buffering around the Chief Roi Mata's Domain site in Vanuatu, the decisions of Melanesian landowners are effectively sovereign. As such, buffer zones are only likely

---

Conservation at Angkor: Questioning the Spatial Regulation of a World Heritage Property' (2012) 18(2) *International Journal of Heritage Studies* 194. For discussion of buffer zones in the Pacific World Heritage context, see Adam M Trau, Chris Ballard, Meredith Wilson, 'Bafa Zon: Localising World Heritage at Chief Roi Mata's Domain, Vanuatu' (2014) 20(1) *International Journal of Heritage Studies* 86.

[72] Ruddle, Hviding and Johannes, above n 4, 268.

[73] Natasha Affolder, 'Democratising or Demonising the World Heritage Convention?' (2007) 39 *University of Wellington Law Review* (2007) 341, 356.

[74] Ibid.

to be effective in Melanesia 'when aligned with or augmenting existing local customary provisions'.[75]

The boundary and buffer zone requirements in the *Operational Guidelines* clearly need to be applied flexibly, to facilitate the listing of sites under customary protection. However, as the East Rennell case demonstrates, the implications of any non-compliance for conservation efforts should be carefully considered when such sites are nominated.

The western boundary of the East Rennell World Heritage site is the border between provincial wards 2 and 3 on Rennell island.[76] As noted in Sect. 5.2.1, it appears that to strictly meet the boundary requirements in the *Operational Guidelines*, the whole of Rennell island should have been listed. However, the nomination was limited to East Rennell, as the West Rennellese people did not consent to the listing of their land. No buffer zone has been created, and it is unlikely that one will be established in the future given the customary tenure of that area.

The World Heritage site is now under threat from activities in West Rennell, including logging, mining, and actions that are facilitating the spread of invasive species. The East Rennellese people have little control over these activities, and the SIG has been reluctant to strongly regulate them, in part because of its reverence for the customary rights of the West Rennellese (see Sect. 5.3.3.2). The preservation of East Rennell is therefore closely tied to the decision-making of the West Rennellese people. However, the West Rennellese are unlikely to voluntarily agree to forgo development opportunities in order to preserve the World Heritage site, particularly if they do not receive any tangible benefits from the site's listing. It is thus imperative that they be included in World Heritage conservation initiatives, including alternative livelihood development projects.

6.5   MANAGEMENT PLANNING FOR THE EAST RENNELL WORLD HERITAGE SITE

A management plan for East Rennell was prepared in 2007 by a consultant with funding from the World Heritage Fund.[77] The plan was endorsed by the East Rennell World Heritage Trust Board, the Rennell Bellona provincial government, and the Solomon Islands National Commission for

[75] Trau, Ballard and Wilson, above n 71, 91.
[76] Wingham, above n 18, 38.
[77] Wein, above n 3, 6.

UNESCO.[78] A revised plan was prepared in 2014,[79] but it has not been finalised or approved by the local communities.

The 2007 East Rennell management plan has barely been implemented.[80] Consequently, the World Heritage Committee has called for the plan to be revised,[81] a measure which is also identified in the *DSOCR*.[82] In the preparation of any new management plan for the site, it is instructive to consider why the existing plan has been ineffective.

The 2007 management plan sets out a vision for the site, management objectives, and a series of actions to achieve those aims. Among other things, it supports banning commercial logging and mining, limiting coconut crab harvesting, and regulating the taking of marine species through the creation of a marine protected area,[83] all of which could assist to protect the site's OUV. However, as the Committee has commented, the management measures lack detail, and the plan lacks a timeline and budget.[84] In addition, the plan does not address the impacts of invasive species or climate change, which currently threaten the site's OUV. These omissions should be rectified in any future plan.

Importantly, the East Rennellese people have little interest in or understanding of the 2007 management plan.[85] This is partly attributable to the plan's scope. The objective of the plan is to ensure that the natural ecosystems of East Rennell, which give the site OUV, are safeguarded.[86] While it recognises the need to support sustainable utilisation of resources by the East Rennellese people,[87] the management actions focus almost exclusively

[78] Ibid., 3.

[79] Price, above n 28.

[80] Tabbasum and Dingwall, above n 57, 9, 19. This finding is consistent with the author's observations from working in East Rennell.

[81] WHC Res 38 COM 7A.29, WHC 38th sess, UN Doc WHC-14/38.COM/16 (7 July 2014) 39, 40; WHC Res 39 COM 7A.16, WHC 39th sess, UN Doc WHC-15/39.COM/19 (8 July 2015) 30, 31; WHC Res 40 COM 7A.49, WHC 40th sess, UN Doc WHC-16/40. COM/19 (15 November 2016) 68, 69.

[82] *State of Conservation of Properties Inscribed on the World Heritage List*, WHC 41st sess, UN Doc WHC/17/41.COM/7A.Add (2 June 2017) 26 (East Rennell, Solomon Islands) 31–32.

[83] Wein, above n 3, 19–20.

[84] WHC Res 31 COM 7B.21, WHC 31st sess, UN Doc WHC-07/31.COM/24 (31 July 2007) 58, 58.

[85] Smith and Turk, above n 68, 28.

[86] Wein, above n 3, 8.

[87] Ibid., 8–10.

on conservation of the site's natural environment.[88] As noted in Chap. 5, the heritage that the East Rennellese are most interested in protecting is linked to their cultural identity, as expressed through their land tenure system, environmental knowledge, traditional resource use, crafts, songs, and dance.[89] In addition, they are extremely concerned about their livelihoods and food security.[90] The 2007 management plan therefore does not align with the key priorities and aspirations of the local communities. The fact that the plan was never translated into East Rennellese is also likely to have contributed to their lack of interest.

Smith has found that there is broad support among the East Rennellese people for the documentation of their cultural values, to provide a framework within which the site's natural heritage could be managed.[91] Such a project could inform the development of a management plan that embedded the measures required for the protection of the site's OUV within a broader plan recognising the customs, values, and aspirations of the East Rennellese people. To the extent possible (given the limited resources available for implementation), the plan could also address issues such as improving communication and transport infrastructure, and ensuring access to sustainable food sources. As these are high priorities for the East Rennellese people, the resulting plan may enjoy greater local support.

The East Rennellese peoples' lack of interest in the 2007 management plan may also be linked to the fact that it has no basis under customary law. The plan does not detail relevant customs of the East Rennellese people, but rather identifies the documentation of land tenure, traditional knowledge, and cultural practices as a future management action.[92] This also appears to be an issue with the management plan for the Rock Islands Southern Lagoon World Heritage site in Palau. That site was found to have OUV based on the remains of stone villages, rock art, cave deposits, and burials, which evidence the development of Pacific Island societies, as

---

[88] Ibid., 19–20.

[89] Smith, above n 64, 605.

[90] Gabrys and Heywood, above n 30, 62; Zikuli and Clothier, above n 65, 13. See also Smith, above n 64, who notes that livelihood issues dominated the meetings with community members that she was involved with to discuss the World Heritage programme: at 598. Similarly, livelihood issues dominated many of the meetings the author attended in East Rennell (in her capacity as legal advisor for a non-government organisation) concerning the protection of the World Heritage site.

[91] Smith, above n 64, 605.

[92] Wein, above n 3, 20.

well as its exceptional marine environment and biodiversity.[93] The site's nomination dossier contends that natural resources are managed through traditional cultural controls, such as marine tenure and *bul* (temporary restrictions imposed by village chiefs on certain activities).[94] However, the site's management plan does not document these practices or explain how they help address threats to the area's OUV.

These management plans can be contrasted with the plan for the Chief Roi Mata's Domain World Heritage site in Vanuatu, which was listed as a cultural landscape in which people's lives are still strongly defined by *kastom*.[95] The management plan for that site describes itself as 'an unprecedented attempt' to document the site's *nafsan natoon* (the local peoples' expression for customary protection).[96] In consultations undertaken to inform the development of the plan, community members discussed how people's behaviours are dictated by *nafsan natoon*. These discussions provided information and strategies that form the basis of the identified management measures.[97] Thus, to a large extent, the management plan reflects the codification and extension of traditional practices.[98]

A further issue with the 2007 East Rennell management plan is that is has no force under State law. The plan was drafted in anticipation that some of the management actions would be strengthened through a provincial ordinance,[99] but no such law has been enacted. In Solomon Islands, a management plan can gain legal effect under the *Protected Areas Act 2010*. However, as explained in Sect. 7.2, there are a raft of legal and practical issues that need to be considered when drafting such a plan and, in any event, East Rennell has not yet been declared a 'protected area' under that Act. As the 2007 management plan has no basis under customary or State law, implementation is entirely voluntary. In the absence of strong community interest in the plan, this makes implementation unlikely.

[93] WHC Res 36 COM 8B.12, WHC 36th sess, UN Doc WHC-12/36.COM/19 (June–July 2012) 165, 165 para 3.

[94] Republic of Palau, *The Rock Islands Southern Lagoon Nomination for Inscription on the World Heritage List* (2012) 109.

[95] WHC Res 32 COM 8B.27, 32nd sess, UN Doc WHC-08/32.COM/24Rev (31 March 2009) 170.

[96] Meredith Wilson, *Plan of Management for Chief Roi Mata's Domain (CRMD)* (2006) 22.

[97] Ibid., 8.

[98] Meredith Wilson, Chris Ballard and Douglas Kalotiti, 'Chief Roi Mata's Domain: Challenges Facing a World Heritage-Nominated Property in Vanuatu' (Paper presented at ICOMOS meeting, Cairns, 21 July 2007) 6.

[99] Wein, above n 3, 17.

The East Rennell management plan has also not been successful because the community associations charged with implementing the plan have lacked the resources and capacity to do so. Responsibility for implementation was originally vested in the East Rennell World Heritage Trust Board,[100] and later in the LTWHSA.[101] As discussed in Sect. 6.3, decisions of the LTWHSA have no force under customary law and there are issues impeding the association's effectiveness.

The 2007 management plan does not make any government entity responsible for implementation. This is in some respects understandable in the Pacific, where heritage protection is not a high priority and government resources are scarce. For example, Denham, Muke, and Genorupa have contended that any management plan for the Kuk Early Agricultural World Heritage site in Papua New Guinea (PNG) needs to be resistant to neglect by the national and provincial governments.[102] They note that in PNG it is both 'unrealistic and inappropriate to burden national or provincial governments with substantial and continuing financial commitments'.[103] Similarly, in Solomon Islands, a management plan that requires a large long-term commitment from the government is unlikely to be successful. However, responsibility for the implementation of a management plan cannot simply be devolved to the local people if they have insufficient capacity and resources to undertake the management measures.

Further work is required to identify the optimal management model for the East Rennell World Heritage site. The *Operational Guidelines* do not prescribe the management approach that should be taken. Indeed, they recognise that the form of management system for a World Heritage site will depend on the characteristics and needs of the site, and that it may incorporate customary practices.[104] However, the *Operational Guidelines* also state that the management system must be documented,[105] and will often include a cycle of planning, implementation, monitoring, evaluation, and feedback; monitoring and assessment of impacts; capacity build-

---

[100] Wein, above n 3, 10, 21.

[101] *Lake Tegano World Heritage Site Association Constitution and Rules* cl 2.2(a).

[102] John Denham, Tim Muke and Vagi Genorupa, 'Nominating and Managing a World Heritage Site in the Highlands of Papua New Guinea' (2007) 39(3) *World Archaeology* 324, 331.

[103] Ibid., 333.

[104] UNESCO, *Operational Guidelines for the Implementation of the World Heritage Convention*, UN Doc WHC.16/01 (26 October 2016) para 110.

[105] Ibid., paras 108, 132(5).

ing; and a description of how the system functions.[106] A site under customary protection may not need such heritage management structures and tools.[107] As such, the *Operational Guidelines* currently suggest a management structure that is not necessarily appropriate for a site under customary protection. It is therefore imperative that the management plan provisions of the *Operational Guidelines* be applied flexibly for such sites.

Around the world, only a 'handful' of places under customary management and protection have been inscribed on the World Heritage List.[108] Most of these are subject to co-management systems, under which the government and local resource users share power and responsibility for the area.[109] For example, the Uluru-Kata Tjuta National Park in Australia is jointly managed by the Director of National Parks and the Uluru-Kata Tjuta Board of Management. Of the 12 members of this Board, eight are Aboriginal members nominated by the traditional owners of the area.[110] In Australia, co-management approaches have emerged to allow Indigenous people greater decision-making powers over their traditional lands,[111] to ensure that their aspirations are incorporated into environmental manage-

---

[106] Ibid., para 111.

[107] Smith and Turk, above n 68, 30.

[108] Ibid., 26.

[109] F Berkes, P J George and R J Preston, 'The Evolution of Theory and Practice of the Joint Administration of Living Resources' (1991) 18(2) *Alternatives* 12, 12; Borrini-Feyerbend, Kothari and Oviedo, above n 35.

[110] Australian Government Department of Environment and Energy, *Park Management* http://www.environment.gov.au/topics/national-parks/uluru-kata-tjuta-national-park/management-and-conservation/park-management. For further discussion of co-management initiatives, see, for example, M Nursey-Bray and P Rist, 'Co-Management and Protected Area Management: Achieving Effective Management of a Contested Site: Lessons from the Great Barrier Reef World Heritage Area (GBRWHA)' (2009) 33(1) *Marine Policy* 118; Tony Corbett, Marcus Lane and Chris Clifford, *Achieving Indigenous Involvement in Management of Protected Areas: Lessons from Recent Australian Experience*, Aboriginal Politics and Public Sector Management Research Paper 5 (Centre for Australian Public Sector Management, 1998); T Bauman, C Haynes and G Lauder, *Pathways to the Co-Management of Protected Areas and Native Title in Australia*, AIATSIS Research Discussion Paper 32 (2013); Melanie Zubra et al, 'Building Co-Management as a Process: Problem Solving Through Partnerships in Aboriginal Country, Australia' (2012) 49 *Environmental Management* 1130; Joseph J Spaeder and Harvey A Feit, 'Co-Management and Indigenous Communities: Barriers and Bridges to Decentralised Resource Management: Introduction' (2005) 47(2) *Anthropologica* 147.

[111] Bauman, Haynes and Lauder, above n 110, 9.

ment initiatives,[112] and as part of the reconciliation process.[113] However, while co-management is an effective approach in some places, it is unlikely to be appropriate for sites such as East Rennell that are owned and occupied by customary owners who rely on the land for their livelihoods, and where the government has limited resources and capacity to participate in site management.

Whatever management approach is taken at East Rennell, it is imperative that the East Rennellese maintain a central role in decision-making concerning the conservation of their land. However, it cannot be assumed that they will be willing and able to implement management measures to conserve the site's OUV, particularly those that have no basis in custom. To date, while local community organisations have received some assistance to implement the 2007 management plan,[114] the funds and expertise available at the local level to dedicate to site management remain very limited. Any future management plans for the site are unlikely to enjoy significant success unless they are accompanied by long-term funding and assistance for implementation.

6.6   Conclusion

The protection regimes of all sites nominated for World Heritage listing should be scrutinised, to help stakeholders agree upon feasible and appropriate conservation objectives, and to anticipate and address any challenges that arise. If the site is under customary protection, that assessment should consider the scope of customary laws, with reference to the current and foreseeable threats to the site's OUV. The level of compliance with relevant customs, and whether they are enforceable against outsiders, also needs to be understood. In addition, the assessment should look at the structure, jurisdiction, and strength of relevant customary governance bodies, and how they will work together if the site comprises land governed by more than one body. Furthermore, the relationship between customary protection and any existing or proposed management plan and heritage protection legislation must be analysed.

---

[112] Nursey-Bray and Rist, above n 110; Corbett, Lane and Clifford, above n 110, 1.

[113] Bauman, Haynes and Lauder, above n 110, 10.

[114] For example, the assistance provided by Gabrys and Heywood: see Gabrys and Heywood, above n 30.

Provisions guiding the assessment of a customary protection regime could be inserted into the *Operational Guidelines*, drawing upon existing best practice guidelines on related topics.[115] Clearly, any provisions concerning these matters would need to be sufficiently broad and flexible to encompass the huge variety of customary legal systems and World Heritage sites that exist around the world. However, appropriately drafted, they could serve as a useful starting point for assessing customary protection in the context of the *World Heritage Convention*.

Although East Rennell was inscribed on the World Heritage List primarily based on its customary protection, there has been little empirical research into how the customary system can contribute to addressing the threats to the site's OUV. The limited available literature suggests that customary practices cannot deal with many of the key threats, and customary governance is not strong. Further interdisciplinary research is needed to document existing practices, and to explore the scope for the customary system to evolve to meet new challenges such as invasive species and climate change.

If a site is well-managed through customary systems, a management plan may not be required to effectively protect the site.[116] However, as customary protection is unable to deal with all threats to the OUV of East Rennell, additional management measures are required. The site's 2007 management plan has not been effective in protecting the site, in part because it lacks any basis under custom or State law. This feature, coupled with the fact that the SIG is not charged with executing the plan, means that implementation relies on the voluntary commitment of the East Rennellese people. The pressing livelihood issues they face, and their limited capacity and resources for heritage protection, make it unlikely that the plan will be implemented.

Implementing any management plan at East Rennell will be challenging. A future plan may however enjoy greater success if management actions to safeguard the site's OUV are embedded in a broader strategy that seeks to preserve culture and support livelihood development. To the extent possible, management actions should be aligned with and/or extend existing customs. This makes the identification and documentation

---

[115] See, for example, Robert Wild and Christopher McLeod (eds), *Sacred Natural Sites: Guidelines for Protected Area Managers*, Best Practice Protected Area Guidelines Series No. 16 (IUCN, 2008); Borrini-Feyerabend, Kothari and Oviedo, above n 35.

[116] Smith and Turk, above n 68, 30.

of relevant customary laws a crucial next step in the protection of the site. As explained in the next chapter, the potential for the management plan to gain legal effect under the *Protected Areas Act 2010* should also be considered. Importantly, the measures must be designed in light of the resource and capacity constraints of the bodies charged with implementation. It is unrealistic to expect that local people will dedicate significant time and resources to World Heritage protection activities unless the proposed management measures closely align with their priorities and/or they are supported to do so.

REFERENCES

ARTICLES, BOOKS AND REPORTS

Affolder, Natasha, 'Democratising or Demonising the World Heritage Convention?' (2007) 39 *University of Wellington Law Review* (2007) 341

Allen, Matthew, Sinclair Dinnen, Daniel Evans, Rebecca Monson, *Justice Delivered Locally: Systems, Challenges and Innovations in Solomon Islands* (World Bank, 2013)

Aswani, Shankar, 'Customary Sea Tenure in Oceania as a Case of Rights-Based Fishery Management: Does it Work?' (2005) 15 *Reviews in Fish Biology and Fisheries* 285

Ballard, Chris and Meredith Wilson, 'Unseen Monuments: Managing Melanesian Cultural Landscapes' in Ken Taylor and Jane L Lennon (eds), *Managing Cultural Landscapes* (Routledge, Oxon 2012) 130

Bauman, T, C Haynes and G Lauder, *Pathways to the Co-Management of Protected Areas and Native Title in Australia*, AIATSIS Research Discussion Paper 32 (2013)

Bennett, Judith, *Wealth of the Solomons: A History of a Pacific Archipelago, 1800–1978* (University of Hawaii Press, 1988)

Berkes, F, P J George and R J Preston, 'The Evolution of Theory and Practice of the Joint Administration of Living Resources' (1991) 18(2) *Alternatives* 12

Birket-Smith, K A J, *An Ethnological Sketch of Rennell Island: A Polynesian Outlier in Melanesia* (Munksgaard, 2nd ed, 1969)

Borrini-Feyerabend, Grazia, Ashish Kothari and Gonzalo Oviedo, *Indigenous and Local Communities and Protected Areas: Towards Equity and Enhanced Conservation* (IUCN, Gland, 2004)

Clarke, Pepe and Charles Taylor Gillespie, *Legal Mechanisms for the Establishment and Management of Terrestrial Protected Areas in Fiji* (IUCN, Suva 2009)

Clarke, Pepe and Stacy D Jupiter, 'Law, Custom and Community-Based Natural Resource Management in Kubulau District (Fiji)' (2010) 37(1) *Environmental Conservation* 98

Corbett, Tony, Marcus Lane and Chris Clifford, *Achieving Indigenous Involvement in Management of Protected Areas: Lessons from Recent Australian Experience*, Aboriginal Politics and Public Sector Management Research Paper 5 (Centre for Australian Public Sector Management, 1998)

Corrin, Jennifer and Don Paterson, *Introduction to South Pacific Law* (Palgrave Macmillan, 3rd ed, 2011)

Corrin Care, Jennifer and Jean G Zorn, 'Legislating for the Application of Customary Law in Solomon Islands' (2005) 34 *Common Law World Review* 144

Denham, John, Tim Muke and Vagi Genorupa, 'Nominating and Managing a World Heritage Site in the Highlands of Papua New Guinea' (2007) 39(3) *World Archaeology* 324

Dingwall, Paul, *Report on the Reactive Monitoring Mission to East Rennell, Solomon Islands, 21–29 October 2012* (IUCN, 2013)

Elbert, Samual H and Torben Monberg, *From the Two Canoes: Oral Traditions of Rennell and Bellona Islands* (Danish National Museum and University of Hawaii Press, 1965)

Gabrys, Kasia and Mike Heywood, 'Community and Governance in the World Heritage Property of East Rennell' in Anita Smith (ed), *World Heritage in a Sea of Islands: Pacific 2009 Programme*, World Heritage Papers 34 (UNESCO, 2012) 60

Gillespie, Josephine, 'Buffering for Conservation at Angkor: Questioning the Spatial Regulation of a World Heritage Property' (2012) 18(2) *International Journal of Heritage Studies* 194

International Heritage Section, Department of Sustainability, Environment, Water, Population and Communities, Australian Government, 'Australian Capacity Building Support for East Rennell World Heritage Area 2007–2013' in Anita Smith (ed), *World Heritage in a Sea of Islands: Pacific 2009 Programme*, World Heritage Papers 34 (UNESCO, 2012) 66

Kuschel, Rolf, 'Early Contacts Between Bellona and Rennell Islands and the Outside World' (1988) 23(2) *Journal of Pacific History* 191

Lindsay, Jonathan M, *Creating Legal Space for Community-Based Management: Principles and Dilemmas* (Food and Agriculture Organization of the United Nations, Rome 1998)

Monberg, Torben, 'Research on Rennell and Bellona: A Preliminary Report' (1960) 2 *Folk* 71

Monberg, Torben, 'Crisis and Mass Conversion on Rennell Island in 1938' (1962) 71(2) *Journal of Polynesian Society* 145

Monberg, Torben, 'Bellona and Rennell Islanders' in Melvin Ember, Carol R Ember and Jan Skoggard (eds), *Encyclopedia of World Cultures Supplement* (Macmillan, 2002) 46

Mumma, Albert, 'Legal Aspects of Cultural Landscape Protection in Africa' in *Cultural Landscapes: The Challenges of Conservation*, World Heritage Papers 7 (UNESCO, 2003) 156

Naitoro, John, 'Mineral Resource Policy in Solomon Islands: The 'Six Feet' Problem' (2000) 15(1) *Pacific Economic Bulletin* 132

Ntumy, Michael A, 'The Dream of a Melanesian Jurisprudence: The Purpose and Limits of Law Reform' in Jonathan Aleck and Jackson Rannells (eds), *Custom at the Crossroads* (University of Papua New Guinea, 1995) 7

Nursey-Bray, M and P Rist, 'Co-Management and Protected Area Management: Achieving Effective Management of a Contested Site: Lessons from the Great Barrier Reef World Heritage Area (GBRWHA) (2009) 33(1) *Marine Policy* 118

Republic of Palau, *The Rock Islands Southern Lagoon Nomination for Inscription on the World Heritage List* (2012)

Ruddle, K, E Hviding and R E Johannes, 'Marine Resources Management in the Context of Customary Tenure' (1992) 7 *Marine Resource Economics* 249

Smith, Anita, 'East Rennell World Heritage Site: Misunderstandings, Inconsistencies and Opportunities in the Implementation of the World Heritage Convention in the Pacific Islands' (2011) 17(6) *International Journal of Heritage Studies* 592

Smith, Anita, 'The World Heritage Pacific 2009 Programme' in Anita Smith (ed), *World Heritage in a Sea of Islands: Pacific 2009 Programme*, World Heritage Papers 34 (UNESCO, Paris 2012) 2

Smith, Anita and Cate Turk, 'Customary Systems of Management and World Heritage in the Pacific Islands' in Sue O'Connor, Denis Byrne and Sally Brockwell (eds), *Transcending the Culture-Nature Divide in Cultural Heritage: Views from the Asia-Pacific Region* (ANU E Press, 2012) 22

Spaeder, Joseph J and Harvey A Feit, 'Co-Management and Indigenous Communities: Barriers and Bridges to Decentralised Resource Management: Introduction' (2005) 47(2) *Anthropologica* 147

Tabbasum, Salamat Ali and Paul Dingwall, *Report on the Mission to East Rennell World Heritage Property and Marovo Lagoon, Solomon Islands, 30 March–10 April 2005* (IUCN and World Heritage Centre, 2005)

Techera, Erika J, 'Samoa: Law, Custom and Conservation' (2006) 10 *New Zealand Journal of Environmental Law* 361

Trau, Adam M, Chris Ballard, Meredith Wilson, 'Bafa Zon: Localising World Heritage at Chief Roi Mata's Domain, Vanuatu' (2014) 20(1) *International Journal of Heritage Studies* 86

Turton, Steve, *East Rennell World Heritage Area: Assessment of the State of Conservation of World Heritage Values. Final Field Report* (James Cook University, 2014)

Veitayaki, Joeli, Akosita D R Nakoro, Tareguci Sigarua and Nanise Bulai, 'On Cultural Factors and Marine Managed Areas in Fiji' in Jolie Liston, Geoffrey

Clark and Dwight Alexander (eds), *Pacific Island Heritage: Archaeology, Identity and Community* (ANU E Press, 2011) 37

Wild, Robert and Christopher McLeod (eds), *Sacred Natural Sites: Guidelines for Protected Area Managers*, Best Practice Protected Area Guidelines Series No. 16 (IUCN, 2008)

Wilson, Meredith, Chris Ballard and Douglas Kalotiti, 'Chief Roi Mata's Domain: Challenges Facing a World Heritage-Nominated Property in Vanuatu' (Paper presented at ICOMOS meeting, Cairns, 21 July 2007)

Wingham, Elspeth J and Ben Devi, 'The Involvement of Local People in the Management of a Proposed World Heritage Site at East Rennell, Solomon Islands' in Hans D Thulstrup (ed), *World Natural Heritage and the Local Community: Case Studies from Asia Pacific, Australia and New Zealand* (UNESCO, 1999) 79

Wingham, Elspeth J, *Nomination of East Rennell, Solomon Islands by the Government of Solomon Islands for Inclusion in the World Heritage List Natural Sites* (1997)

Wolff, T, 'The Fauna of Rennell and Bellona, Solomon Islands' (1969) 255(800) *Philosophical Transactions of the Royal Society of London, Series B, Biological Sciences* 321

Zikuli, Jacob and Hazel Clothier, *Community Attitudes and Perceptions Towards the East Rennell World Heritage Programme* (Live and Learn Environmental Education, 2008)

Zubra, Melanie, Helen Ross, Arturu Izurieta, Phillip Ris, Ellie Bock and Fikret Berkes, 'Building Co-Management as a Process: Problem Solving Through Partnerships in Aboriginal Country, Australia' (2012) 49 *Environmental Management* 1130

LEGISLATION AND BILLS: SOLOMON ISLANDS

Charitable Trusts Act (Cap. 55)
Forest Resources and Timber Utilisation Act (Cap. 40)
Local Courts Act (Cap. 19)
Mines and Minerals Act (Cap. 42)
Protected Areas Act (2010)

LEGISLATION AND BILLS: SAMOA

Village Fono Act 1990

CONVENTIONS

Convention Concerning the Protection of the World Cultural and Natural Heritage, opened for signature 16 November 1972, 1037 UNTS 151 (entered into force 17 December 1975)

UNITED NATIONS DOCUMENTS

State of Conservation of Properties Inscribed on the World Heritage List, WHC 41st sess, UN Doc WHC/17/41.COM/7A.Add (2 June 2017) 26 (East Rennell, Solomon Islands)

UNESCO, *Operational Guidelines for the Implementation of the World Heritage Convention*, UN Doc WHC.16/01 (26 October 2016)

WHC Res 28 COM 15B.12, WHC 28th sess, UN Doc WHC-04/28.COM/26 (29 October 2004) 84

WHC Res 31 COM 7B.21, WHC 31st sess, UN Doc WHC-07/31.COM/24 (31 July 2007) 58

WHC Res 32 COM 8B.27, 32nd sess, UN Doc WHC-08/32.COM/24Rev (31 March 2009) 170

WHC Res 36 COM 8B.12, WHC 36th sess, UN Doc WHC-12/36.COM/19 (June–July 2012) 165

WHC Res 38 COM 7A.29, WHC 38th sess, UN Doc WHC-14/38.COM/16 (7 July 2014) 39

WHC Res 39 COM 7A.16, WHC 39th sess, UN Doc WHC-15/39.COM/19 (8 July 2015) 30

WHC Res 40 COM 7A.49, WHC 40th sess, UN Doc WHC-16/40.COM/19 (15 November 2016) 68

WHC Res 41 COM 7A.19, 41st sess, UN Doc WHC-17/41.COM/18 (12 July 2017) 35

INTERNET MATERIALS

Australian Government Department of Environment and Energy, *Park Management* http://www.environment.gov.au/topics/national-parks/uluru-kata-tjuta-national-park/management-and-conservation/park-management

OTHER

Lake Tegano World Heritage Site Association Constitution and Rules (2009)
Pacific World Heritage Action Plan 2016–2020 (2016)
Price, Anna, *(Draft) Management Plan – East Rennell, Solomon Islands* (2014)
Wilson, Meredith, *Plan of Management for Chief Roi Mata's Domain* (CRMD) (2006)
Wein, Laurie, *East Rennell World Heritage Site Management Plan* (Solomon Islands National Commission for UNESCO, 2007)

Protecting the East Rennell World Heritage Site: Legislation

7.1 Introduction

The World Heritage nomination dossier for East Rennell stated that the Solomon Islands government (SIG) would enact a *World Heritage Protection Act* to supplement the site's customary protection.[1] However, that never occurred, and today the site does not enjoy broad protection under any legislation. Consequently, the World Heritage Committee has repeatedly called upon the SIG to strengthen the legal protection of the site.[2] Substantial legislative reform for heritage protection is unlikely in Solomon Islands (at least in the short term) given the country's economic and political situation. This chapter therefore considers the extent to which existing legislation could be utilised to address the threats to

---

[1] Elspeth J Wingham, *Nomination of East Rennell, Solomon Islands by the Government of Solomon Islands for Inclusion in the World Heritage List Natural Sites* (1997) 38. The customary protection of East Rennell was explored in Chap. 6.

[2] WHC Res 29 COM 7B.10, WHC 29th sess, UN Doc WHC-05/29.COM/22 (9 September 2005) 45, 45; WHC Res 31 COM 7B.21, WHC 31st sess, UN Doc WHC-07/31.COM/24 (31 July 2007) 58, 58; WHC Res 33 COM 7B.19, WHC 33rd sess, UN Doc WHC-09/33.COM/20 (20 July 2009) 68, 68; WHC Res 34 COM 7B.17, WHC 34th sess, UN Doc WHC-10/34.COM/20 (3 September 2010) 71, 71; WHC Res 37 COM 7B.14, WHC 37th sess, UN Doc WHC-13/37.COM/20 (5 July 2013) 68, 68; WHC Res 38 COM 7A.29, WHC 38th sess, UN Doc WHC-14/38.COM/16 (7 July 2014) 39, 40.

© The Author(s) 2018
S. C. Price, *World Heritage Conservation in the Pacific*,
Palgrave Series in Asia and Pacific Studies,
https://doi.org/10.1007/978-981-13-0602-0_7

233

East Rennell's outstanding universal value (OUV).[3] The chapter analyses relevant legislative provisions, and issues affecting the ability and willingness of the SIG and customary landowners to use these laws to protect World Heritage.

The chapter begins by exploring the implementation of the *Protected Areas Act 2010* (the *PA Act*) at East Rennell (Sect. 7.2). It explains that the approach to conservation facilitated by the *PA Act* is appropriate for Solomon Islands, where most land is under customary tenure, many people rely on natural resources to support their subsistence livelihoods, and the government's capacity to enforce legislation is limited. There are however significant limitations to the protection provided by the Act. In addition, as the Act regulates matters traditionally dealt with under customary law, the relationship between the legislation and customary legal systems must be carefully considered in the preparation of the site's management plan and the selection of its management committee.

The chapter then analyses other legislation that could be used to protect East Rennell, focusing on the regulation of logging and mining, which are arguably the most pressing threats to the site's OUV. There appears to be uncertainty among some people working for the SIG regarding the scope of the government's power to regulate these activities.[4] However, as will be explained, under existing legislation, SIG decision-makers could refuse to approve developments that may impact East Rennell's OUV. This legislation also gives customary landowners the right to dictate whether logging or mining occurs on their land. Notwithstanding this, for various economic, social, political, and legal reasons, operations that threaten the site may continue to be approved. The East Rennellese must be supported to exercise their rights under logging and mining laws, if they are to successfully maintain their opposition to such developments (Sect. 7.3.1).

The chapter then briefly considers legislation that could be utilised to address the over-harvesting of certain animals (Sect. 7.3.2) and invasive species (Sect. 7.3.3), and highlights the need for climate change legisla

---

[3] These threats were explained in Sect. 5.3.1.

[4] See Solomon Islands Government, *State Party Report on the State of Conservation of the East Rennell World Heritage Area (Solomon Islands)* (SIG, 2013) 6; Solomon Islands Government, *State Party Report on the State of Conservation of the East Rennell World Heritage Area (Solomon Islands)* (SIG, 2014) 3. The authors of these reports contend that the SIG has limited power to prevent logging or mining occurring on customary land.

tion in Solomon Islands (Sect. 7.3.4). It also comments on the absence of laws specifically dealing with the protection of cultural and intangible heritage (Sect. 7.4).

Overall, the chapter demonstrates that while legislation exists that could help address the threats to East Rennell, ensuring compliance with those laws is challenging. Suggestions for legislative amendments and other steps that could strengthen the site's protection are provided. However, given the inherent challenges that exist, East Rennell's future is uncertain.

7.2   Protecting East Rennell Under the *Protected Areas Act 2010*

One of the measures identified in the *Desired State of Conservation for the Removal of East Rennell from the List of World Heritage in Danger* (*DSOCR*) is the declaration of the site under the *PA Act*.[5] That Act empowers the responsible Minister[6] to declare an area of land or sea to be a 'protected area'.[7] If that occurs, activities undertaken within the site are regulated under the *Protected Areas Regulations 2012* (the *PA Regulations*).

As explained further in the sections below, in most circumstances, the Minister cannot declare a protected area unless the landowners or a non-government organisation managing the area submit an application to the

---

[5] *State of Conservation of Properties Inscribed on the World Heritage List*, WHC 41st sess, UN Doc WHC/17/41.COM/7A.Add (2 June 2017) 26 (East Rennell, Solomon Islands) 31. The *DSOCR* is explained in Sect. 5.3.2. For analysis of the *PA Act* generally, see Stephanie Price et al, *Environmental Law in Solomon Islands* (Public Solicitor's Office, Solomon Islands Government, 2015) ch 9. The *DSOCR* also calls for the implementation of the Rennell Bellona *Lake Tegano Natural Heritage Park Ordinance* at the World Heritage site. This ordinance provides for the establishment of a protected area at East Rennell. A draft of the ordinance was prepared in 2009. However, the ordinance has not been passed by the Rennell Bellona provincial assembly, and it remains in draft. There are several legal issues associated with the draft that should be considered before the ordinance is passed. For analysis of the draft, see Stephanie Clair Price, *Strengthening World Heritage Protection in the Pacific: An Exploration of Solomon Islands' Implementation of the World Heritage Convention* (PhD Thesis, University of Western Australia, 2017) ch 8.

[6] Currently the Minister of Environment, Climate Change, Disaster Management, and Meteorology.

[7] *PA Act* s 10(1).

Director of Environment.[8] The Director then assesses the application and makes a recommendation to the Minister.[9] The Minister must be satisfied that several requirements are met before he or she can make a declaration, including that the people with 'rights or interests in the area' consent.[10] An appropriate management plan must also be in place.[11] This plan contains rules regulating the use of the site, and addresses matters such as research, training, public awareness, and monitoring.[12] Once a protected area is declared, the national Protected Areas Advisory Committee appoints a management committee for the site,[13] which is responsible for overseeing the implementation and periodic review of the site's management plan.[14]

The SIG recently indicated that it is committed to implementing the *PA Act* at East Rennell, but a 'thorough and cautious approach is required to ensure community ownership of the decision'.[15] While some work has been done towards preparing a protected area application for the site,[16] at this stage it is unclear whether the East Rennellese people will consent to a declaration. The SIG has noted that before the *PA Act* is implemented at the site, a comprehensive and detailed roadmap needs to be developed to address the governance, management, and technical issues that may arise.[17] Key issues that need to be considered if and when the Act is implemented at East Rennell (and which could inform the development of such a roadmap) are highlighted below. Many of these concern the relationship between the *PA Act* regime and customary law.

---

[8] Ibid., s 10(4). The exception is that in some circumstances the Minster for Forests or the Minister for Fisheries may recommend a site be declared as a protected area: s 11(2).

[9] Ibid., s 10(2).

[10] Ibid., s 10(7) (c).

[11] Ibid., s 10(7) (d).

[12] *PA Regulations* reg 23.

[13] *PA Act* s 12(1).

[14] Ibid., s 12(3); *PA Regulations* reg 29(1).

[15] Solomon Islands Government, *State Party Report on the State of Conservation of East Rennell (Solomon Islands)* (SIG, 2017) 1.

[16] As legal adviser for a non-government organisation, in 2013 the author participated in meetings in the East Rennell communities to discuss the process for, and implications of, establishing a protected area at the World Heritage site. In the same year, a draft protected area management plan for the site was prepared. However, the plan has not yet been finalised or approved by the local communities.

[17] Solomon Islands Government, above n 15.

7.2.1   Addressing the Threats to the Outstanding Universal Value of East Rennell Under the Protected Areas Act

If East Rennell was declared under the *PA Act*, the activities undertaken in the site would be regulated under the *PA Regulations* and the site's management plan. Some rules prescribed in the *PA Regulations* apply to all declared sites. For example, the industrial and commercial extraction of timber, round logs, and minerals from protected areas is prohibited.[18] Consequently, the declaration of East Rennell under the Act would make logging and mining in the site unlawful (which is an outcome the World Heritage Committee is seeking).[19] The *PA Regulations* also restrict certain activities in Marine Protected Areas (MPAs) such as fishing within spawning aggregations or during spawning seasons, and the use of drag nets.[20] These provisions could help address the over exploitation of marine species at East Rennell.

The *PA Regulations* also provide a mechanism for the development of site-specific rules, as some rules in the Regulations can be modified by a site's management plan and/or a decision of the site's management committee. The rules in the *PA Regulations* are very broad, so most activities in a protected area must be expressly authorised to be lawful. This authorisation process could be used to tailor the rules to address the threats to East Rennell's OUV. For example, pursuant to the *PA Regulations*, it is an offence to take any organism from a protected area without authorisation under the management plan or by the management committee.[21] The management plan for East Rennell could state that no person may take a coconut crab that is carrying eggs or is less than 90 mm long. If a person failed to comply, he or she could be fined for taking an organism without authorisation in contravention of the *PA Regulations*.

A limitation of the *PA Act* regime is that it provides little protection against activities outside the declared site.[22] This is significant for the protection of East Rennell, which is threatened by activities occurring in

---

[18] *PA Regulations* reg 61(1). The terms 'commercial' and 'industrial' are not defined in the *PA Regulations*, and their precise meaning is unclear. For discussion of this issue and its implications, see Price et al, above n 5, 269–70.

[19] See, for example, WHC Res 41 COM 7A.19, WHC 41st sess, UN Doc WHC/17/41. COM/18 (12 July 2017) 35.

[20] *PA Regulations* regs 50(1), 52(1).

[21] Ibid., reg 62.

[22] The exceptions are: (1) activities on land that may be harmful or destructive to an MPA are prohibited unless Ministerial approval has been obtained (*PA Regulations* reg 54(1)); (2)

West Rennell (see Sect. 5.3.1). Importantly, the declaration of the site would not prevent further logging and mining operations in West Rennell being approved. The regulation of those activities would still be subject to the *Forest Resources and Timber Utilisation Act (Cap. 40) (FRTU Act)*, the *Mines and Minerals Act (Cap. 42) (MM Act)*, and the *Environment Act 1998* (see Sect. 7.3.1). Furthermore, the *PA Act* could not be used to establish rules regulating the harvesting of species or requiring the implementation of biosecurity measures outside the declared protected area. As such, notwithstanding the declaration of East Rennell, legislation such as the *Fisheries Management Act 2015* and the *Biosecurity Act 2013* would still be relevant (see Sects. 7.3.2 and 7.3.3).

7.2.2   Landowner Consent for the Declaration of a Protected Area

The *PA Regulations* prescribe a process which must be implemented before a protected area application for customary land can be submitted. In summary, that process involves the 'landowning tribe' holding a meeting to discuss the submission of an application.[23] If the tribe reaches a consensus or makes a resolution in support of an application, the tribe's leaders must document this in a written agreement, which must also be signed by the leaders of 'neighbouring tribes'.[24]

As noted previously, the Minister for Environment cannot declare a site to be protected without the consent of the people with 'rights or interests in the area'[25] (referred to here for convenience as 'Landowners'[26]). To help the Minister determine whether the Landowners have consented, the *PA Act* requires the Director to verify who the Landowners are and discuss the application with them.[27] It is however unlikely that the Director will strictly comply with this requirement in all instances. Given the Ministry of Environment's resource constraints, and the difficulties involved with determining who has rights to customary land, in practice, the Director

---

logging and mining in a buffer zone of up to 1 km around every protected area is prohibited (*PA Regulations* reg 61(1)).

[23] Ibid., reg 44(1) (a).

[24] Ibid., reg 44(1) (d), (e).

[25] *PA Act* s 10(7) (c).

[26] It is acknowledged that the term 'Landowner' over-simplifies the nature of the rights and obligations that characterise most customary tenure systems. People who have the right to occupy and/or use customary land do not 'own' that land in the Western sense of that word. The term 'Landowner' is used here for convenience only.

[27] *PA Act* s 10(2) (a), (d).

(and the Minister) are likely to rely on the documents submitted with the application as proof of Landowner consent. These include the documents arising from the consent process prescribed in the *PA Regulations*, such as the minutes of the meeting at which the landowning tribe consented to the application, and the agreement signed by the tribal leaders. This heightens the importance of the prescribed consent process.

Implementing that process may however be problematic because of its potential inconsistency with customary law. Unlike many other places in Solomon Islands, on Rennell, customary land rights are held individually by male members of the lineage.[28] Thus, in Rennell, if a decision to apply for a protected area was made on a tribal basis, an individual landowner might not consider himself bound by that decision. Furthermore, implementing the process could cause or exacerbate disputes about land rights (as has been the experience with logging and mining agreements entered into by tribal leaders—see Sect. 7.3.1.2).

It is well accepted that local people should be involved in decisions concerning the conservation of their land.[29] However, questions remain concerning what processes should be implemented to ensure that this occurs.[30] The most appropriate process to be undertaken at East Rennell (to ensure widespread community support for a protected area, and to satisfy the requirements of the *PA Regulations*) is yet to be determined. If a protected area is to be pursued at the site, a consent process needs to be designed through extensive discussions with the local communities and their leaders.

In any event, it is not yet clear whether the East Rennell Landowners will consent to the declaration of their land under the *PA Act*. As discussed in Sect. 5.3.3.3, Solomon Islanders are often not interested in participating in conservation programmes that are not accompanied by a real promise of alternative development,[31] which can make establishing protected areas

[28] Samuel H Elbert and Torben Monberg, *From the Two Canoes: Oral Traditions of Rennell and Bellona Islands* (Danish National Museum and University of Hawaii Press, 1965) 10.

[29] See, for example, *Rio Declaration on Environment and Development*, Report of the United Nations Conference on Environment and Development, UN Doc A/CONF.151/6/Rev.1 (1992) principles 10, 22; Barbara Lausche, *Guidelines for Protected Area Legislation* (IUCN, 2011) 75, 45; *Operational Guidelines for the Implementation of the World Heritage Convention*, UN Doc WHC.16/01 (26 October 2016) para 123.

[30] Robert James Hales et al, 'Indigenous Free Prior Informed Consent: A Case for Self Determination in World Heritage Nomination Processes' (2013) 19(3) *International Journal of Heritage Studies* 270, 273.

[31] Martha Macintyre and Simon Foale, 'Global Imperatives and Local Desires: Competing Economic and Environmental Interests in Melanesian Communities' in Victoria Lockwood

challenging. The East Rennellese may therefore consider they have insufficient incentive to support a protected area declaration, unless it is accompanied by (for example) support for livelihood projects. Significantly, in 2018 the Tuhunui Tribe of East Rennell (which claims that it owns a significant portion of the World Heritage site) wrote to the World Heritage Centre stating that it opposes the declaration of their land as a protected area under the Act.[32] If that opposition is maintained, the PA Act cannot be implemented at East Rennell.

If the East Rennellese people do consent to a declaration, ways to maintain the support of the East Rennellese people also need to be explored. Protected area legislation commonly incorporates mechanisms for securing long-term landowner support through binding agreements and instruments such as easements and covenants registered against the title of the land.[33] However, written agreements often carry little weight among Solomon Islanders, particularly agreements that impact on issues traditionally governed through customary law.[34] Thus, future generations may not feel bound by an agreement entered into by tribal leaders to support a protected area declaration. Furthermore, they could not be bound by the agreement through an easement or covenant as such instruments cannot apply to customary land. Maintaining the support of the East Rennellese people will likely require ongoing consultations and negotiations, and the provision of assistance to enable the local communities to improve their livelihoods.

7.2.3   Customary Law and Protected Area Management Plans

In a legally plural society, the incorporation of customary law into a State law will often be an effective approach to heritage protection.[35] The PA Act in effect enables this because the rules in a protected area management

---

(ed), *Globalisation and Culture Change in the Pacific Islands* (Pearson Prentice Hall, 2004) 149, 161.

[32] *State of Conservation of the Properties Inscribed on the List of World Heritage in Danger*, WHC 42nd sess, UN Doc WHC/18/42.COM/7A.Add.2 (15 June 2018) 17 (East Rennell, Solomon Islands).

[33] Lausche, above n 29, 100–102.

[34] Elspeth J Wingham and Ben Devi, 'The Involvement of Local People in the Management of a Proposed World Heritage Site at East Rennell, Solomon Islands' in Hans D Thulstrup (ed), *World Natural Heritage and the Local Community: Case Studies from Asia Pacific, Australia and New Zealand* (UNESCO, 1999) 79, 83.

[35] See, for example, *Barbados Programme of Action for the Sustainable Development of Small Island Developing States*, UN Doc A/CONF.167/9 (October 1994), part I annex II

plan can be based on customary laws. Such a plan could acknowledge and allow for the continuation of relevant customary practices. This may improve the plan's effectiveness, as substantial inconsistency between custom and management measures may reduce compliance with both. It could also make the customs enforceable through the State legal system (if the rule authorised an activity that would otherwise be prohibited under the *PA Regulations*—see Sect. 7.2.1). While enforcing the *PA Act* through the State system is likely to be challenging, it could be more effective than the customary system if the offender is an outsider or if customary governance is weak.

The use of customary laws as the basis for a protected area management plan could however be problematic. For example, customary laws addressing the threats to a World Heritage site may not exist, or customs being practised may be inconsistent with conservation.[36] In addition, many customs are inherently flexible and they may lose this characteristic through codification in State law.[37] Thus, the incorporation of customary laws into a management plan could render the customs static. It is notable however that a protected area management plan can be amended by the area's

para 79; Albert Mumma, 'The Link Between Traditional and Formal Legal Systems' in Webber Ndoro and Gilbert Pwiti (eds), *Legal Frameworks for the Protection of Immoveable Cultural Heritage in Africa*, ICCROM Conservation Studies 5 (ICCROM, 2005) 22, 24; Catherine Giraud-Kinley, 'The Effectiveness of International Law: Sustainable Development in the South Pacific Region' (1999–2000) 12 *Georgetown Environmental Law Review* 125, 159; Erika Techera, *Local Approaches to the Protection of Biological Diversity: The Role of Customary Law in Community Based Conservation in the South Pacific*, Macquarie Law Working Paper 2007-2 (2007).

[36] As noted in Sect. 6.2, it is questionable whether customary rules regulating the taking of species at East Rennell are sufficient to ensure harvesting levels are sustainable.

[37] See, for example, Tom Graham, 'Flexibility and the Codification of Traditional Fisheries Management Systems', *SPC Traditional Marine Resource Management and Knowledge Information Bulletin* 3 (1994) 2, 2; K Ruddle, E Hviding and R E Johannes, 'Marine Resources Management in the Context of Customary Tenure' (1992) 7 *Marine Resource Economics* 249, 267; Blaise Kuemlangan, *Creating Legal Space for Community-Based Fisheries and Customary Marine Tenure in the Pacific: Issues and Opportunities* (Food and Agriculture Organisation of the United Nations, 2004) 36–37; Kenneth Ruddle, 'The Context of Policy Design for Existing Community-Based Fisheries Management Systems in the Pacific Islands' (1998) 40 *Ocean and Coastal Management* 105, 113; Jean Zorn and Jennifer Corrin Care, '"Barava Tru": Judicial Approaches to the Pleading and Proof of Custom in the South Pacific' (2002) 51 *International and Comparative Law Quarterly* 612, 635; Anita Smith, 'East Rennell World Heritage Site: Misunderstandings, Inconsistencies and Opportunities in the Implementation of the World Heritage Convention in the Pacific Islands' (2011) 17(6) *International Journal of Heritage Studies* 592, 601.

management committee,[38] without the involvement of the Director or the Minister for Environment. This reduces the risk of a custom being rendered static by its incorporation into a management plan.

A further challenge is that customary laws are often broad principles, rather than unambiguous rules.[39] Consequently, a custom may need to be altered to gain the certainty required to become an enforceable management plan rule. Furthermore, the incorporation of customary laws into a management plan could lead community members to have less respect for laws not codified in the plan,[40] and for customary governance.[41] For these reasons, the extent to which East Rennell's management plan can and should incorporate customary laws needs to be carefully considered.

7.2.4   Customary and Protected Area Governance

Under the *PA Act*, the national Protected Areas Advisory Committee must appoint a management committee for each declared protected area. Anyone (including local community members) can be appointed to a management committee.[42] The committee could be a new organisation established specifically for the purposes of the *PA Act*, or an existing body (such as a Council of Chiefs).[43] Whatever approach is taken, the relationship between the management committee and customary governance bodies needs to be considered.

The *PA Act* regulates rights to lands and resources, which are issues traditionally governed under customary law. Therefore, in exercising its functions under the Act, the management committee is likely to be regulating some issues within the jurisdiction of a customary governance body. Legally, certain decisions made by the management committee

---

[38] *PA Act* s 12(1).

[39] See, for example, Miranda Forsyth, 'Beyond Case Law: *Kastom* and Courts in Vanuatu' (2004) 35 *Victoria University of Wellington Law Review* 427.

[40] See, for example, Pampa Mukherjee, 'Community Rights and Statutory Laws: Politics of Forest Use in Uttrakhand Himalayas' (2004) 50 *Journal of Legal Pluralism and Unofficial Law* 161.

[41] See, for example, R E Johannes and F R Hickey, *Evolution of Village-Based Marine Resource Management in Vanuatu Between 1993 and 2001*, Coastal Region and Small Island Papers 15 (UNESCO, 2004) 35. Johannes and Hickey make this point in relation to the creation of community conservation areas under the *Environmental Management and Conservation Act (Cap. 283)* (Vanuatu).

[42] *PA Act* s 12(1)–(2).

[43] *PA Regulations* reg 28(1).

under the *PA Act* prevail over customary law.[44] However, as the East Rennell case demonstrates, in practice, the relationship between such decisions and custom is more complex.

The appointment of the East Rennell Council of Chiefs as the management committee would allow the Council to retain its status as the key local decision-making body in the area. However, it raises several questions. For example, would this erode local peoples' respect for the Council's decisions on issues of custom? In addition, would the Council be able to manage its affairs so that people could distinguish between the decisions it makes in its different capacities? This would sometimes be necessary because decisions regarding protected area matters could have legal implications under the *PA Act*, whereas decisions concerning custom could not.

Implementing the provisions of the *PA Regulations* concerning the decision-making procedures of management committees could also be challenging. Among other things, these deal with the frequency of management committee meetings, quorums, and how decisions are made.[45] These procedures are unlikely to be consistent with the Council of Chiefs' customary processes. Thus, to comply with the *PA Regulations*, when dealing with protected area matters, the Council would have to adopt procedures that differ from custom. This could be logistically very difficult. In addition, it could result in the creation of a hybrid body that has little legitimacy among the local people.[46]

If the East Rennell Council of Chiefs was not adopted as the management committee, a non-customary organisation would take that role (e.g. the committee of the existing Lake Tegano World Heritage Site Association, which was discussed in Sect. 6.3). Would the appointment of a non-customary management committee with statutory powers erode the status of the Council of Chiefs among the local people? How would any overlap between the jurisdiction of the Council and the management committee be managed to ensure both work effectively?[47] Would the power dynamic among chiefs be

---

[44] *Solomon Islands Independence Order 1978*, sch (*Constitution of Solomon Islands*), sch 3 para 3(2).

[45] *PA Regulations* reg 27(6), sch 3.

[46] For discussion of this point generally, see Jonathan M Lindsay, *Creating Legal Space for Community-Based Management: Principles and Dilemmas* (Food and Agriculture Organisation of the United Nations, 1998) 7; Lausche, above n 29, 161.

[47] For discussion of this point generally, see Albert Mumma, 'Legal Aspects of Cultural Landscape Protection in Africa' in *Cultural Landscapes: The Challenges of Conservation*, World Heritage Papers 7 (UNESCO, 2003) 156, 156.

affected if some but not all were appointed to the management committee? These questions warrant further investigation before a decision on the composition of a management committee for East Rennell is made.

7.2.5   Financing Protected Area Management

Although the *PA Act* facilitates local people to carry out local governance and enforcement roles, with one exception, it makes no provision for such people to be paid.[48] Chiefs involved with local community governance commonly contend that they should be paid by the State for their services, as it diverts them from livelihood activities.[49] Similarly, local people may require financial support before they will participate in protected area governance and enforcement. This was an observation made by Heywood and Gabrys during their time in East Rennell.[50] The SIG is not currently willing and able to provide such support. As Joe Horokou (the Director of Environment and Conservation Division of the Ministry of Environment) has commented:

> There is a perception from the [East Rennellese] people that we [the SIG] should be funding the site. They are asking the Ministry [of Environment] to employ locals as rangers and managers. The government is faced with financial difficulties and human resource constraint. The best we can do is facilitate.[51]

The *PA Act* also does not specifically provide for management committees and rangers to be provided with any resources to enable them to carry out their functions.[52] The *PA Act* establishes a Protected Areas Fund, to

---

[48] The exception is that a member of a management committee is entitled to $SBD60 for attending a meeting: *PA Act* s 12(5); *PA Regulations* reg 27(5).

[49] Matthew Allen et al, *Justice Delivered Locally: Systems, Challenges and Innovations in Solomon Islands* (World Bank, 2013) 69.

[50] See, for example, Mike Heywood and Kasia Gabrys, *Evaluation Report on Training in Community-Based Natural Resources Management* (2009). Heywood and Gabrys trained East Rennellese people in bird and tilapia monitoring, to collect baseline data against which future changes could be assessed. They concluded that while the East Rennellese people are interested in natural resources management, they are 'generally not willing to contribute voluntarily and expect monitory remuneration for their services': at 4.

[51] Interview by the author with Joe Horokou, Director of the Environment and Conservation Division of the Ministry of Environment (Honiara, 15 August 2013).

[52] In this respect, the *PA Act* and the *PA Regulations* are not consistent with best practice. IUCN's guidelines for protected area legislation state that such laws should elaborate on the

assist with the establishment and management of protected areas.[53] However, it remains to be seen whether the East Rennellese will receive any money through this Fund. They will therefore likely need financial assistance from external sources to enable them to fulfil their protected area governance and enforcement roles. They may also require technical assistance and training to help them understand and perform the tasks required of them under the *PA Act*. Without such support, it is unlikely that the protected area will be effective.

7.3   ADDRESSING THE THREATS TO THE OUTSTANDING UNIVERSAL VALUE OF EAST RENNELL UNDER OTHER LEGISLATION

7.3.1   *Logging and Mining*

The principal national laws regulating logging and mining in Solomon Islands are the *FRTU Act*, the *MM Act*, and the *Environment Act*. Under this legislation, a person wishing to undertake logging or mining on customary land[54] requires

- the consent of the customary landowners[55];
- for logging, a felling licence granted by the Commissioner for Forests under the *FRTU Act*[56];
- for mining,[57] a mining tenement (namely a prospecting licence or mining lease) granted by the Minister for Mines under the *MM Act*[58]; and

---

kinds of assistance the protected areas authority should provide to support communities and individuals in managing their conserved lands: Lausche, above n 29, 138.

[53] *PA Act* ss 13, 15.

[54] This analysis is restricted to considering logging and mining on customary land because most land in Solomon Islands is under customary tenure (see Sect. 2.3.5).

[55] *FRTU Act* s 5; *MM Act* ss 21, 36(a).

[56] *FRTU Act* ss 4(1) (d), 5. A licence is not required if the Minister exempts the applicant from this requirement (s 4(1) (c)), if the trees are felled for use as firewood or unmilled timber (s 4(1) (a)), or if the trees are felled to supply logs to a licenced mill (s 4(1) (b)).

[57] The term 'mining' is used here to mean prospecting or mining, unless the context dictates otherwise. The other stage of the mining process regulated under Solomon Islands' law (reconnaissance) is not considered here, because it is not commonly undertaken in Solomon Islands: Price et al, above n 5, 127.

[58] *MM Act* ss 20(1), 36. The Minister does however have the power to compulsorily acquire land for mining (as distinct from prospecting) (*MM Act* s 33(1)).

- a development consent granted by the Director of Environment under the *Environment Act*.[59]

As explained below, this legislation gives both the SIG and customary landowners a role in the regulation of logging and mining.

7.3.1.1 Regulation of Logging and Mining by the Solomon Islands Government

The *DSOCR* calls for logging and mining in East Rennell to be banned.[60] In accordance with this, a Cabinet paper adopted by the SIG in 2016 directed the Commissioner of Forest to 'revoke and/or refuse granting of felling licences within the World Heritage site'.[61] The *DSOCR* also calls for the implementation of measures to ensure that any extractive industries undertaken in West Rennell are sustainable and do not impact the OUV of East Rennell.[62]

Some people working within the SIG have contended that under existing laws the government lacks the power to prevent logging and mining from impacting the World Heritage site.[63] However, an analysis of the relevant legislation reveals that SIG decision-makers do have this power:

- The Commissioner for Forests could refuse to grant a felling licence on the grounds that the proposed logging development will have an unacceptable impact on a heritage site.[64]
- The Minister for Mines could refuse to grant a mining tenement on the grounds that the proposed mining development will have an unacceptable impact on a heritage site.[65]

---

[59] *Environment Act 1998* s 19(1) (b). The Director can grant an exemption from this requirement (s 19(1) (c)).

[60] *State of Conservation of Properties Inscribed on the World Heritage List*, UN Doc WHC/17/41.COM/7A.Add, 29.

[61] Ibid., 27.

[62] Ibid., 29.

[63] See above n 4.

[64] While the *FRTU Act* does not expressly refer to natural or cultural heritage, several provisions indicate that the likely impact of a logging proposal on a heritage site is a relevant consideration. See, for example, *FRTU Act* s 5(2) (c) (iii)–(iv); *Forest Resources and Timber Utilisation (Felling Licences) Regulations 2005* regs 10(f), 13(1) (b), 13(1) (d). Thus, the Commissioner for Forests could refuse to grant a felling licence under Section 5(1).

[65] Environmental and cultural impacts are relevant to the Minister for Mine's decision as to whether to grant a mining tenement under the *MM Act*. See, for example, *MM Act* ss 4(2) (a), 31(1) (h), 36(b) (ii); *Mines and Minerals Regulations 1996* regs 18(b), 18(f).

- The Director of Environment could refuse to grant a development consent for a logging or mining project that will impact a heritage site.[66] Indeed the Director is obliged to do so, if approval of the project would lead Solomon Islands to be in breach of its obligations under an international law such as the *World Heritage Convention*.[67]
- These decision-makers could impose conditions on any approvals granted, which are designed to minimise the impact of a proposed operation on a heritage site.[68]
- These decision-makers could cancel approvals if the operator was in breach of the relevant legislation or the conditions of the approval.[69]

SIG decision-makers therefore already have powers which could be exercised to regulate logging and mining in order to protect the East Rennell World Heritage site. It remains to be seen however whether such powers will be exercised. Indeed, several logging and mining developments have been approved in West Rennell even though their environmental impact assessments (EIAs) identified the loss of East Rennell's OUV as a potential impact.[70]

There is clearly an economic rationale behind the decision-making of SIG officials concerning logging and mining on Rennell and elsewhere. Logging has been a major revenue earner for Solomon Islands since the 1980s.[71] While there has been little mineral sector development in Solomon Islands to date, the industry has accelerated rapidly over the last

---

[66] *Environment Act* s 15.

[67] *Environment Regulations 2008* reg 14(1) (d).

[68] *FRTU Act* s 5(2); *MM Act* ss 22(h), 38(1) (e); *Environment Act* ss 22(3), 24(3); *Interpretation and General Provisions Act (Cap. 85)* s 30(1) (c).

[69] *FRTU Act* s 39; *MM Act* s 71(1); *Interpretation and General Provisions Act (Cap. 85)* s 30(1) (e).

[70] For example, the EIA for Asia Pacific Investment and Development Ltd's mining proposal at West Rennell identified the loss of OUV of East Rennell as one of the main potential adverse impacts of the operation: Asia Pacific Investment Development Ltd, *Rennell Island Bauxite Project, Renbel Province: Environment Impact Statement* (2014) 4. Similarly, the EIA for PT Mega Bintang Borneo's proposed mining operation in central Rennell found that the World Heritage site is likely to be affected by the development: PT Mega Bintang Borneo Ltd, *Environment Impact Statement: Central Rennell Bauxite Mining Project* (2014) 68.

[71] See, for example, Daniel Gay (ed), *Solomon Islands Diagnostic Trade Integration Study 2009 Report* (Solomon Islands Government, 2009) 48; Morgan Wairiu, 'History of the Forestry Industry in Solomon Islands: The Case of Guadalcanal' (2007) 42(2) *Journal of Pacific History* 233, 243.

few years. With tenements now covering large tracts of terrestrial and marine areas in Solomon Islands, mining could become a significant contributor to the State's economy in coming years.[72] The contribution of logging and mining to Solomon Islands' (albeit limited) economic growth creates a disincentive for SIG officials to reign in the industries. As a conservation officer in the Ministry of Environment has commented:

> It is too big an ask of the international community to [ask SIG to] ban logging [on Rennell]. Although it is destructive it is a source of revenue for government as well as the communities.[73]

The close connection between the SIG and these industries is also influential. It is well known that there is widespread corruption within the forestry industry.[74] Many State officials, including politicians, have been directly involved in logging operations, or have benefited from bribes and inducements paid by foreign companies to influence government policy and evade regulatory requirements.[75] Logging companies in Solomon Islands are renowned for utilising their connections with the government to bend the rules in their favour, and indications are that many mining companies are likely to operate in a similar manner.[76] The 'big men' style of politics and leadership prevalent in Melanesia, and the social norms of reciprocity and obligation that underlie Solomon Island culture, also help

---

[72] See, for example, Gay (ed), above n 71, 54; Tubagus Feridhanusetyawan and Shanaka J Peiris, *Solomon Islands: Selected Issues*, IMF Country Report 11/360 (International Monetary Fund, 2011) 8, 10.

[73] Interview by the author with a conservation officer in the Ministry of Environment (Honiara, 2 August 2013).

[74] See, for example, Solomon Islands Office of the Auditor General, *An Auditor-General's Insights into Corruption in Solomon Islands Government*, National Parliament Paper 48 (2007) 10.

[75] See, for example, Sinclair Dinnen, 'State-Building in a Post-Colonial Society: The Case of Solomon Islands' (2008) 9 *Chicago Journal of International Law* 51, 59; Matthew Allen, 'The Political Economy of Logging in Solomon Islands' in Ron Duncan (ed), *The Political Economy of Economic Reform in the Pacific* (Asian Development Bank, 2011) 277, 289–90.

[76] See, for example, Graham Baines, *Solomon Islands is Unprepared to Manage a Minerals-Based Economy*, State, Society and Governance in Melanesia Discussion Paper 2015/6 (Australian National University, 2015); Tony Hughes and Ali Tuhanuku, *Logging and Mining in Rennell: Lessons for Solomon Islands. Report to the World Bank and Solomon Islands Government* (2015).

explain why the logging industry is poorly regulated and susceptible to corruption.[77]

Government decision-makers are also reluctant to more strongly regulate logging and mining because of their reverence for the rights of customary landowners. In Solomon Islands, State law prevails to the extent of any inconsistency with customary law.[78] Therefore, the government can regulate access to land and resources notwithstanding any customary rights. However, many Solomon Islanders believe differently. Before Europeans arrived in their region, it was a foreign idea to Pacific islanders that anyone other than the landowners could have rights to the resources on or under that land.[79] Despite Solomon Islands becoming a protectorate and then an independent State, landowners commonly claim ownership over minerals and trees pursuant to their customary laws.[80] Many consider the State has no authority to control how customary land and resources are used.[81] For example, when asked to comment on the World Heritage Committee's requests for SIG to ban logging on Rennell, Joe Horokou (Director of the Environment) stated:

The resource is owned by the people and they make decisions about how to use it, especially the forest. While government can work with people to look after the lake [Tegano] it would be difficult to stop logging on the whole island ... To me there is some contradiction between requirements of the [World Heritage] Convention and customary law.[82]

Similarly, an officer in the Ministry of Culture has stated:

---

[77] Allen, above n 75. Allen notes that 'big man' societies are those where leaders achieve their status largely because of their ability to generate and distribute wealth: 280.

[78] *Solomon Islands Independence Order 1978*, sch (*Constitution of Solomon Islands*), sch 3 para 3(2).

[79] Glenn Banks, 'Mining' in Moshe Rapaport (ed), *The Pacific Islands: Environment and Society* (University of Hawai'i Press, 2013) 379, 383.

[80] Phillip Iro Tagini, *The Search for King Solomon's Gold: An Examination of the Policy and Regulatory Framework for Mining in Solomon Islands* (PhD Thesis, The Australian National University, 2007) 261.

[81] Jan McDonald, *Marine Resource Management and Conservation in Solomon Islands: Roles, Responsibilities and Opportunities* (Griffith Law School, 2010) 2.

[82] Interview by the author with Joe Horokou, Director of the Environment and Conservation Division of the Ministry of Environment (Honiara, 15 August 2013).

Yes, we should stop logging because as a World Heritage site, East Rennell puts us [Solomon Islands] on the map, it's universal, it's for the good of humanity. But the man on the ground does not see it like this.[83]

Further limiting the SIG's capacity to regulate logging and mining is the issue of compliance. It is well recognised that logging commonly occurs outside licenced areas, in contravention of the *FRTU Act*.[84] Literature on the implementation of the *MM Act* is more limited, but it suggests that breaches are common.[85] Furthermore, many developments occur without approval under the *Environment Act*.[86]

This lack of compliance is partly a result of the government's failure to reform the industries and strengthen regulation, as discussed above. It is also due to a lack of staff and resources within the relevant Ministries, which impedes their ability to carry out their statutory duties.[87] A lack of coordination between the relevant Ministries further hampers the effective implementation of legislation.[88] There is often little incentive for the SIG to enforce regulatory provisions, and it is very difficult for landowners to seek enforcement through the court system because of their limited access to legal services.[89] In this context, compliance with the law has effectively been left to the whim of the logging and mining companies.[90] For example,

---

[83] Interview by the author with an officer in the Ministry of Culture (Honiara, 26 July 2013).

[84] See, for example, Gay, above n 71, 212, 218; Allen, above n 75, 287; Price et al, above n 5, 119–20; Baines, above n 76, 2; Laurence Cordonnery, 'Environmental Law Issues in the South Pacific and the Quest for Sustainable Development and Good Governance' in Anita Jowitt and Tess Newton Cain (eds), *Passage of Change: Law, Society and Governance in the Pacific* (ANU Press, 2010) 233, 235; Douglas Hou, Elaine Johnson and Stephanie Price, 'Defending the Forest in the Clouds: Public Interest Law in Solomon Islands' (2013) 15 *Asia Pacific Journal of Environmental Law* 167, 170.

[85] See, for example, Baines, above n 76; Hughes and Tuhanuku, above n 76.

[86] Price et al, above n 5, 220.

[87] See, for example, Allen, above n 75, 287; Tagini, above n 80, 149; Baines, above n 76, 2, Ian Frazer, 'The Struggle for Control of Solomon Island Forests' (1997) 9(1) *Contemporary Pacific* 39, 47–8.

[88] Tagini, above n 80, 382.

[89] For example, landowners have commenced very few cases to enforce the *Environment Act 1998* against resource companies, despite the prevalence of breaches by such companies. For discussion of one such case, see Hou, Johnson and Price, above n 84.

[90] Ben Boer, 'Solomon Islands' in Ben Boer (ed), *Environmental Law in the South Pacific: Consolidated Report of the Reviews of Environmental Law in the Cook Islands, Federated States*

Hughes and Tuhanuku have said that the regulation of the industries on Rennell is 'weak and haphazard', creating a situation where 'the commercial players have been making their own rules and getting away with it'.[91] These issues impact both the willingness and the ability of the SIG to protect World Heritage against the impacts of logging and mining. It is beyond the scope of this work to explore the full suite of reforms required to address these issues, which have been noted elsewhere.[92] In terms of World Heritage protection, the *FRTU Act* and the *MM Act* should be amended to make the likely impact of a logging or mining project on heritage an express relevant consideration. This would reduce any ambiguity that exists concerning the scope of the powers decision-makers have under these laws. The development of a national World Heritage policy could also assist with this.

7.3.1.2   *Customary Landowner Involvement in the Regulation of Logging and Mining*

Except in limited circumstances, under the *FRTU Act* and the *MM Act*, a person who wishes to undertake logging and mining on customary land must first obtain the consent of the landowners.[93] On the face of it, this requirement gives the East Rennellese a powerful tool to prevent logging or mining occurring on their land. However, in practice, these developments often occur without the consent of all customary landowners. As explained below, key reasons for this include inadequacies in the drafting of the legislation and a lack of government oversight of the consent process. As a result, it is uncertain whether the East Rennellese will be able to protect the World Heritage site from the impacts of logging and mining.

of Micronesia, Kingdom of Tonga, Republic of the Marshall Islands and Solomon Islands (South Pacific Regional Environment Programme and IUCN Environmental Law Centre, 1996) 189, 224. Boer made this point in relation to logging companies, but it equally applies to mining companies.

[91] Hughes and Tuhanuku, above n 76, 8, 10.

[92] See, for example, Siobhan McDonnell, Joseph Foukana and Alice Pollard, *Building a Pathway for Successful Land Reform in Solomon Islands* (2015); Ben Boer, *Solomon Islands: Review of Environmental Law* (SPREP, 1993), in particular 96–8; Baines, above n 76; Hughes and Tuhanuku, above n 76.

[93] *FRTU Act* s 5; *MM Act* ss 21, 36(a). The Minister for Forests can however exempt a person from requiring a felling licence (*FRTU Act* s 4(1) (c)), in which case the applicant does not need to obtain the consent of the landowners for logging (confirmed by the High Court of Solomon Islands in *Alevangana v Kegu* [2012] SBHC 1). The Minister for Mines can compulsorily acquire land for mining (as distinct from prospecting) (*MM Act* s 33(1)), which would mean that landowner consent for that operation is not required.

Under the *FRTU Act*, landowner consent for logging proposals is given through the 'timber rights process'.[94] That process involves the provincial executive holding a 'timber rights meeting', and making a determination about whether the landowners wish to grant 'timber rights' to the licence applicant.[95] If they do, the landowners and the logging company must then enter into a 'timber rights agreement', after which the Commissioner for Forests can grant the company a felling licence.[96]

The 'timber rights process' was inserted into the *FRTU Act* in 1977,[97] when the logging industry expanded from land owned or leased by the government onto customary land.[98] However, the purpose of these provisions was not to protect landowners, but to give logging companies someone with whom to make an agreement.[99] Consequently, the legislation does not incorporate sufficient checks and balances to ensure that logging agreements are only made with the consent of all relevant landowners. In practice, these provisions are often manipulated by powerful individuals within landowning groups who declare themselves entitled to grant 'timber rights' notwithstanding the true customary position.[100] Logging companies are generally happy to enter into agreements with such persons to facilitate the development of the land. As a result, the *FRTU Act* has effectively enabled 'people with tenuous claims, or even no claims at all, to become the principal beneficiaries' of logging operations.[101]

There is no equivalent to the 'timber rights process' under the *MM Act*. Under that law, the tenement applicant is responsible for identifying the people with customary rights in the area. The applicant must enter into a 'surface access agreement' with those people before it can obtain a mining tenement from the Minister for Mines.[102] Implementation of the *MM Act*

---

[94] For detailed analysis of this process, see Price et al, above n 5, 76–97.

[95] *FRTU Act* s 8(3).

[96] Ibid., s 8(4).

[97] Pursuant to the *Forest and Timber Amendment Act 1977*.

[98] Frazer, above n 87, 48. For comprehensive analysis of the history of the regulation of logging in Solomon Islands, see, for example, Judith Bennett, 'Forestry, Public Land, and the Colonial Legacy in Solomon Islands' (1995) 7(2) *Contemporary Pacific* 243; Judith Bennett, *Pacific Forest: A History of Resource Control and Contest in Solomon Islands, c 1800–1997* (Brill Academic Publishers Inc, 2000); Wairiu, above n 71.

[99] *Tovua v Meki* [1989] SBHC 3; [1988–1989] SILR 74.

[100] See, for example, Wairiu, above n 71; Baines, above n 76, 1.

[101] *Tovua v Meki* [1989] SBHC 3; [1988–1989] SILR 74.

[102] *MM Act* ss 21, 36(a).

has been analysed less than the *FRTU Act*[103] (in part because until recently relatively little mining had occurred in Solomon Islands).[104] However, it appears that the approval processes for many mining projects have been plagued by difficulties.[105]

Some problems with the landowner consent provisions in the *FRTU Act* and the *MM Act* stem from how these laws refer to customary rights holders. Both laws use inconsistent and potentially inappropriate terminology when referring to the persons whose consent is required for logging and mining projects.[106] As Corrin notes, 'This has set up a serious dilemma in Solomon Islands where the legislation may permit those with a restricted interest in land to dispose of the most valuable fruit of the land.'[107]

Implementing the landowner consent provisions in the *FRTU Act* and the *MM Act* is also problematic because of their inconsistency with some customary decision-making processes.[108] This issue is particularly pertinent in Rennell, where customary land tenure differs from many other places in Solomon Islands. In accordance with the *FRTU Act* and the *MM Act*, logging and mining agreements are generally signed by a community member purporting to act on behalf of a landowning group such as a tribe. While this may reflect the customary land tenure system in some areas, land ownership on Rennell is more individualised than elsewhere in Solomon Islands. As Hughes and Tuhanuku have explained:

> In Rennell, these [land] rights are held at the family level, grouped geographically on a tribal basis but jealously guarded at the level of the family,

---

[103] A significant exception to this is Tagini's doctoral thesis: see Tagini, above n 80. See also Joe Fardin, *Mining Law and Agreement Making in Solomon Islands* (Public Solicitor's Office, Solomon Islands Government, 2011).

[104] For history of the regulation of mining in Solomon Islands, see generally Tagini, above n 80, ch 2.

[105] See, for example, Baines, above n 76. The saga involving Sumitomo Metal Mining Solomons Ltd obtaining approval to conduct mining on Isabel is a key example of this. See, for example, *SMM Solomon Ltd v Attorney General; Bogotu Minerals Ltd v Attorney General* [2014] SBHC 91.

[106] For full analysis of this issue, see Jennifer Corrin, 'Customary Land and the Language of the Common Law' (2008) 37 *Common Law World Review* 305, 320–1; Price et al, above n 5, 97–8; 143–5.

[107] Corrin, above n 106, 320.

[108] Tagini, above n 80, 221.

even on occasion setting brother against brother. Family boundaries are well known and defended.[109]

Consequently, while customary laws in other parts of Solomon Islands may authorise certain individuals to make decisions on behalf of their tribe, on Rennell

[i]ndividual families can hold out against the wishes of the tribe, including the chief or chiefs. This feature of land rights has led to inter-family rows, delays in reaching decisions about adjoining land areas, and family dissatisfaction with agreements about logging and mining entered into on a 'tribal' basis.[110]

Consent for much of the logging and mining that is now occurring in West Rennell was given by the heads of only a few of the families who own land within the relevant area, which has led to conflict.[111] As a result, Hughes and Tuhanuku contend that the Rennellese peoples' experience with logging and mining to date has been 'unhappy and divisive'.[112]

A lack of government oversight over the landowner consent processes has enabled many of the problems referred to above to occur. In general, the SIG has tended to look after the interests of investors over those of landowners.[113] Some landowners do not learn about proposals for their land until a company representative arrives to persuade them to sign an agreement,[114] or even until the operations begin.[115] Even if landowners are notified of an application, they often lack the information they require to properly assess the proposal and make an informed decision.[116] In addition, landowner agreements are commonly signed and negotiated by a few people within a landowning community, without input from all people with customary rights in the area.[117] These are rarely scrutinised by the government to ensure they have been signed by the people who have the right under custom to make deci-

---

[109] Hughes and Tuhanuku, above n 76, 12.

[110] Ibid.

[111] Ibid., 18.

[112] Ibid., 13.

[113] See, for example, McDonnell, Foukana and Pollard, above n 92, 62.

[114] See, for example, Baines, above n 76, 5.

[115] See, for example, J C Corrin, 'Abrogation of the Rights of Customary Land Owners by the Forest Resources and Timber Utilisation Act' (1992) 3 *Queensland University of Technology Law Journal* 131, 136; Baines, above n 76, 11.

[116] See, for example, Tagini, above n 80, 147.

[117] Price et al, above n 5, 90.

sions with respect to the land.[118] This lack of government oversight makes the consent processes highly vulnerable to exploitation.[119] There are also inadequate processes in place for the resolution of disputes concerning land-owner approval for logging and mining. In West Rennell, for example, 'the combination of physical remoteness, lack of understanding of issues and pos-sibilities, and capture of the regulators by the loggers and miners, has deprived the people ... of orthodox avenues of complaint'.[120]

In practice therefore, Solomon Islanders often have little power to pre-vent logging and mining from occurring on their land, which reduces their ability to protect World Heritage. Legislative reform to address the issues referred to above is long overdue. In particular, logging and mining laws should be amended to incorporate new mechanisms for identifying customary landowners and resolving disputes.[121] There is also a critical need for landowners to have greater access to information, advice, and representation concerning logging and mining proposals, to enable them to make informed decisions about the development of their land, and to challenge approvals they consider are unlawful.

It is also notable that while the *FRTU Act* and the *MM Act* give land-owners the right to refuse consent to logging and mining operations, they have no power to halt operations that have already been approved. The prescribed form of a 'timber rights agreement' does not give landowners a right to terminate, and these agreements do not come to an end merely because a felling licence has expired.[122] Although there is no prescribed form for a mining 'surface access agreement' under the *MM Act*, it is unlikely that a company would enter into an agreement that gives the landowners a broad right to terminate. Consequently, if the East Rennellese ever approved logging or mining within the World Heritage site, it is unlikely that they could subsequently retract their authorisation.[123]

---

[118] See, for example, McDonnell, Foukana and Pollard, above n 92, 7, 62; Price et al, above n 5, 83; Baines, above n 76, 6.

[119] See, for example, Baines, above n 76, 9.

[120] Hughes and Tuhanuku, above n 76, 19.

[121] McDonnell, Foukana and Pollard, above n 92, 11.

[122] *Linear Perspective Ltd v Attorney General* [2011] SBHC 18. For further discussion, see Price et al, above n 5, 93–94.

[123] Although the landowners are unlikely to have grounds for ending the agreement based on breach of contract, there may be equitable causes of action open to them such as uncon-scionable conduct or duress. Pursuing such a case through the courts would however be very difficult for most East Rennellese people.

Similarly, even if the landowners of West Rennell wanted the logging and mining of their land to cease, they have little capacity to achieve this.

Landowners also have little power to prevent logging and mining from occurring on neighbouring land. The *FRTU Act* and the *MM Act* do not give third parties the right to participate in the approval process. Landowners could object to the grant of a 'development consent' under the *Environment Act* if the proposed operations may affect them.[124] However, this presupposes that the landowners are aware of their rights under that law, and have the necessary skills and resources to review the applicant's EIA and prepare an objection. The isolation of East Rennell and the limited resources available there mean it would be difficult for most East Rennellese people to exercise their right to object under the *Environment Act*. As such, the East Rennellese have little capacity to protect the World Heritage site from the impacts of logging and mining occurring in West Rennell. Safeguarding the site must therefore involve supporting and encouraging the West Rennellese people to oppose any further such developments.

It must also be recognised that while the East Rennellese people have not yet consented to the logging or mining of their land, it appears some community members support such developments.[125] As noted in Sect. 5.3.3.3, there are limited alternative development opportunities in East Rennell, and local communities are increasingly concerned about their livelihoods and food security. In time, these concerns may manifest into support for logging and mining.[126]

Whether the East Rennellese would in practice benefit from such developments is of course debatable. The history of logging in Solomon Islands

---

[124] *Environment Act* ss 22(2), 24(2).

[125] Environment and Conservation Division (Solomon Islands Ministry of Environment, Climate Change, Disaster Management, and Meteorology) *Lake Tegano World Heritage Site, East Rennell, Rennell-Bellona Province: A Report on Community Consultation Visit on the Status of East Rennell World Heritage Site, 5–12 October 2011* (SIG, 2012); John Marnell, 'Concerns Raised Over East Rennell Logging Application', *Sunday Isles*, 25 March 2012, 9; Hughes and Tuhanuku, above n 76, 12; Teddy Kafo, 'Proposed logging threatens World Heritage Lake Tegano', *The Solomon Star*, 24 February 2015; Paul Dingwall, *Report on the Reactive Monitoring Mission to East Rennell, Solomon Islands, 21–29 October 2012* (IUCN, 2013) 18.

[126] For analysis of factors that lead many Solomon Islanders to support logging projects, see Michelle Dyer, 'Eating money: Narratives of equality on customary land in the context of natural resource extraction in Solomon Islands' (2017) 28 *The Australian Journal of Anthropology* 88.

shows that community members often receive little from the sale of their timber rights.[127] Logging companies commonly under-report their takings to minimise royalty payments,[128] and fail to deliver on promises to construct local infrastructure.[129] In addition, royalties are frequently horded by the individual landowners who signed the agreement with the logging company, rather than being distributed to all community members[130] or invested.[131]

Logging can also have negative social consequences for communities. It can degrade water sources and destroy gardens that local people rely upon for their livelihoods. It also commonly causes or exacerbates land disputes,[132] and contributes to problems such as the loss of community pride and respect for leadership structures, and increased instances of alcoholism and prostitution.[133] Mining projects have had similar consequences for local communities.[134]

---

[127] See, for example, Frazer, above n 87, 39; Bennett, *Pacific Forest*, above n 98, 319–38; Pacific Horizon Consultancy Group, *Solomon Islands State of Environment Report* (Solomon Islands Government, 2008) 52; Debra McDougall, 'Church, Company, Committee, Chief: Emergent Collectivities in Rural Solomon Islands' in Mary Patterson and Martha Macintyre (eds), *Managing Modernity in the Western Pacific* (University of Queensland Press, 2011) 121, 139; Sue Farran, 'Timber Extraction in Solomon Islands: Too Much, Too Fast; Too Little, Too Late' in Emma Gilberthrope and Gavin Hilson (eds), *National Resource Extraction and Indigenous Livelihoods: Development Challenges in an Era of Globalisation* (Routledge, 2014) 179, 179.

[128] See, for example, Gay (ed), above n 71, 218; Sinclair Dinnen, 'The Solomon Islands Intervention and the Instabilities of the Post-Colonial State' (2008) 20(3) *Global Change, Peace and Security* (formerly *Pacific Review: Peace, Security and Global Change*) 338, 351.

[129] See, for example, Price et al, above n 5, 191.

[130] See, for example, Allen et al, above n 49, 21; Judith Bennett, *Roots of Conflict in Solomon Islands – Though Much is Taken, Much Abides: Legacies of Tradition and Colonialism*, State, Society and Governance in Melanesia Discussion Paper (Australian National University, 2002) 13; Chris Brown, *Regional Study: The South Pacific*, Asia-Pacific Forestry Sector Outlook Study (Food and Agriculture Organisation of the United Nations, 1997) 4.

[131] See, for example, Gay (ed), above n 71, 218.

[132] See, for example, Allen et al, above n 49, 21; Hughes and Tuhanuku, above n 76, 8; Tarcisius Tara Kabutaulaka, 'Rumble in the Jungle: Land, Culture and (Un)sustainable Logging in Solomon Islands' in Antony Hooper (ed), *Culture and Sustainable Development in the Pacific* (ANU E Press and Asia Pacific Press, 2005) 88, 92.

[133] See, for example, Greenpeace Pacific, *Caught Between Two Worlds: A Social Impact Study of Large and Small Scale Development in Marovo Lagoon, Solomon Islands* (2001) 13, 16.

[134] See, for example, Baines, above n 76; Tagini, above n 80, chs 7–8; Daniel Evans, 'Tensions at the Gold Ridge Mine, Guadalcanal, Solomon Islands' (2010) 25(3) *Pacific Economic Bulletin* 121, 129–130.

Although operations in West Rennell are following this sorry pattern,[135] some community members still support such developments. A West Rennellese community leader who consented to the logging of his land has said that those who support World Heritage are 'dreamers' while those who support logging are 'doers'.[136] In the absence of alternative economic opportunities, the East Rennellese people may follow in his footsteps and allow logging and mining companies to operate within the World Heritage site. Indeed, East Rennell's paramount chief has said that logging will be the only option if assistance is not provided to help the communities meet their livelihood needs.[137]

7.3.2   The Over-harvesting of Species

The *DSOCR* calls on the SIG to ensure that species harvesting at East Rennell is sustainable. In particular, it states that the taking of coconut crabs should be regulated through establishing no-take zones and seasonal restrictions.[138] Commercially valuable invertebrates including beche de mer and trochus are also potentially being over-harvested at East Rennell (see Sect. 5.3.1).

Like many Pacific Island States, Solomon Islands has no broad threatened species legislation.[139] However, existing laws could be utilised to impose harvesting restrictions. If East Rennell was declared to be a protected area under the *PA Act*, the taking of species in the site would be regulated under the site's management plan and the *PA Regulations* (see Sect. 7.2). The *Fisheries Management Act 2015* also provides several mechanisms for regulating the harvesting of marine species.

The *Fisheries Management Act* empowers the Minister for Fisheries and the Director for Fisheries to make regulations, declarations, and orders regulating fishing. For example, if certain requirements are met, the Minister can declare a species as protected or endangered.[140] If such a declaration is made, the taking of the species is prohibited.[141] The Minister also has a

---

[135] Hughes and Tuhanuku, above n 76, 9. 13.

[136] Dingwall, above n 125, 18.

[137] Environment and Conservation Division, above n 125.

[138] *State of Conservation of Properties Inscribed on the World Heritage List*, UN Doc WHC/17/41.COM/7A.Add, 31.

[139] Cf *Endangered Species Act 1975* (Marshall Islands).

[140] *Fisheries Management Act 2015* s 31(1).

[141] Ibid., s 31(2).

broad power to make regulations,[142] which could contain rules restricting harvesting.[143] In addition, the Director is empowered to make orders regulating matters such as when fishing for a particular species can occur, specifications and quantity of fish that can be taken, and what gear and vessels can be used.[144] The Director can also introduce management measures through the development of a Fisheries Management Plan, which has legal effect when published in the Government Gazette.[145] The Director and the Minister therefore have ample power under the *Fisheries Management Act* to restrict the taking of marine species that are under threat at East Rennell.

Ensuring compliance with these restrictions will be an ongoing challenge. Indeed, the harvesting of coconut crabs, trochus, and beche de mer was until recently regulated under the *Fisheries Regulations 1972*.[146] Thus, controls on the harvesting of these species (which the World Heritage Committee has called on Solomon Islands to introduce[147]) have existed for many years. However, compliance and enforcement has been very poor.

One reason for this is that in Solomon Islands, the State legal system is of marginal significance to much of the population.[148] Consequently, many Solomon Islanders would be unaware of the existence of laws regulating harvesting and the restrictions imposed under them. This is particularly the case for people living in rural areas such as Rennell, where the national government is effectively absent. The laws that have enjoyed greatest success relate to 'high-profile' species. For example, the prohibition on the capture of dolphins for sale or export[149] has been an important step

---

[142] Ibid., s 129.

[143] Regulations concerning inshore fisheries were expected to be approved in 2017 (see Solomon Islands Government, *Fisheries Acts and Supporting Regulations* http://www.fisheries.gov.sb/fisheries-acts) but appear to have been delayed.

[144] Ibid., s 22(3).

[145] Ibid., s 17.

[146] These Regulations were made under Section 20 of the *Fisheries Ordinance 1972*. They continued to have effect (until repealed) pursuant to Section 61(2) of the *Fisheries Act 1998* and then Section 130(2) of the *Fisheries Management Act 2015*. They were repealed pursuant to Regulation 70(1) (a) of the *Fisheries Management Regulations 2017*. For a summary of the *Fisheries Regulations 1972*, see Price et al, above n 5, 176–8.

[147] WHC Res 37 COM 7B.14, WHC 37th sess, UN Doc WHC-13/37.COM/20 (5 July 2013) 68.

[148] See, for example, Allen et al, above n 49, 45. For further discussion of the (ir)relevance of the State legal system in Solomon Islands, see Sect. 2.5.1.2.

[149] *Fisheries (Prohibition of Export of Dolphins) Regulations 2013* reg 3. These regulations were recently repealed: *Fisheries Management Regulations 2017* reg 70(1) (e). Restrictions on the capture of dolphins are now found in the *Fisheries Management Regulations 2017* reg 20.

towards reducing this practice in Solomon Islands.[150] However, in East Rennell at least, State legal rules regulating harvesting are not well known among much of the population.

A further issue is that some laws regulating the taking of commercially valuable species (including beche de mer) have been amended several times.[151] Since independence, Solomon Islands has experienced significant political instability, with government leaders and Ministers frequently changing. As harvesting laws are not based on well-established policy, they have been amenable to decisions of the government of the day. This decreases their effectiveness.

Importantly, people often have little incentive to comply with laws regulating the taking of species, particularly laws that are inconsistent with customary harvesting rights. Any such inconsistency may also reduce the likelihood of the State enforcing the restriction. For example, an SIG employee formerly working within the National Commission for the United Nations Educational, Scientific and Cultural Organisation (UNESCO) has commented:

> On Rennell, if the government makes a rule that says that people can't take coconut crab, and a person from there wants to take coconut crab, how do we [the government] tell them that they can't?

A further disincentive for compliance is that some Solomon Islanders rely on resource harvesting for their livelihoods. Coconut crabs taken by the East Rennellese are often sold in Honiara, or to loggers and miners in West Rennell, providing locals with a rare income opportunity. Joe Horokou (Director of the Environment) has thus stated:

> It is not practical [for the government] to deny people from harvesting some of the things they require from the environment. It's their livelihood.[152]

---

[150] See, for example, Francis Pituvaka, 'Dolphins Freed After Raid', *The Solomon Star* (online), 1 November 2016 http://www.solomonstarnews.com/news/national/11645-dolphins-freed-after-raid. Pituvaka writes about the release of dolphins that were captured and held in contravention of the *Fisheries (Prohibition of Export of Dolphins) Regulations 2013*.

[151] Price et al, above n 5, 191.

[152] Interview by the author with Joe Horokou, Director of the Environment and Conservation Division of the Ministry of Environment (Honiara, 15 August 2013).

The penalties for non-compliance are often insufficient to promote compliance,[153] and monitoring and enforcement by the State is difficult, particularly in remote areas. Hence, on Rennell, laws regulating fishing have rarely, if ever, been enforced.[154]

The compilation of consolidated and up-to-date versions of the relevant legislation, and the creation and distribution of copies of the rules in a format readily understandable by those involved with harvesting may be beneficial. Substantially increasing the resources available to the relevant Ministries to monitor and enforce the laws would also assist, but is unlikely without funding from donors or other States. Increasing the penalties for non-compliance, and assisting people involved with harvesting to undertake alternative and sustainable livelihood activities, could also reduce over-harvesting.

The use of such laws to protect World Heritage will however always be difficult, given their potential inconsistency with customary rights and the reliance of local people on the resources for their livelihoods. As such, the *PA Act* may be a better approach to addressing this threat. Regardless of the approach taken, information on the harvesting that is occurring at East Rennell should be collected. At present there is limited data on what species are being taken, by whom, using what methods, and for what purpose. Collecting this data may be difficult, due to the financial and human resources required, and the potential sensitivity of the information. However, the data could help inform appropriate management responses, and ensure that the limited resources available for addressing this threat are used efficiently.

7.3.3   Invasive Species

The *DSOCR* calls for biosecurity measures to be implemented to address the threats associated with the black ship rat (*Rattus rattus*) and the giant African snail (*Achatina* spp.) and to prevent the introduction of new invasive species.[155] Measures such as baiting and trapping around log loading and storage sites, and vehicle washdowns have been recom-

---

[153] For example, the maximum fine for taking undersized trochus in contravention of the *Fisheries Regulations 1972* was $SBD100: Regulation 6. It is notable that the fines for non-compliance under the more recent *Fisheries Management Act 2015* are significantly higher.

[154] Dingwall, above n 125, 21.

[155] *State of Conservation of Properties Inscribed on the World Heritage List*, UN Doc WHC/17/41.COM/7A.Add, 30–31.

mended.[156] As explained below, the *Biosecurity Act 2013* could provide a legal basis for such measures.

The *Biosecurity Act* provides the SIG with several legal mechanisms that could help prevent the introduction of invasive species. For example, the Act requires the master of every incoming vessel to take steps to prevent any animals on board the vessel from coming to shore.[157] Incoming vessels must be taken to a biosecurity port holding area so that they can be searched.[158] No crew or cargo from the vessel can be landed unless and until landing clearance is granted by an SIG biosecurity officer.[159] Used logging vehicles and machinery will only obtain such clearance if they are free of soil, pests, seeds, and other plant and animal matter.[160] It is an offence to fail to comply with these requirements, and persons found guilty of non-compliance can be subjected to fines and/or imprisonment.[161] If these requirements are strictly enforced, they could reduce the risk of further invasive species being introduced to Rennell.

The *Biosecurity Act* also gives the Minister for Agriculture the power to take various steps to control the spread of invasive species in an area, which could help address the threats posed to East Rennell by ship rats and giant snails. For example, the Minister could declare Rennell or part of it as a biosecurity controlled area.[162] The Director for Agriculture could then require that biosecurity measures such as baiting be taken within that area.[163]

The *Biosecurity Act* is a significant addition to Solomon Islands' legislative framework for World Heritage protection. However, it remains to be seen if and how it will be implemented on Rennell. The SIG will require substantial resources to set up the administrative structures needed to implement the Act. Furthermore, enforcing the legislation, particularly in a remote place such as Rennell, will no doubt be a challenge.

---

[156] Steve Turton, *East Rennell World Heritage Area: Assessment of the State of Conservation of World Heritage Values. Final Field Report* (James Cook University, 2014) 16, 18–19.

[157] *Biosecurity Act* s 21(1).

[158] Ibid., s 15(1); *Biosecurity Regulations 2015* Regulation 4.

[159] *Biosecurity Act* s 15(3).

[160] *Biosecurity Regulations* Regulation 36.

[161] *Biosecurity Act* sch.

[162] *Biosecurity Act* s 62.

[163] Ibid., s 63.

7.3.4 Climate Change

The World Heritage Committee has urged Solomon Islands to revise East Rennell's management plan to include climate change adaptation and mitigation measures.[164] Recommended measures include monitoring tilapia[165] populations to assess the impacts of increasing salinity in the lake, and introducing new species of fish, taro, or coconut that are tolerant to changing climatic conditions.[166] Lakeside areas also need to be replanted to mitigate the impacts of flooding.[167]

Implementing these measures does not require legislation, but legislative reform in this area would be beneficial. Solomon Islands does not currently have any climate change legislation. Its national climate change policy states that such a law will be enacted, to give a legal mandate to the agency responsible for climate change[168] and to facilitate the planning, implementation, and evaluation of adaptation and mitigation actions.[169] However, this has not yet happened.

Furthermore, there is no express requirement under any legislation for climate change to be considered in administrative decision-making, such as the determination of development approvals under the *Environment Act*, the *FRTU Act*, or the *MM Act*. Sustainable forest management is often essential for climate change adaptation.[170] Currently, logging and mining developments are commonly approved despite their impacts on ecosystems. These impacts will grow as the effects of climate change are increasingly felt. Legislative amendment to require decision-makers to consider whether a proposed development may increase vulnerability to the predicted impacts of climate change would be beneficial.[171]

---

[164] WHC Res 37 COM 7B.14, WHC 37th sess, UN Doc WHC-13/37.COM/20 (5 July 2013) 68, 68.

[165] Tilapia (*Tilapia mozambica*) were introduced into Lake Tegano in the 1950s as a food source for the local people.

[166] Turton, above n 156, 15.

[167] Dingwall, above n 125, 6.

[168] The Climate Change Division of the Ministry of Environment, Climate Change, Disaster Management, and Meteorology.

[169] *Solomon Islands National Climate Change Policy: 2012–2017* (2012) cl 8.1.1(1) (b).

[170] Ben Boer and Pepe Clarke, *Legal Frameworks for Ecosystem-Based Adaptation to Climate Change in the Pacific Islands* (SPREP, 2012) 14.

[171] Ibid., 21.

7.4   CONSERVING CULTURAL AND INTANGIBLE HERITAGE

Although East Rennell was inscribed on the World Heritage List on the basis of its natural heritage values, the heritage that the East Rennellese people are most concerned about protecting is linked to their cultural identity, as expressed through their land tenure system, environmental knowledge, traditional resource use, crafts, songs, and dance.[172] It is therefore worthwhile mentioning here that Solomon Islands has no national cultural heritage[173] or intangible heritage laws.

Cultural heritage legislation in many other jurisdictions provides for the establishment of registers of built heritage sites, and imposes restrictions on the ownership, use, and development of such places. That type of legislation may be appropriate for some Pacific Island States. For example, a *Heritage Act* creating such a regime has been proposed for Fiji,[174] and if passed will be utilised to help protect the Levuka Historical Port World Heritage site. However, such a law would be of limited benefit in Solomon Islands, where heritage sites generally comprise places evidencing the connection between people and their environment, and associated traditions, knowledge, stories, and songs.[175] Furthermore, the *PA Act* already establishes a protective regime for landscapes and seascapes of natural and cultural significance. Amending Solomon Islands' existing laws to strengthen the protection they offer to sites of cultural significance could therefore be a more efficient and effective approach than enacting new legislation. New laws would however be required to establish a comprehensive protection regime for intangible heritage, as anticipated by the 2003 *Convention for the Safeguarding of Intangible Cultural Heritage.*[176]

---

[172] Smith, above n 37, 605.

[173] Although Solomon Islands lacks any comprehensive cultural heritage legislation, some legislation does contain provisions relating to the protection of cultural heritage, including the *Protection of Wrecks and War Relics Act (Cap. 150)*, the *Town and Country Planning Act (Cap. 154)*, the *Land and Titles Act (Cap. 133)*, and some provincial ordinances such as the *Choiseul Province Preservation of Culture Ordinance 1999* and the *Makira Preservation of Culture and Wildlife Ordinance 1985*.

[174] *Heritage Bill 2016* (No. 10 of 2016) (Fiji).

[175] Anita Smith, 'The World Heritage Pacific 2009 Programme' in Anita Smith (ed), *World Heritage in a Sea of Islands: Pacific 2009 Programme*, World Heritage Papers 34 (UNESCO, 2012) 2, 4.

[176] *Convention on the Safeguarding of the Intangible Cultural Heritage*, opened for signature 17 October 2003, 2368 UNTS 3 (entered into force 20 April 2006). Solomon Islands ratified the Convention in May 2018.

7.5   Conclusion

If East Rennell is to be removed from the List of World Heritage in Danger, its protection under law must be strengthened. This chapter demonstrated that while legislative reform would be beneficial, to some extent this could be achieved utilising existing laws.

The declaration of East Rennell under the *PA Act* could help address some of the threats to the site's OUV. However, the Act cannot be implemented there without the East Rennellese peoples' consent. At present, it is unclear whether they will agree, given the restrictions it will place on their activities and the lack of alternative development opportunities on the island. They are more likely to consent if the management plan closely aligns with their priorities and aspirations, and if they supported to carry out the governance and enforcement roles available to them under the *PA Act*. Importantly, the relationship between the *PA Act* regime and customary law must be carefully considered if and when the protected area is established. Issues to be investigated include the extent to which the management plan can and should incorporate customary laws, and the relationship between the site's management committee and the Council of Chiefs.

The *FRTU Act*, the *MM Act*, and the *Environment Act* give both customary landowners and SIG officials a role in decision-making concerning logging and mining. However, various economic, social, political, and legal issues influence their ability and willingness to protect World Heritage from the impacts of these activities. Steps could be taken to mitigate some of these issues, such as addressing the inadequacies in legislation and improving monitoring and enforcement. Others however are deeply rooted in Solomon Islander culture, such as peoples' reverence for customary rights and the social factors influencing SIG decision-making. Given this, it is critical that the Rennellese people (from both East and West Rennell) are encouraged and supported to oppose logging and mining. This should include ensuring they are informed of resource development proposals, helping them exercise their rights under relevant legislation, supporting alternative livelihood development projects, and assisting with *PA Act* applications.

To date, legislation restricting the taking of species such as beche de mer, trochus, and coconut crabs has been relatively ineffective in East Rennell. Actions that could improve compliance include increasing the penalties for breaches, raising peoples' awareness of the laws (e.g. by translating them into local languages), and providing more resources for monitoring and enforcement. However, implementing such laws will always be challenging, given the geography of the country, the lack of relevance of State laws to many Solomon Islanders, and the widespread belief in the pre-eminence of customary harvesting rights. As such, the *PA Act* is likely to be a more effective approach to regulating harvesting.

Addressing the threats posed by invasive species and climate change is also a major challenge on Rennell. The *Biosecurity Act* provides a range of regulatory mechanisms for controlling invasive species, but it is not yet clear whether it will be implemented and enforced at Rennell. Biosecurity measures could also be incorporated into the site's management plan, which could have some legal effect if the site is declared under the *PA Act*. This plan could also incorporate adaptation and mitigation measures designed to help the East Rennellese people cope with the impacts of climate change.

REFERENCES

ARTICLES, BOOKS AND REPORTS

Allen, Matthew, 'The Political Economy of Logging in Solomon Islands' in Ron Duncan (ed), *The Political Economy of Economic Reform in the Pacific* (Asian Development Bank, 2011) 277

Allen, Matthew, Sinclair Dinnen, Daniel Evans and Rebecca Monson, *Justice Delivered Locally: Systems, Challenges and Innovations in Solomon Islands* (World Bank, 2013)

Asia Pacific Investment Development Ltd, *Rennell Island Bauxite Project, Renbel Province: Environment Impact Statement* (2014)

Baines, Graham, *Solomon Islands is Unprepared to Manage a Minerals-Based Economy*, State, Society and Governance in Melanesia Discussion Paper 2015/6 (Australian National University, 2015)

Banks, Glenn, 'Mining' in Moshe Rapaport (ed), *The Pacific Islands: Environment and Society* (University of Hawai'i Press, 2013) 379

Bennett, Judith, 'Forestry, Public Land, and the Colonial Legacy in Solomon Islands' (1995) 7(2) *Contemporary Pacific* 243

Bennett, Judith, *Pacific Forest: A History of Resource Control and Contest in Solomon Islands, c 1800–1997* (Brill Academic Publishers Inc, 2000)

Bennett, Judith, *Roots of Conflict in Solomon Islands – Though Much is Taken, Much Abides: Legacies of Tradition and Colonialism*, State, Society and Governance in Melanesia Discussion Paper (Australian National University, 2002)

Boer, Ben, *Solomon Islands: Review of Environmental Law* (SPREP, 1993)

Boer, Ben, 'Solomon Islands' in Ben Boer (ed), *Environmental Law in the South Pacific: Consolidated Report of the Reviews of Environmental Law in the Cook Islands, Federated States of Micronesia, Kingdom of Tonga, Republic of the Marshall Islands and Solomon Islands* (South Pacific Regional Environment Programme and IUCN Environmental Law Centre, 1996) 189

Boer, Ben and Pepe Clarke, *Legal Frameworks for Ecosystem-Based Adaptation to Climate Change in the Pacific Islands* (SPREP, 2012)

Brown, Chris, *Regional Study: The South Pacific*, Asia-Pacific Forestry Sector Outlook Study (Food and Agriculture Organisation of the United Nations, 1997)

Cordonnery, Laurence, 'Environmental Law Issues in the South Pacific and the Quest for Sustainable Development and Good Governance' in Anita Jowitt and Tess Newton Cain (eds), *Passage of Change: Law, Society and Governance in the Pacific* (ANU Press, 2010) 233

Corrin, J C, 'Abrogation of the Rights of Customary Land Owners by the Forest Resources and Timber Utilisation Act' (1992) 3 *Queensland University of Technology Law Journal* 131

Corrin, Jennifer, 'Customary Land and the Language of the Common Law' (2008) 37 *Common Law World Review* 305

Dingwall, Paul, *Report on the Reactive Monitoring Mission to East Rennell, Solomon Islands, 21–29 October 2012* (IUCN, 2013)

Dinnen, Sinclair, 'State-Building in a Post-Colonial Society: The Case of Solomon Islands' (2008) 9 *Chicago Journal of International Law* 51

Dinnen, Sinclair, 'The Solomon Islands Intervention and the Instabilities of the Post-Colonial State' (2008) 20(3) *Global Change, Peace and Security* (formerly *Pacific Review: Peace, Security and Global Change*) 338

Dyer, Michelle, 'Eating money: Narratives of equality on customary land in the context of natural resource extraction in Solomon Islands' (2017) 28 *The Australian Journal of Anthropology* 88

Elbert, Samuel H and Torben Monberg, *From the Two Canoes: Oral Traditions of Rennell and Bellona Islands* (Danish National Museum and University of Hawaii Press, 1965)

Environment and Conservation Division (Solomon Islands Ministry of Environment, Climate Change, Disaster Management, and Meteorology) *Lake Tegano World Heritage Site, East Rennell, Rennell-Bellona Province: A Report on Community Consultation Visit on the Status of East Rennell World Heritage Site, 5–12 October 2011* (SIG, 2012)

Evans, Daniel, 'Tensions at the Gold Ridge Mine, Guadalcanal, Solomon Islands' (2010) 25(3) *Pacific Economic Bulletin* 121

Fardin, Joe, *Mining Law and Agreement Making in Solomon Islands* (Public Solicitor's Office, Solomon Islands Government, 2011)

Farran, Sue, 'Timber Extraction in Solomon Islands: Too Much, Too Fast; Too Little, Too Late' in Emma Gilberthrope and Gavin Hilson (eds), *National Resource Extraction and Indigenous Livelihoods: Development Challenges in an Era of Globalisation* (Routledge, 2014) 179

Feridhanusetyawan, Tubagus and Shanaka J Peiris, *Solomon Islands: Selected Issues,* IMF Country Report 11/360 (International Monetary Fund, 2011)

Forsyth, Miranda, 'Beyond Case Law: *Kastom* and Courts in Vanuatu' (2004) 35 *Victoria University of Wellington Law Review* 427

Frazer, Ian, 'The Struggle for Control of Solomon Island Forests' (1997) 9(1) *Contemporary Pacific* 39

Gay, Daniel (ed), *Solomon Islands Diagnostic Trade Integration Study 2009 Report* (Solomon Islands Government, 2009) 48

Giraud-Kinley, Catherine, 'The Effectiveness of International Law: Sustainable Development in the South Pacific Region' (1999–2000) 12 *Georgetown Environmental Law Review* 125

Graham, Tom, 'Flexibility and the Codification of Traditional Fisheries Management Systems', *SPC Traditional Marine Resource Management and Knowledge Information Bulletin* 3 (1994) 2

Greenpeace Pacific, *Caught Between Two Worlds: A Social Impact Study of Large and Small Scale Development in Marovo Lagoon, Solomon Islands* (2001)

Hales, Robert James, John Rynne, Cathy Howlett, Jay Devine and Vivian Hauser, 'Indigenous Free Prior Informed Consent: A Case for Self Determination in World Heritage Nomination Processes' (2013) 19(3) *International Journal of Heritage Studies* 270

Heywood, Mike and Kasia Gabrys, *Evaluation Report on Training in Community-Based Natural Resources Management* (2009)

Hou, Douglas, Elaine Johnson and Stephanie Price, 'Defending the Forest in the Clouds: Public Interest Law in Solomon Islands' (2013) 15 *Asia Pacific Journal of Environmental Law* 167

Hughes, Tony and Ali Tuhanuku, *Logging and Mining in Rennell: Lessons for Solomon Islands. Report to the World Bank and Solomon Islands Government* (2015)

Johannes, R E and F R Hickey, *Evolution of Village-Based Marine Resource Management in Vanuatu Between 1993 and 2001*, Coastal Region and Small Island Papers 15 (UNESCO, 2004)

Kabutaulaka, Tarcisius Tara, 'Rumble in the Jungle: Land, Culture and (Un)sustainable Logging in Solomon Islands' in Antony Hooper (ed), *Culture and Sustainable Development in the Pacific* (ANU E Press and Asia Pacific Press, 2005) 88

Kafo, Teddy, 'Proposed logging threatens World Heritage Lake Tegano', *The Solomon Star*, 24 February 2015

Kuemlangan, Blaise, *Creating Legal Space for Community-Based Fisheries and Customary Marine Tenure in the Pacific: Issues and Opportunities* (Food and Agriculture Organisation of the United Nations, 2004)

Lausche, Barbara, *Guidelines for Protected Area Legislation* (IUCN, 2011)

Lindsay, Jonathan M, *Creating Legal Space for Community-Based Management: Principles and Dilemmas* (Food and Agriculture Organisation of the United Nations, 1998)

Macintyre, Martha and Simon Foale, 'Global Imperatives and Local Desires: Competing Economic and Environmental Interests in Melanesian Communities' in Victoria Lockwood (ed), *Globalisation and Culture Change in the Pacific Islands* (Pearson Prentice Hall, 2004) 149

Marnell, John, 'Concerns Raised Over East Rennell Logging Application', *Sunday Isles*, 25 March 2012, 9

McDonald, Jan, *Marine Resource Management and Conservation in Solomon Islands: Roles, Responsibilities and Opportunities* (Griffith Law School, 2010)

McDonnell, Siobhan, Joseph Foukana and Alice Pollard, *Building a Pathway for Successful Land Reform in Solomon Islands* (2015)

McDougall, Debra, 'Church, Company, Committee, Chief: Emergent Collectivities in Rural Solomon Islands' in Mary Patterson and Martha Macintyre (eds), *Managing Modernity in the Western Pacific* (University of Queensland Press, 2011) 121

Mukherjee, Pampa, 'Community Rights and Statutory Laws: Politics of Forest Use in Uttrakhand Himalayas' (2004) 50 *Journal of Legal Pluralism and Unofficial Law* 161

Mumma, Albert, 'Legal Aspects of Cultural Landscape Protection in Africa' in *Cultural Landscapes: The Challenges of Conservation*, World Heritage Papers 7 (UNESCO, 2003) 156

Mumma, Albert, 'The Link Between Traditional and Formal Legal Systems' in Webber Ndoro and Gilbert Pwiti (eds), *Legal Frameworks for the Protection of Immoveable Cultural Heritage in Africa*, ICCROM Conservation Studies 5 (ICCROM, 2005) 22

Pacific Horizon Consultancy Group, *Solomon Islands State of Environment Report* (Solomon Islands Government, 2008)

Pituvaka, Francis, 'Dolphins Freed After Raid', *The Solomon Star* (online), 1 November 2016 http://www.solomonstarnews.com/news/national/11645-dolphins-freed-after-raid

Price, Stephanie, Adam Beeson, Joe Fardin and Jennifer Radford, *Environmental Law in Solomon Islands* (Public Solicitor's Office, Solomon Islands Government, 2015)

PT Mega Bintang Borneo Ltd, *Environment Impact Statement: Central Rennell Bauxite Mining Project* (2014)

Ruddle, K, E Hviding and R E Johannes, 'Marine Resources Management in the Context of Customary Tenure' (1992) 7 *Marine Resource Economics* 249

Ruddle, Kenneth, 'The Context of Policy Design for Existing Community-Based Fisheries Management Systems in the Pacific Islands' (1998) 40 *Ocean and Coastal Management* 105

Smith, Anita, 'East Rennell World Heritage Site: Misunderstandings, Inconsistencies and Opportunities in the Implementation of the World Heritage Convention in the Pacific Islands' (2011) 17(6) *International Journal of Heritage Studies* 592

Smith, Anita, 'The World Heritage Pacific 2009 Programme' in Anita Smith (ed), *World Heritage in a Sea of Islands: Pacific 2009 Programme*, World Heritage Papers 34 (UNESCO, 2012) 2

Solomon Islands Government, *State Party Report on the State of Conservation of the East Rennell World Heritage Area (Solomon Islands)* (SIG, 2013)

Solomon Islands Government, *State Party Report on the State of Conservation of the East Rennell World Heritage Area (Solomon Islands)* (SIG, 2014)

Solomon Islands Government, *State Party Report on the State of Conservation of East Rennell (Solomon Islands)* (SIG, 2017)

Solomon Islands Office of the Auditor General, *An Auditor-General's Insights into Corruption in Solomon Islands Government*, National Parliament Paper 48 (2007)

Techera, Erika, *Local Approaches to the Protection of Biological Diversity: The Role of Customary Law in Community Based Conservation in the South Pacific*, Macquarie Law Working Paper 2007-2 (2007)

Turton, Steve, *East Rennell World Heritage Area: Assessment of the State of Conservation of World Heritage Values. Final Field Report* (James Cook University, 2014)

Wairiu, Morgan, 'History of the Forestry Industry in Solomon Islands: The Case of Guadalcanal' (2007) 42(2) *Journal of Pacific History* 233

Wingham, Elspeth J, *Nomination of East Rennell, Solomon Islands by the Government of Solomon Islands for Inclusion in the World Heritage List Natural Sites* (1997)

Wingham, Elspeth J and Ben Devi, 'The Involvement of Local People in the Management of a Proposed World Heritage Site at East Rennell, Solomon Islands' in Hans D Thulstrup (ed), *World Natural Heritage and the Local Community: Case Studies from Asia Pacific, Australia and New Zealand* (UNESCO, 1999) 79

Zorn, Jean and Jennifer Corrin Care, "Barava Tru': Judicial Approaches to the Pleading and Proof of Custom in the South Pacific' (2002) 51 *International and Comparative Law Quarterly* 612

CASES

Alevangana v Kegu [2012] SBHC 1
Linear Perspective Ltd v Attorney General [2011] SBHC 18
SMM Solomon Ltd v Attorney General; Bogotu Minerals Ltd v Attorney General
[2014] SBHC 91
Tovua v Meki [1989] SBHC 3; [1988–1989] SILR 74

LEGISLATION AND BILLS: FIJI

Heritage Bill 2016 (No. 10 of 2016)

LEGISLATION AND BILLS: MARSHALL ISLANDS

Endangered Species Act 1975

LEGISLATION AND BILLS: SOLOMON ISLANDS

Biosecurity Act 2013
Biosecurity Regulations 2015
Choiseul Province Preservation of Culture Ordinance 1999
Environment Act 1998
Environment Regulations 2008
Fisheries Act 1998
Fisheries Management Act 2015
Fisheries Management Regulations 2017
Fisheries Ordinance 1972
Fisheries (Prohibition of Export of Dolphins) Regulations 2013
Fisheries Regulations 1972
Forest and Timber Amendment Act 1977
Forest Resources and Timber Utilisation Act (Cap. 40)
Forest Resources and Timber Utilisation (Felling Licences) Regulations 2005
Interpretation and General Provisions Act (Cap. 85)
Land and Titles Act (Cap. 133)
Lake Tegano Natural Heritage Park Ordinance 2009 (draft)
Makira Preservation of Culture and Wildlife Ordinance 1985
Mines and Minerals Act (Cap. 42)
Mines and Minerals Regulations 1996
Protected Areas Act 2010
Protected Areas Regulations 2012
Protection of Wrecks and War Relics Act (Cap. 150)
Solomon Islands Independence Order 1978, sch (*Constitution of Solomon Islands*)
Town and Country Planning Act (Cap. 154)

LEGISLATION AND BILLS: VANUATU

Environmental Management and Conservation Act (Cap. 283)

CONVENTIONS

Convention on the Safeguarding of the Intangible Cultural Heritage, opened for signature 17 October 2003, 2368 UNTS 3 (entered into force 20 April 2006)

UNITED NATIONS DOCUMENTS

Barbados Programme of Action for the Sustainable Development of Small Island Developing States, UN Doc A/CONF.167/9 (October 1994)

Operational Guidelines for the Implementation of the World Heritage Convention, UN Doc WHC.16/01 (26 October 2016)

Rio Declaration on Environment and Development, Report of the United Nations Conference on Environment and Development, UN Doc A/CONF.151/6/Rev.1 (1992)

State of Conservation of Properties Inscribed on the World Heritage List, WHC 41st sess, UN Doc WHC/17/41.COM/7A.Add (2 June 2017) 26 (East Rennell, Solomon Islands)

State of Conservation of the Properties Inscribed on the List of World Heritage in Danger, WHC 42nd sess, UN Doc WHC/18/42.COM/7A.Add.2 (15 June 2018) 17 (East Rennell, Solomon Islands)

WHCRes 29 COM 7B.10, WHC 29th sess, UN Doc WHC-05/29.COM/22 (9 September 2005) 45

WHCRes 31 COM 7B.21, WHC 31st sess, UN Doc WHC-07/31.COM/24 (31 July 2007) 58

WHC Res 33 COM 7B.19, WHC 33rd sess, UN Doc WHC-09/33.COM/20 (20 July 2009) 68

WHC Res 34 COM 7B.17, WHC 34th sess, UN Doc WHC-10/34.COM/20 (3 September 2010) 71

WHC Res 37 COM 7B.14, WHC 37th sess, UN Doc WHC-13/37.COM/20 (5 July 2013) 68

WHC Res 38 COM 7A.29, WHC 38th sess, UN Doc WHC-14/38.COM/16 (7 July 2014) 39

WHC Res 41 COM 7A.19, WHC 41st sess, UN Doc WHC-17/41.COM/18 (12 July 2017) 35

THESES

Price, Stephanie Clair, *Strengthening World Heritage Protection in the Pacific: An Exploration of Solomon Islands' Implementation of the World Heritage Convention* (PhD Thesis, University of Western Australia, 2017)
Tagini, Phillip Iro, *The Search for King Solomon's Gold: An Examination of the Policy and Regulatory Framework for Mining in Solomon Islands* (PhD Thesis, The Australian National University, 2007)

INTERNET MATERIALS

Solomon Islands Government, *Fisheries Acts and Supporting Regulations* http://www.fisheries.gov.sb/fisheries-acts

INTERVIEWS

Interview by the author with an officer in the Ministry of Culture (Honiara, 26 July 2013)
Interview by the author with a conservation officer in the Ministry of Environment (Honiara, 2 August 2013)
Interview by the author with Joe Horokou, Director of the Environment and Conservation Division of the Ministry of Environment (Honiara, 15 August 2013)

OTHER

Solomon Islands National Climate Change Policy: 2012–2017 (2012)

PART IV

Conclusion

Strengthening World Heritage Protection in the Pacific: Lessons from Solomon Islands

8.1   Introduction

This book explored the opportunities and challenges associated with the protection of World Heritage in the Pacific by analysing the implementation of the *World Heritage Convention*[1] at two scales.

Firstly, these issues were considered at the Pacific level (Part II). Chapter 2 set the scene by exploring the types of heritage sites prevalent in the Pacific, and the key characteristics of Pacific Island States and their legal systems. Two key aspects of the *World Heritage Convention* were then critically analysed: the origins and interpretation of the concept of 'World Heritage' (Chap. 3), and the protection regime established by the *Convention* (Chap. 4). The book demonstrated that many opportunities and challenges stem from the nature of the region's heritage, land tenure, and legal systems, while others are attributable to characteristics of the *Convention* and the World Heritage Committee's approach to heritage and its protection.

Secondly, the implementation of the *World Heritage Convention* in Solomon Islands was assessed (Part III). This Part began by analysing the inscription of East Rennell on the World Heritage List, and discussing the context for World Heritage conservation in Solomon Islands (Chap. 5).

---

[1] *Convention Concerning the Protection of the World Cultural and Natural Heritage*, opened for signature 16 November 1972, 1037 UNTS 151 (entered into force 17 December 1975) ('*World Heritage Convention*').

© The Author(s) 2018   277
S. C. Price, *World Heritage Conservation in the Pacific*,
Palgrave Series in Asia and Pacific Studies,
https://doi.org/10.1007/978-981-13-0602-0_8

East Rennell's protection under customary law, management plans (Chap. 6), and State legislation (Chap. 7) were then explored. The book demonstrated that the involvement of the East Rennellese people in World Heritage protection is critical. However, greater State intervention (including through the implementation and enforcement of legislation) is also necessary to deal with some of the threats to the site's outstanding universal value (OUV).

Like elsewhere, in Solomon Islands, the implementation of the *World Heritage Convention* is influenced by a range of economic, social, political, and cultural issues. The country's low economic growth, political instability, governance issues, and the close connection between many politicians and resource development industries reduce the ability and willingness of the Solomon Islands government (SIG) to participate in World Heritage protection. Increasing concern over food security and livelihoods is leading some local community members to support developments that are detrimental to conservation efforts. Additionally, forces such as globalisation, urbanisation, and migration are degrading many customary legal systems, impeding the ability of community leaders to effectively manage their land and resources.

Addressing or mitigating these issues will require efforts at a much broader scale than is possible or appropriate through the implementation of the *World Heritage Convention*. The purpose of this book was therefore not to find 'the solution' to World Heritage in the Solomon Islands, which would not only be inappropriate for a single, foreign scholar, but also impossible, given the nature and complexity of the challenges that exist. It is however instructive to consider the lessons that can be learned from Solomon Islands' experience, for the protection of East Rennell and other places sharing similar characteristics.

Drawing upon the findings in previous chapters, this chapter therefore identifies some key lessons from Solomon Islands and the Pacific more broadly. They concern the involvement of Pacific Island States in the *World Heritage Convention* regime (Sect. 8.2), the nomination of Pacific sites (Sect. 8.3), and the protection of Pacific World Heritage (Sect. 8.4). Throughout this chapter, options for addressing some of the challenges associated with implementing the *Convention* are identified. Each of these options might only lead to small, incremental improvements, and obtaining funding and assistance to enable their implementation will always be challenging. However, in time, they could assist Pacific Island governments, customary landowners, and others to safeguard the region's impressive heritage sites.

8.2   THE INVOLVEMENT OF PACIFIC ISLAND STATES IN THE *WORLD HERITAGE CONVENTION* REGIME

8.2.1   *Ensuring the Pacific Voice Continues to Be Heard*

Over the past three decades, the *World Heritage Convention* regime has evolved. The World Heritage Committee has broadened its interpretation of the concept of 'World Heritage', in recognition of the great diversity of heritage places that exist around the world. The Committee is also now more open to different forms of heritage protection, including that offered by customary law. These changes have increased the scope for Pacific Island States to effectively use the *Convention* to protect their heritage (discussed in Sects. 3.4 and 4.3).

This evolutionary process will no doubt continue, and as it does, the *Convention* bodies (the Committee and the Advisory Bodies) must ensure that the needs and aspirations of Pacific Island States are taken into account. Pacific representation on the Committee would be beneficial, but it is uncertain whether any Pacific Island State has the capacity, resources, and willingness to effectively serve in that role. These States should however be encouraged and supported to engage with the *Convention* regime in other ways.

Although State parties have a duty to report to the World Heritage Committee on their implementation of the *Convention*,[2] this obligation is not always complied with. For example, the Committee has repeatedly called upon Solomon Islands to submit reports,[3] but few such requests

---

[2] *World Heritage Convention* art 29; UNESCO, *Operational Guidelines for the Implementation of the World Heritage Convention*, UN Doc WHC.16/01 (26 October 2016) (*'Operational Guidelines 2016'*) para 199.

[3] WHC Res 29 COM 7B.10, WHC 29th sess, UN Doc WHC-05/29.COM/22 (9 September 2005) 45, 46; WHC Res 31 COM 7B.21, WHC 31st sess, UN Doc WHC-07/31.COM/24 (31 July 2007) 58, 58; WHC Res 33 COM 7B.19, WHC 33rd sess, UN Doc WHC-09/33.COM/20 (20 July 2009) 68, 68; WHC Res 34 COM 7B.17, WHC 34th sess, UN Doc WHC-10/34.COM/20 (3 September 2010) 71, 71; WHC Res 36 COM 7B.15, WHC 36th sess, UN Doc WHC-12/36.COM/19 (June–July 2012) 63, 64; WHC Res 37 COM 7B.14, WHC 37th sess, UN Doc WHC-13/37.COM/20 (5 July 2013) 68, 69; WHC Res 38 COM 7A.29, WHC 38th sess, UN Doc WHC-14/38.COM/16 (7 July 2014) 39, 40; WHC Res 39 COM 7A.16, WHC 39th sess, UN Doc WHC-15/39.COM/19 (8 July 2015) 30, 31; WHC Res 40 COM 7A.49, WHC 40th sess, UN Doc WHC-16/40.COM/19 (15 November 2016) 68, 69.

have been complied with.[4] The SIG's failure to comply is somewhat understandable, given the limited funds and personnel it has to dedicate to World Heritage matters. It is however also regrettable. State parties can use these reports to not only record their compliance with the *Convention*, but also inform the Committee of their broader views concerning World Heritage.

It is also understandable that Solomon Islands is not always represented at World Heritage Committee annual meetings.[5] As noted in the *Pacific World Heritage Action Plan 2016–2020*, the isolation and resource constraints of Pacific Island States impede their ability to participate in global forums, particularly those held in the northern hemisphere.[6] However, a consequence is that the Committee often has limited information about the SIG's perspective when making its decision. This may have contributed to the gulf that exists between the actions that the Committee is seeking from the SIG and those that the State party is willing and able to take.[7]

Regional World Heritage meetings are therefore critical, as they provide Pacific Islanders with an opportunity to discuss common issues, develop strategic plans,[8] and formulate shared visions which can be articulated to the *Convention* bodies.[9] Mechanisms to allow Pacific Island States to participate in Committee meetings without having to physically attend could also be explored. For example, a regional meeting could potentially be held simultaneously with the Committee meeting, with a video link between the two venues.

---

[4] Solomon Islands only submitted State party reports in 2012, 2013, 2014, and 2017. See Solomon Islands Government, *State Party Report on the State of Conservation of the East Rennell World Heritage Area (Solomon Islands)* (SIG, 2012); Solomon Islands Government, *State Party Report on the State of Conservation of the East Rennell World Heritage Area (Solomon Islands)* (SIG, 2013); Solomon Islands Government, *State Party Report on the State of Conservation of the East Rennell World Heritage Area (Solomon Islands)* (SIG, 2014); Solomon Islands Government, *State Party Report on the State of Conservation of the East Rennell World Heritage Site* (SIG, 2017).

[5] For example, no representative of Solomon Islands attended the World Heritage Committee annual meetings in 2014, 2015, or 2016.

[6] *Pacific World Heritage Action Plan 2016–2020* (2016) 3.

[7] Attendance at World Heritage Committee meetings may also influence representation on the World Heritage List. For discussion of this point, see generally Lynn Meskell, Claudia Liuzza and Nicholas Brown, 'World Heritage Regionalism: UNESCO from Europe to Asia' (2015) 22 *International Journal of Cultural Property* 437.

[8] For example, the *Pacific World Heritage Action Plan 2016–2020* (2016).

[9] For example, *Presentation of the World Heritage Programme for the Pacific*, WHC 31st sess, UN Doc WHC-07/31.COM/11C (10 May 2007) annex I (Appeal to the World Heritage Committee from the Pacific Island State Parties) (the '*Pacific Appeal*').

More broadly, the development of a consortium of Pacific Island States should be considered. This approach has enjoyed some success in other contexts. A notable example is the Alliance of Small Island States (AOSIS), an intergovernmental organisation of low-lying coastal and island countries that functions as an 'ad hoc lobby and negotiating voice'.[10] AOSIS has been relatively successful in articulating the views of member nations to the international community, particularly on the issue of global warming. A formal consortium of Pacific Island States (or a larger organisation encompassing other nations facing similar challenges, such as the Small Island Developing States) could provide a channel for such States to influence the implementation and evolution of the *Convention* regime.

8.2.2 Recognising the Pacific Perspective

The *Pacific Appeal* (which was presented to the World Heritage Committee by representatives of the Pacific Island States in 2007) clearly articulated the vision of Pacific Islanders concerning their heritage and the *World Heritage Convention*. It noted that the implementation of the *Convention* must be considered in the context of the types of sites prevalent in the region, including 'spectacular and highly powerful spiritually-valued natural features and cultural places'.[11] Furthermore, the protection of that heritage must be based on respect for 'traditional cultural practices, indigenous knowledge and systems of land and sea tenure'.[12] The *Convention* bodies' contemporary approach to World Heritage and its protection has, to some extent, been influenced by statements such as this. If the *Convention* is to become more relevant and effective in the region, the Pacific perspective must continue to be taken into account.

One issue that resonated strongly from this research is the reverence that many Solomon Islanders have for the rights of customary landowners. As was explained in Sect. 4.2.3, State parties to the *Convention* have an obligation to implement the measures, including the laws, required to protect World Heritage. The SIG has the power under its Constitution to comply with this obligation, notwithstanding any customary rights.[13]

---

[10] Alliance of Small Island States, *About AOSIS* http://aosis.org/about/.
[11] *Pacific Appeal*, UN Doc WHC-07/31.COM/11C, annex I para 11.
[12] Ibid., annex I para 13.
[13] *Solomon Islands Independence Order 1978*, sch ('*Constitution of Solomon Islands*') sch 3 para 3.

However, in practice, it is a fallacy to consider State law at the top of the legal hierarchy in Solomon Islands. Many Solomon Islanders (including those in government) believe that customary landowners have complete rights to their land and resources, and the State has no authority to decide how they are used.[14] Thus, people working in the SIG commonly consider that the government's role in the protection of World Heritage is only to facilitate conservation, rather than to dictate any measures to be taken (see Sect. 5.3.3.2).

It is unhelpful to advocate for the SIG to undertake measures that fundamentally diverge from the views of Solomon Islanders concerning customary rights. If that is done, it will likely exacerbate the perception that the *Convention* is an ill fit in the region. That perception was reflected in a statement made by Malchoir Mataki (Permanent Secretary of the Ministry of Environment) when asked to comment on the World Heritage Committee's request for the SIG to ban logging in West Rennell:

> They [the Committee] are making that suggestion without any clue as to how things operate in this country.[15]

Similarly, Joe Horokou (Director of the Environment and Conservation Division) has said:

> To me there is some contradiction between requirements of the [World Heritage] Convention and customary law.[16]

World Heritage protection in Solomon Islands will only be achieved if conservation measures can accommodate both the SIG's international obligations and its reverence for the rights of customary landowners. Of course, identifying approaches that achieve this will be an ongoing challenge, particularly given the economic aspirations of many East Rennellese people and the limited development opportunities available on the island.

---

[14] Jan McDonald, *Marine Resource Management and Conservation in Solomon Islands: Roles, Responsibilities and Opportunities* (Griffith Law School, 2010) 2; Phillip Iro Tagini, *The Search for King Solomon's Gold: An Examination of the Policy and Regulatory Framework for Mining in Solomon Islands* (PhD Thesis, Australian National University, 2007) 261.

[15] Interview by the author with Malchoir Mataki, Permanent Secretary of the Ministry of Environment (Honiara, 1 October 2013).

[16] Interview by the author with Joe Horokou, Director of the Environment and Conservation Division of the Ministry of Environment (Honiara, 15 August 2013).

It is however clear that conservation efforts will not succeed without the broad support of the site's customary owners. As such, it is imperative that they are supported to protect World Heritage (see Sect. 8.4.4).

8.3 THE NOMINATION OF PACIFIC SITES FOR WORLD HERITAGE LISTING

The *Pacific World Heritage Action Plan* specifies a range of regional- and national-level measures designed to help Pacific Island governments nominate sites for World Heritage listing. These include thematic studies to identify appropriate sites, and capacity building programmes.[17] These initiatives should clearly be supported. Furthermore, lessons learned from existing World Heritage sites should be taken into account when decisions are made about nominations. Three key lessons from Solomon Islands' experience are highlighted below.

8.3.1 Recognising the Implications of Any Disconnect Between the Global and Local Significance of the Site

When assessing whether a Pacific site should be nominated, and the criteria upon which it should be nominated, a key issue to consider is the extent and implications of any disconnect between the global and local significance of the place. All places exist within a hierarchy of spatial scales, and the value of a place may vary considerably at different levels within that hierarchy.[18] By defining World Heritage to be heritage of 'outstanding universal value', the *Convention* 'manufactures history and heritage at a global scale'.[19] However, the OUV of a World Heritage site may not coincide with the local population's view of why the place is significant.[20] While the potential for such a disconnect exists at sites

---

[17] *Pacific World Heritage Action Plan 2016–2020* (2016) 5.

[18] Brian Graham, Gregory J Ashworth and John E Tunbridge, *A Geography of Heritage: Power, Culture and Economy* (Arnold, 2000) 4.

[19] Steve Brown, 'Poetics and Politics: Bikini Atoll and World Heritage Listing' in Sue O'Connor, Denis Byrne and Sally Brockwell (eds), *Transcending the Culture-Nature Divide in Cultural Heritage: Views from the Asia-Pacific Region* (ANU E Press, 2012) 35, 48.

[20] See, for example, William Logan, 'Cultural Diversity, Cultural Heritage and Human Rights: Towards Heritage Management as Human Rights-Based Cultural Practice' (2012) 18(3) *International Journal of Heritage Studies* 231, 237–239; Naomi Deegan, 'The Local-Global Nexus in the Politics of World Heritage: Space for Community Development?' in

around the world, its implications may be more significant in regions such as the Pacific, where the involvement of local people in heritage protection is particularly critical.

This issue is relevant to both cultural and natural World Heritage sites. Many of the listed cultural sites in the Pacific have been recognised as having OUV as 'expressions of a global narrative' rather than because of the values attributed to them by Pacific islanders.[21] Similarly, the two listed natural World Heritage sites in the region were listed because of their outstanding environmental features as opposed to their local cultural significance. UNESCO's management manuals state that a World Heritage site should be managed to conserve all its heritage values, not just those that give the site OUV.[22] However, in practice, the Committee is most concerned about the preservation of a World Heritage site's OUV. Consequently, at East Rennell, for example, the Committee's focus is on the preservation of the site's forest and marine ecosystems, while the East Rennellese people are more concerned about conserving their cultural identity. This situation is not conducive to the creation of a cooperative approach to World Heritage protection.

The listing of further natural World Heritage sites in the Pacific region should not be ruled out. Many such places would not qualify as cultural or mixed sites. As such, to preclude their listing as natural sites would significantly reduce the potential for the *Convention* to be utilised in the region. This issue must however be explored when sites are considered for nomination, including investigating the implications it will have for the site's conservation.

Successful World Heritage management often requires that conflicting interests at different levels be reconciled.[23] Deegan refers to this as finding the local-global nexus, 'where forces from diverging dimensions of scale ...

Marie-Theres Albert, Marielle Richon, Marie José Viñals and Andrea Witcomb (eds), *Community Development through World Heritage*, World Heritage Papers 31 (UNESCO, 2012) 77, 80.

[21] Anita Smith, 'World Heritage and Outstanding Universal Value in the Pacific Islands' (2015) 21(2) *International Journal of Heritage Studies* 177.

[22] See, for example, UNESCO et al, *Managing Natural World Heritage*, World Heritage Resource Manual (UNESCO, 2012) 37.

[23] Rick van der Ploeg, 'Welcome Address by the Chair of the conference' in Eléonore de Merode, Rieks Smeets and Carol Westrik (eds), *Linking Universal and Local Values: Managing a Sustainable Future for World Heritage*, World Heritage Papers 13 (UNESCO, 2004) 24, 24.

interconnect and interpenetrate'.[24] However, this does not always require the complete alignment of international, national, and local perceptions of World Heritage. For example, Trau (who has worked at and researched the Chief Roi Mata's Domain World Heritage site in Vanuatu) writes about the 'glocalisation' of the concept of World Heritage at that site. Like the East Rennellese, the Lelema people (the customary owners of Chief Roi Mata's Domain) consider income generation for education, health, and transport as the overwhelming priority.[25] However, unlike at East Rennell, World Heritage is becoming increasingly understood and valued at the Vanuatu site. This is occurring not because the Lelema communities have 'absorbed the global doctrine' of World Heritage, but because they are adapting and applying global and local principles of development and conservation to meet their own knowledge and aspirations.[26] This local adaptation of the concept of World Heritage has become integral to the ongoing management and protection of the site by the Vanuatu government and the Lelema people.[27]

This 'glocalisation' process has not occurred in Solomon Islands, partly because neither the SIG nor the East Rennellese people have enjoyed economic benefits from World Heritage. In addition, the cultural heritage of the East Rennellese people was not recognised in the site's World Heritage listing (unlike Chief Roi Mata's Domain, which was listed as a 'cultural landscape'). In this context, World Heritage is a low priority for the SIG and a source of misunderstanding and disenchantment among the local people. Mechanisms for strengthening the protection of East Rennell must involve finding and capitalising the local-global nexus. This will likely require broadening heritage conservation efforts to encompass the preservation of East Rennellese culture, and supporting the local communities to improve their livelihoods (see Sect. 8.4.4).

[24] Deegan, above n 20, 81.
[25] Adam M Trau, 'The Glocalisation of World Heritage at Chief Roi Mata's Domain, Vanuatu' (2012) 24(3) *Historic Environment* 4, 7. See also Adam M Trau, *World Heritage at Chief Roi Mata's Domain: The Global-Local Nexus of Community Heritage Conservation and Tourism Development in Vanuatu* (PhD Thesis, University of Western Sydney, 2013).
[26] Trau, above n 25, 'The Glocalisation of World Heritage', 4.
[27] Other reasons for the relative success of the Chief Roi Mata's Domain site include the accessibility of the site from Port Vila (which has facilitated tourism) and the level of government support. See generally Trau, *World Heritage at Chief Roi Mata's Domain*, above n 25; Meredith Wilson, Chris Ballard, Richard Matanik and Topie Warry, 'Community as the First C: Conservation and Development through Tourism at Chief Roi Mata's Domain, Vanuatu' in Anita Smith (ed), *World Heritage in a Sea of Islands: Pacific 2009 Programme*, World Heritage Papers 34 (UNESCO, 2012) 68.

8.3.2   Understanding the Potential and Limitations of Customary Protection of World Heritage Sites

As we saw in Sect. 4.3.3, the Committee's decision to allow sites protected under customary mechanisms to be listed was important for the Pacific, where land and resources have been managed by Pacific Islanders for millennia. Indeed, that decision enabled the listing of East Rennell, which had little protection under State law when nominated. However, perhaps reflecting the Committee's desire to improve the balance of the World Heritage List, it appears that East Rennell was listed despite a lack of clarity concerning its protection regime (see Sect. 5.2). As explained below, in the future, when a place under customary protection is nominated, the scope and strength of that protection, and its relationship to any proposed management plan and State legislation should be more closely examined.

8.3.2.1   Customary Laws and Governance

If a site is to be nominated based on its customary protection, the relevant customary legal system/s should first be researched and documented.[28] The scope of customary laws should be assessed with reference to the site's World Heritage values. As noted previously, the motivation behind the development of customary laws in some parts of the Pacific was the sustainable use of resources; however, in other places, population densities were too low for a conservation ethic to develop[29] (see Sect. 2.5.2). Thus, whether customary laws support the conservation of a site's OUV needs to be verified. In addition, customary laws need to be examined in light of the current and foreseeable threats to the site. If it is evident that additional protection measures such as legislation will be required, research should consider if and how they will interact with custom.

Customary governance should also be researched, to understand who has authority to make decisions and how those decisions are made. Most

---

[28] The need to research and document customary legal systems before assuming they will form part of an effective heritage or resource management regime has been recognised elsewhere: see, for example, Joseph Eboreime, 'Nigeria's Customary Laws and Practices in the Protection of Cultural Heritage with Special Reference to the Benin Kingdom' in Webber Ndoro and Gilbert Pwiti (eds), *Legal Frameworks for the Protection of Immoveable Cultural Heritage in Africa*, ICCROM Conservation Studies 5 (ICCROM, 2005) 9, 11; Shankar Aswani, 'Customary Sea Tenure in Oceania as a Case of Rights-Based Fishery Management: Does it Work?' (2005) 15 *Reviews in Fish Biology and Fisheries* 285, 304–305.

[29] See, for example, Simon Foale et al 'Tenure and Taboos: Origins and Implications for Fisheries in the Pacific' (2011) 12 *Fish and Fisheries* 357, 357.

customary governance bodies in the Pacific have changed substantially since pre-colonial times, and many are weakening under modern pressures such as the introduction of the cash economy, migration, and globalisation (see Sect. 2.4.4). Therefore, their contemporary role needs to be assessed, including their legitimacy among the landowning communities and the extent to which they can ensure compliance with custom. Such a study may reveal that the World Heritage values of the area are being well managed, and there is little need for intervention. Alternatively, it may reveal that customary governance needs to be strengthened and/or supplemented (for example, by the establishment of another local governance structure). If a new structure is to be established, its relationship to any customary governance bodies needs to be understood so that all have clear mandates and can work cooperatively together.

8.3.2.2 Boundaries and Buffer Zones

The provisions of the *Operational Guidelines for the Implementation of the World Heritage Convention* concerning the delineation of World Heritage site boundaries should be carefully considered when a site under customary protection is nominated. These provisions require that all attributes necessary to convey a site's OUV be within the site's boundaries.[30] As customary land tenure in some parts of the Pacific (particularly Melanesia) is highly fragmented, compliance with this requirement may result in the site encompassing land owned by more than one group. Coordinating the management of the site by the various customary landowning groups may be challenging. Thus, it may be appropriate to advocate for the boundary requirements to be relaxed, to allow the delineation of a site that can be effectively protected under one customary legal system, rather than creating a large site under fragmented ownership.

The *Operational Guidelines* also state that a buffer zone around a World Heritage site should be established where necessary to protect the site.[31] The feasibility of creating such a buffer zone needs to be assessed, particularly if the land surrounding the site is owned by a different customary group. For example, the owners of the buffer zone may not accept restrictions on the use and development of their land, particularly if they receive no tangible benefits from the World Heritage listing.

---

[30] *Operational Guidelines 2016*, UN Doc WHC.16/01, paras 100–102.
[31] Ibid., paras 104–107.

While the boundary and buffer zone provisions may need to be applied flexibly for sites under customary protection, the consequences of any non-compliance should also be considered. As explained in Sect. 5.2, East Rennell did not strictly comply with the boundary requirements for listing because the forests across the island are intrinsically linked. In addition, no buffer zone has been established, possibly because of land tenure issues. Today, many of the threats to the World Heritage values of the site arise from activities in West Rennell, which the East Rennellese have little power to control, and which the SIG has been unwilling to strongly regulate. The site's deviation from the boundary and buffer zone requirements thus continues to have significant implications for conservation efforts.

8.3.2.3 The Relationship Between Customary Protection, Management Plans, and State Legislation

Customary protection will rarely, in itself, be sufficient to protect a World Heritage site from all modern threats.[32] Consequently, when a site under customary protection is nominated for World Heritage listing, the extent to which that protection needs to be strengthened and supplemented through other mechanisms, such as a management plan and/or State legislation, should be determined. The relationship between such mechanisms and custom also has to be clearly understood. Numerous issues concerning such relationships were revealed through the analysis in Chaps. 6 and 7. For example, how will any inconsistencies between the management plan provisions and customary law be resolved? Will State legislation incorporate aspects of custom, and if so will that affect (positively or negatively) compliance with those customs? If a new governance body will be established, what will be the composition of that body? And how will its jurisdiction relate to that of customary governance bodies? Understanding issues such as these is crucial if the additional management measures are to be effective.

The role that the State is likely to play in the protection of the site also needs to be examined. For example, there is a reluctance among people working within SIG to implement any measures that are not widely supported by the East Rennellese people, reflecting their reverence for the rights of customary owners, and recognition of the peoples' reliance on the land for their livelihoods (see Sect. 5.3.3.2). It therefore cannot be assumed

---

[32] Chris Ballard and Meredith Wilson, 'Unseen Monuments: Managing Melanesian Cultural Landscapes' in Ken Taylor and Jane L Lennon (eds), *Managing Cultural Landscapes* (Routledge, 2012) 130, 132, 149; Anita Smith, 'The World Heritage Pacific 2009 Programme' in Anita Smith (ed), *World Heritage in a Sea of Islands: Pacific 2009 Programme*, World Heritage Papers 34 (UNESCO, 2012) 2, 5; Pepe Clarke and Charles Taylor Gillespie, *Legal Mechanisms for the Establishment and Management of Terrestrial Protected Areas in Fiji* (IUCN, 2009) 2.

that a State party will be willing to do all it takes to strictly protect the OUV of a site under customary protection, despite its *Convention* obligations.

A thorough assessment of customary protection at the nomination stage would provide a more realistic picture of the strength of the site's protection regime, allowing all stakeholders to agree upon feasible conservation objectives. It might also help the Pacific State parties and the *Convention* bodies to anticipate and address issues concerning the conservation of the site. The inclusion of provisions in the *Operational Guidelines* to guide such an assessment may be beneficial.

8.3.3   Shifting the Focus to World Heritage Protection

The *Global Strategy for a Representative, Balanced and Credible World Heritage List* was adopted by the Committee in 1994 on the basis that the List would only remain credible if it better reflected the diversity of heritage sites around the world. It therefore supported activities such as encouraging States from under-represented regions to nominate sites, and broadening the interpretation of the notion of cultural heritage.[33] However, to date, less attention has been paid to improving World Heritage conservation, including addressing non-compliance with Committee decisions.[34] Consequently, Anderson has argued:

> To maintain credibility, a shift in focus from quantity to quality must take place. This means that sites put forward for nomination should be clearly identified as gaps in the World Heritage List and receive advice from the earliest stages on how to meet the standards of the Convention. It also means that the management of existing World Heritage sites should be central to the Convention's focus.[35]

The increased focus on conservation advocated for by Anderson must occur in the Pacific.

Unless the challenges associated with the conservation of Pacific World Heritage are addressed, not only will such places remain at risk of being damaged or destroyed, but the representation of the region on the World Heritage List is unlikely to substantially increase.

---

[33] *Operational Guidelines 2016*, UN Doc WHC.16/01, para 60.

[34] *Reports of the Advisory Bodies*, WHC 39th sess, UN Doc WHC-15/39.COM/5B (15 May 2015) 9 para 63.

[35] Inger Anderson, 'Today Defines Tomorrow: World Heritage as Litmus Test for Action on Agreements' (2016) 79 *World Heritage* 4, 9.

8.4   THE PROTECTION OF PACIFIC WORLD HERITAGE SITES

8.4.1   Prioritising World Heritage Conservation

The extent to which the protection of World Heritage should be prioritised over other places needs to be discussed. The importance of this issue elsewhere has been well recognised. For example, in relation to Africa, Breen has stated:

> World Heritage inscription lays undue emphasis on single sites in a national context, diverts resources and expertise from the broader context of state services and national heritage provision.[36]

Similarly, Mumma has said:

> There is a danger that, by prioritizing action in support of those places at the highest level, elements of the wider resources may not be properly considered and this may result in detriment to the heritage.[37]

Likewise in the Pacific, World Heritage protection stretches the very limited resources of governments and other institutions. While World Heritage sites have been internationally recognised as having OUV, unlisted sites may be just as significant (both internationally and locally) and thus warrant the same level of protection.

Whether States should develop conservation measures (such as legislation) that apply solely to World Heritage sites or to heritage places more broadly needs to be considered on a case-by-case basis. Some World Heritage sites may be sufficiently unique to justify specific measures. For example, Kiribati's listed site, the Phoenix Islands Protected Area, was established and is protected under the *Phoenix Islands Protected Area Regulations 2008* (Kiribati).[38] That site is the largest marine protected area

---

[36] Colin Breen, 'Advocacy, International Development and World Heritage Sites in Sub-Saharan Africa' (2007) 39(3) *World Archaeology* 355, 365.

[37] Albert Mumma, 'Framework for Legislation on Immoveable Cultural Heritage in Africa' in Webber Ndoro, Albert Mumma and George Abungu (eds), *Cultural Heritage and the Law: Protecting Immoveable Heritage in English-Speaking Countries of Sub-Saharan Africa*, ICCROM Conservation Studies 8 (ICCROM, 2008) 97, 98.

[38] Made under the *Environment Act 1999* (Kiribati).

in the Pacific, and hosts a range of marine environments and incredible biodiversity. Its unique characteristics arguably warrant site-specific legislation.[39] In contrast, it is arguable that East Rennell is not so dissimilar from other places in Solomon Islands as to justify such a law, at least not at the national level. It was therefore reasonable for the SIG to enact the *Protected Areas Act 2010* (the *PA Act*) rather than a *World Heritage Protection Act* as envisaged in East Rennell's nomination dossier.[40]

The corollary is that the *PA Act* is not specifically designed for East Rennell, and there are many issues that require careful consideration if and when the Act is implemented there (see Sect. 7.2). Furthermore, unlike the *PA Act*, specific World Heritage legislation could provide a broad framework for decision-making concerning all aspects of the *Convention*. For example, it could address the nomination of sites for World Heritage listing (including landowner consultation and/or consent requirements), site management plans (including the process for their development, review, and approval), administrative decision-making concerning World Heritage matters, the financing of World Heritage protection, and income sharing.[41] Additionally, such legislation could be drafted to apply to sites with both cultural and natural heritage values, which may not fit well under protected area or cultural heritage legislation. These benefits may make specific World Heritage legislation the appropriate choice for some Pacific Island States, if they are willing to commit the human and financial resources needed to implement and administer such a law.

The views of the broader population towards World Heritage may also be influential. In some parts of the world, sites nominated for World Heritage listing have often already been 'reterritorialized from a local scale to the national and been interpreted as representations of the nation and nationalism'.[42] In such places, the inscription of a site on the World Heritage List may engender a sense of national pride, which translates into

---

[39] This is also the approach taken in relation to several World Heritage sites in Australia, such as the Great Barrier Reef Marine Park.

[40] One reason why the *World Heritage Protection Act* was not pursued in Solomon Islands is because it would only apply to World Heritage sites, as opposed to important heritage places more broadly: Interview by the author with an officer in the Ministry of Education, who was formerly the focal point for World Heritage within the Solomon Islands National Commission for UNESCO (Honiara, 28 July 2013).

[41] Legislation providing such a framework for implementing the *Convention* is proposed for Fiji. See *Heritage Bill 2016* (Bill no. 10 of 2016) (Fiji), in particular Parts 5–6.

[42] Deegan, above n 20, 80.

the site's protection being prioritised. In contrast, in many Pacific Island States (including Solomon Islands), people's main affiliation rests with their clan or tribe, as opposed to their nation. In the absence of a strong sense of national unity, it is less likely that a place will gain national significance, even if it is inscribed on the World Heritage List. Consequently, the idea that East Rennell warrants protection more than other places in Solomon Islands is not necessarily one that resonates widely among Solomon Islanders. This reinforces the argument that broad protected area legislation was a more appropriate choice for Solomon Islands than a *World Heritage Protection Act*.

8.4.2   Achieving Sustainable Development and Respecting the Rights and Roles of Local Communities in the Protection of World Heritage

The *World Heritage Convention* gives State parties discretion to adopt legal measures appropriate to their circumstances. However, as explored in Chap. 4, for many years the World Heritage Committee favoured a 'fortress'-style approach to World Heritage protection, which is often inappropriate in the Pacific. In recent years, the Committee's views on the conservation of World Heritage have evolved. An important milestone in this regard was the Committee's resolution that rights recognised under the *United Nations Declaration on the Rights of Indigenous People*[43] (*UNDRIP*) must be respected in the implementation of the *Convention*. The adoption of the *World Heritage Sustainable Development Policy* by the General Assembly of State parties[44] (which followed the endorsement of a similar document by the Committee[45]) was also significant, as it demonstrated broad acknowledgement of the need to pursue heritage protection through the framework of sustainable development. While these are prom-

---

[43] *United Nations Declaration on the Rights of Indigenous Peoples*, GA Res 61/295, UN GAOR, 61st sess, 107th plen mtg, Supp No 49, UN Doc A/RES/61/295 (13 September 2007).

[44] *Policy for the Integration of a Sustainable Development Perspective into the Processes of the World Heritage Convention*, WHC GA Res 20 GA 13, 20th sess, UN Doc WHC-15/20. GA/15 (20 November 2015) 7.

[45] WHC Res 39 COM 5D, WHC 39th sess, UN Doc WHC-15/39.COM/19 (8 July 2015) 7; *World Heritage and Sustainable Development*, WHC 39th sess, UN Doc WHC-15/39.COM/5B (15 May 2015) annex.

ising developments, further work is required to translate them into practice. The Committee could assist by amending the *Operational Guidelines* to fully reflect the modern approach to heritage protection. For example, while the guidelines note that World Heritage sites may be subject to sustainable use,[46] they fall short of the call in the *Sustainable Development Policy* for State parties to balance conservation, sustainability, and development, so that World Heritage protection activities can contribute to the development and quality of life of communities.[47] The *Operational Guidelines* also do not guarantee compliance with *UNDRIP*, as States are merely encouraged, not required, to involve local communities in the preparation of site nominations and the protection of World Heritage sites.[48] In 2011, the Committee adopted a four-year cycle for the amendment of the *Operational Guidelines*,[49] with the next revision due in 2019.[50] Before then, the amendments required to align the *Operational Guidelines* with the modern principles of heritage protection should be identified.

The Committee must also ensure that its resolutions concerning specific World Heritage sites reflect these principles in practice. As noted in Chap. 5, it has repeatedly requested that Solomon Islands address the threats to East Rennell by banning logging and mining on the island, regulating the taking of species, developing a new management plan, and implementing heritage protection legislation. There has been little acknowledgement in its decisions of the critical role of local people in decision-making concerning World Heritage protection. For example, in 2013 the Committee called upon Solomon Islands to apply the *PA Act* to East Rennell 'to ensure full and strict legal protection of the property'.[51] This request fails to recognise an important feature of the Act, namely that

---

[46] *Operational Guidelines 2016*, UN Doc WHC.16/01, para 119.

[47] *Policy for the Integration of a Sustainable Development Perspective into the Processes of the World Heritage Convention*, WHC GA Res 20 GA 13, 20th sess, UN Doc WHC-15/20. GA/15 (20 November 2015) 7, para 1.

[48] *Operational Guidelines 2016*, UN Doc WHC.16/01, para 123.

[49] WHC Res 35 COM 12B, WHC 35th sess, UN Doc WHC-11/35.COM/20 (7 July 2011) 266.

[50] The last review of the *Operational Guidelines* was done in 2015. However, amendments were made in 2016 after the review of certain provisions of the *Guidelines* on an exceptional basis.

[51] WHC Res 37 COM 7B.14, WHC 37th sess, UN Doc WHC-13/37.COM/20 (5 July 2013) 68.

the Minister for Environment cannot declare a protected area under this law without landowner consent.[52] It would be more appropriate, and more consistent with the *Sustainable Development Policy*, for the Committee to request that Solomon Islands encourage and support the landowners to apply for a protected area declaration. Such a request may engender more support among the SIG, because it accurately reflects the scope of its legal authority under the *PA Act*.

Until recently, the Committee's decisions have also not expressly recognised the intrinsic link between local economic development and conservation at East Rennell. In 2016 however, the Committee called upon to Solomon Islands to 'develop an Action Plan which would prioritise local communities and alternative income generating mechanisms that derive benefits from the conservation of the property's Outstanding Universal Value (OUV)'.[53] In 2017, it requested that the international community support the State party in its efforts to develop sustainable livelihoods for the East Rennellese people.[54] This is perhaps evidence that the Committee is shifting towards an approach that more strongly reflects the principles of sustainable development. If the gap between the positions of the Committee and the SIG concerning the protection of East Rennell is to be narrowed, that shift must continue.

8.4.3   Supporting the Development and Implementation of Legislation for the Protection of World Heritage

The *Pacific World Heritage Action Plan* aims to ensure that Pacific heritage places are effectively protected and managed, and specifies regional- and national-level actions designed to help achieve that goal. The regional-level actions include capacity building, holding regular regional meetings, and establishing a cultural heritage database and a register of cultural heritage legal experts.[55] National-level activities vary from State to State, and include increasing cooperation between relevant Ministries, capacity building, information sharing, and improving the effectiveness

---

[52] *PA Act* s 10(7) (c).
[53] WHC Res 40 COM 7A.49, WHC 40th sess, UN Doc WHC/16/40.COM/19 (15 November 2016) 68.
[54] WHC Res 41 COM 7A.19, WHC 41st sess, UN Doc WHC/17/41.COM/18 (12 July 2017) 35.
[55] *Pacific World Heritage Action Plan 2016–2020* (2016) 5, 9, 10.

and coordination of heritage policy and legislation.[56] If and how these proposed actions will be implemented in practice remains to be seen. Drawing upon the *Action Plan* and the findings of this research, some observations about supporting the protection of World Heritage are made below.

8.4.3.1 Regional-Level Activities

Databases and Registers
Academic scholarship and practical experience concerning the protection of Pacific World Heritage is amassing (albeit slowly), so the creation of a comprehensive repository for such information would be beneficial. The logical host of the database would be the Pacific Heritage Hub, a World Heritage facility for Pacific Island States established in 2013 at the University of the South Pacific.[57] The scope of the proposal in the *Action Plan* should be expanded from cultural heritage sites to include all World Heritage places. Similarly, the proposal to create a register of cultural heritage legal experts could be expanded to include people with expertise in natural World Heritage sites.

Importantly, the database should be sufficiently broad to encompass information concerning laws relevant to World Heritage protection. Currently, there is no central location where such information can be found. While UNESCO hosts a database of cultural heritage laws,[58] it is incomplete.[59] Some Pacific legislation can be obtained through Paclii,[60] but that site is also not always up to date.[61] Furthermore, the Paclii site is a database of legislation on all topics, potentially making it difficult to find laws relevant to a particular site. The proposed database should also encompass all key legislation relevant to World Heritage protection, not simply laws specifically aimed at heritage conservation. Including links to

---

[56] Ibid., 7.

[57] Pacific Heritage Hub, *Who We Are* http://www.pacificheritagehub.org/about-us/who-we-are/. The Pacific Heritage Hub is now a section of the Oceania Centre for Arts, Culture and Pacific Studies at the University of the South Pacific.

[58] UNESCO, *UNESCO Database of National Cultural Heritage Laws* http://www.unesco.org/culture/natlaws/index.php.

[59] Of the Pacific Island States, only Cook Islands, Fiji, Niue, Palau, Samoa, and Tonga are covered.

[60] Pacific Islands Legal Information Institute http://www.paclii.org.

[61] For example, the *Phoenix Islands Protected Area Regulations 2008* (Kiribati) are not on the Paclii website.

information about the implementation of the relevant laws (such as the relationship of the laws to custom, and enforcement issues) would also enhance the database's usefulness.

Model Management Plans and Model Laws

It is notable that while the *Action Plan* supports the development of model management plans for World Heritage sites and places on Tentative Lists,[62] it makes no reference to model laws. The merits of developing model laws for the protection of Pacific heritage should however be investigated.

A *Model for a National Act on the Protection of Cultural Heritage* already exists,[63] but it is principally concerned with underwater and moveable heritage, and is global in scope. A regional model would be preferable, as it could be better tailored to the Pacific context. A model law for the protection of cultural heritage has already been developed for the Caribbean, and lessons could potentially be learned from that process for the Pacific. Furthermore, the Pacific region already has a *Model Law for the Protection of Traditional Ecological Knowledge, Innovations and Practices*.[64] The experiences of Pacific Island States in utilising that document could be drawn upon in the development of any model for the protection of World Heritage.

Model laws have many benefits, including allowing for the pooling of expertise in legislative drafting. This is particularly pertinent in the Pacific, where the number of people with the requisite skills is somewhat limited.[65] There is however a risk that model laws can fail to accommodate the diverse characteristics of the relevant States. This risk is exacerbated in the World Heritage context by the diversity of Pacific heritage, which means that no one piece of legislation will be appropriate for all sites.

---

[62] *Pacific World Heritage Action Plan 2016–2020* (2016) 11.

[63] UNESCO, *Model for a National Act on the Protection of Cultural Heritage* (2013) http://www.unesco.org/new/fileadmin/MULTIMEDIA/HQ/CLT/pdf/UNESCO_MODEL_UNDERWATER_ACT_2013.pdf.

[64] *Model Law for the Protection of Traditional Ecological Knowledge, Innovations and Practices* http://www.grain.org/system/old/brl_files/brl-model-law-pacific-en.pdf.

[65] Craig Forrest and Jennifer Corrin, 'A Model Law to Implement the Convention on the Protection of the Underwater Cultural Heritage and its Possible Application in Plural Legal Regimes in Pacific Small Island States: A Case Study of Solomon Islands' (Paper presented at Solomon Islands National University Workshop, Honiara, December 2014) http://www.themua.org/collections/files/original/602a7962da5dd01ceafc413b8ec2d8fe.pdf 4.

Notwithstanding this, the development of a series of options for World Heritage protection legislation may still be useful. The drafting process would have to be led by Pacific Islanders (including representatives of regional Pacific organisations and Pacific Island governments, and customary owners) to ensure that the model enjoys wide support. The model would also need to be culturally and institutionally appropriate for the Pacific context. Importantly, it would need to reflect the diversity of customary legal systems that exist across the region and, in some cases, within States.

8.4.3.2 National-Level Activities for Solomon Islands

The national-level activities in the *Pacific World Heritage Action Plan* vary from State to State. Thus, reflecting the focus of this book, the discussion here is limited to those activities identified for Solomon Islands. These include implementing sustainable income-generating mechanisms for the local communities, strengthening local governance, and banning logging and mining within the World Heritage site.[66] Similar measures are contained in the *Desired State of Conservation for the Removal of East Rennell from the List of World Heritage in Danger*[67] (*DSOCR*—discussed in Sect. 5.3.2). Comments about key measures are set out below.

Prohibition on Logging and Mining Within the World Heritage Site

The *Action Plan* and the *DSOCR* call for logging and mining to be prohibited in East Rennell. As explained in Chap. 7, this could be achieved through the declaration of the site under the *PA Act* (see Sect. 7.2). Even if that did not occur, under the *Forest Resources and Timber Utilisation Act (Cap. 40)*, the *Mines and Minerals Act (Cap. 42)*, and the *Environment Act 1998*, the Commissioner for Forests, the Minister for Mines, and the Director of the Environment have the power to refuse to approve operations within the World Heritage site (see Sect. 7.3.1). However, whether these decision-makers will exercise this power remains to be seen.

---

[66] *Pacific World Heritage Action Plan 2016–2020* (2016) 16–17.
[67] WHC Res 41 COM 7A.19, WHC 41st sess, UN Doc WHC/17/41.COM/18 (12 July 2017) 35.

Regulation of Logging and Mining in West Rennell

The *Action Plan* and the *DSOCR* call for the establishment of legal mechanisms to ensure that logging and mining in West Rennell do not negatively impact the OUV of East Rennell. As demonstrated in Sect. 7.3.1, under existing legislation, the Commissioner for Forests, the Minister for Mines, and the Director for Environment could refuse to approve projects in West Rennell if they may degrade the OUV of the World Heritage site. They could also revoke existing approvals, if the operators are in breach of relevant laws or conditions. Given the history of resource development in Solomon Islands, it is likely that most if not all operators are in breach, so logging and mining occurring in West Rennell could probably be lawfully halted.

Despite this, logging and mining in West Rennell are likely to continue. While this partly reflects Solomon Islanders' reverence for the rights of customary owners (discussed in Sect. 8.2.2), it also reflects the weakness of the regulatory regimes for these industries in Solomon Islands. The need for substantial reform of these regimes has been well recognised elsewhere.[68] Fundamental changes are required to stamp out corruption, protect landowners' rights, and ensure that the industries are sustainable (to the extent that this is possible). While it is beyond the scope of this

---

[68] See, for example, Judith Bennett, *Roots of Conflict in Solomon Islands – Though Much is Taken, Much Abides: Legacies of Tradition and Colonialism*, State, Society and Governance in Melanesia Discussion Paper (Australian National University, 2002); Judith Bennett, *Pacific Forest: A History of Resource Control and Contest in Solomon Islands, c 1800–1997* (Brill Academic Publishers Inc, 2000); Judith Bennett, 'Forestry, Public Land, and the Colonial Legacy in Solomon Islands' (1995) 7(2) *Contemporary Pacific* 243; Daniel Gay (ed), *Solomon Islands Diagnostic Trade Integration Study 2009 Report* (Solomon Islands Government, 2009); Tarcisius Tara Kabutaulaka, 'Rumble in the Jungle: Land, Culture and (Un)sustainable Logging in Solomon Islands' in Antony Hooper (ed), *Culture and Sustainable Development in the Pacific* (ANU E Press and Asia Pacific Press, 2005) 88; Siobhan McDonnell, Joseph Foukana and Alice Pollard, *Building a Pathway for Successful Land Reform in Solomon Islands* (2015); Graham Baines, *Solomon Islands is Unprepared to Manage a Minerals-Based Economy*, State, Society and Governance in Melanesia Discussion Paper 2015/6 (Australian National University, 2015); Tony Hughes and Ali Tuhanuku, *Logging and Mining in Rennell: Lessons for Solomon Islands. Report to the World Bank and Solomon Islands Government* (2015); Ian Frazer, 'The Struggle for Control of Solomon Island Forests' (1997) 9(1) *Contemporary Pacific* 39; Solomon Islands Office of the Auditor General, *An Auditor-General's Insights into Corruption in Solomon Islands Government*, National Parliament Paper 48 (2007).

book to detail the full suite of required reforms,[69] specific changes that could improve World Heritage protection include the amendment of the *Forest Resources and Timber Utilisation Act* and the *Mines and Minerals Act* to make the impact of a logging or mining proposal on heritage an express relevant consideration, and to confirm that approvals cannot be granted over sites declared under the *PA Act*. The development of a national World Heritage policy may also be beneficial. Other suggested reforms are referred to in Sect. 8.4.4.3.

Implementation of Biosecurity Measures

The biosecurity measures called for in the *DSOCR* and the *Action Plan* could be introduced through the *Biosecurity Act 2013* (see Sect. 7.3.3). Among other things, this law requires incoming ships to obtain biosecurity clearance before docking. Additionally, ship captains must try to prevent certain animals from reaching the islands. If enforced, these requirements could minimise the chance of further invasive species being introduced to Rennell. The *Biosecurity Act* also empowers the Minister for Agriculture to declare Rennell or part of it to be a biosecurity controlled area, which would then allow the Director to mandate measures such as baiting. Of course enforcing the legislation, particularly on a remote island such as Rennell, will require significant human and financial resources.

Establishment of Sustainable Harvesting Limits

The *DSOCR* calls for the SIG to ensure species are harvested in a sustainable manner based on traditional resource use regimes. As noted in Sect. 6.2, existing literature raises doubts as to whether these regimes support sustainable harvesting. Further work is needed to research and document relevant customary laws, to ascertain whether the measure in the *DSOCR* is achievable.

The Minister for Fisheries and the Director for Fisheries have ample powers under the *Fisheries Management Act 2015* to regulate the taking of marine species under threat at East Rennell (see Sect. 7.3.2). A study of the harvesting that is occurring at East Rennell should be undertaken, to ascertain what species are being taken, by whom, using what methods, and for what purpose. That information could help inform the appropriate management response, and ensure that the limited

---

[69] For recommendations concerning legislative amendment of forestry laws, see Ben Boer, *Solomon Islands: Review of Environmental Law* (SPREP, 1993), in particular 96–8. See also Hughes and Tuhanuku, above n 68; McDonnell, Foukana and Pollard, above n 68.

resources available for monitoring and enforcement are utilised efficiently. The compilation of consolidated and up-to-date versions of all relevant laws, and the creation and distribution of copies of the rules in a format readily understandable by the East Rennellese people may also have some impact. The use of such laws to protect East Rennell will however always be challenging, particularly given their potential inconsistency with customary rights and the reliance of local people on the resources for their livelihoods. As such, the *PA Act* may be a more effective approach, at least in relation to harvesting undertaken by the East Rennellese people themselves.

Development of a Revised Management Plan Enforceable Through the *Protected Areas Act*

Lessons learned from the 2007 East Rennell management plan should be heeded in the preparation of any new plan for the site (see Sect. 6.5). Ideally, management measures to protect East Rennell's World Heritage values should be incorporated into a broader strategy that addresses the East Rennellese peoples' desire to improve their livelihoods and preserve their cultural identity. Other ways to make the plan understandable and relevant to the local communities should also be investigated, such as translating it into their language.

The implementation of the *PA Act* at East Rennell should also be supported (see Sect. 7.2). The declaration of the site under that Act would make logging and mining within the World Heritage site illegal. In addition, rules addressing issues such as the harvesting of species and biosecurity could be included in the site's management plan. The management plan could also provide the framework for climate change adaptation and mitigation measures. Importantly, the *PA Act* allows local community members to play a lead role in the governance and enforcement of the protected area. The limitations of the *PA Act* must however be recognised. The declaration of East Rennell as a protected area would not prevent the approval of further logging or mining developments in West Rennell. It could also not be used to mandate biosecurity measures or harvesting restrictions outside the boundaries of the site.

8.4.4   Supporting Pacific Islanders to Protect World Heritage

8.4.4.1   Strengthening Customary Protection of Pacific World Heritage Sites

The *Pacific World Heritage Action Plan* reinforces that the protection of Pacific heritage 'must be based on respect for and understanding and maintenance of the traditional cultural practices, indigenous knowledge and systems of land and sea tenure' in the region.[70] It also aims to ensure that Pacific communities are actively engaged in conserving their heritage, and promotes activities such as awareness-raising among communities, and capacity building for local heritage management.[71] Yet, in many parts of the region, customary practices and systems are weakening. Therefore, the absence of specific activities in the *Action Plan* for strengthening customary protection appears to be a significant omission.[72]

While no comprehensive empirical research concerning customary protection at East Rennell has been conducted, recent literature suggests that it is weak (see Sects. 6.2 and 6.3). If East Rennell is to be safeguarded, there is a critical need for further work to explore if and how customary protection can be strengthened. This should involve assessing if and how the East Rennellese people can be supported to utilise their customary system to protect the site's OUV, including the extent to which customs can evolve and adapt to meet new challenges, such as invasive species and climate change. This work should also explore whether the legitimacy of the chiefs within the communities can be improved. For example, following field work in Solomon Islands, Allen et al. reported that many local community members would support external assistance to strengthen customary systems, including more training and awareness work among chiefs, the establishment of a code of conduct for chiefs, support from the police and State courts to back the resolutions of customary governance bodies, and the payment of chiefs for their services.[73] Whether these or

---

[70] *Pacific World Heritage Action Plan 2016–2020* (2016) 1.

[71] Ibid., 4, 5, 7.

[72] The *Action Plan* does refer to the need for awareness raising and capacity building in communities (at 5), which is arguably broad enough to encompass strengthening customary protection. However, the only specific reference to strengthening customary protection is in a national-level activity for Papua New Guinea. That activity is 'promoting respect for customary practices and decision making in heritage protection and management': 7.

[73] Matthew Allen et al, *Justice Delivered Locally: Systems, Challenges and Innovations in Solomon Islands* (World Bank, 2013) 69.

other initiatives would assist the East Rennellese to strengthen their customary protection warrants further investigation.

8.4.4.2 Recognising the Diversity and Fluidity of Views Held by Customary Landowners

Land and marine tenure in Solomon Islands is highly fragmented, so a World Heritage site will rarely be owned by one landowner group. Even if the site is under one system of customary land tenure, it cannot be assumed that all landowners will agree on its future.[74] Long-standing rivalries and tensions between and within such groups may contribute to them holding diverse views.[75] Written agreements that community leaders make concerning World Heritage will not necessarily hold significant weight, as there is no guarantee that future (or even present) generations will feel bound by them.

While there was broad support among the East Rennellese people for World Heritage listing when the site was nominated, available information suggests that many are now disappointed with the World Heritage programme,[76] and some support the logging of the area.[77] The level of

---

[74] See, for example, Marianne Pederson, *Conservation Complexities: Conservationists' and Local Landowners' Different Perceptions of Development and Conservation in Dandaun Province, Papua New Guinea*, State, Society and Governance in Melanesia Discussion Paper 7 (The Australian National University, 2013); Joeli Veitayaki et al, 'On Cultural Factors and Marine Managed Areas in Fiji' in Jolie Liston, Geoffrey Clark and Dwight Alexander (eds), *Pacific Island Heritage: Archaeology, Identity and Community* (ANU E Press, 2011) 37, 45; Adam M Trau, Chris Ballard, Meredith Wilson, 'Bafa Zon: Localising World Heritage at Chief Roi Mata's Domain, Vanuatu' (2014) 20(1) *International Journal of Heritage Studies* 86, 98; Paige West and Dan Brockington, 'An Anthropological Perspective on Some Unexpected Consequences of Protected Areas' (2006) 20(3) *Conservation Biology* 609, 614; Simon Foale, 'Where's Our Development? Landowner Aspirations and Environmentalist Agendas in Western Solomon Islands' (2001) 2(2) *Asia Pacific Journal of Anthropology* 44, 45; Jonathan M Lindsay, *Creating Legal Space for Community-Based Management: Principles and Dilemmas* (Food and Agriculture Organisation of the United Nations, 1998) 8.

[75] Veitayaki et al, above n 74, 45.

[76] Anita Smith, 'East Rennell World Heritage Site: Misunderstandings, Inconsistencies and Opportunities in the Implementation of the World Heritage Convention in the Pacific Islands' (2011) 17(6) *International Journal of Heritage Studies* 592; *State of Conservation of the Properties Inscribed on the List of World Heritage in Danger*, WHC 42nd sess, UN Doc WHC/18/42.COM/7A.Add.2 (15 June 2018) 17 (East Rennell, Solomon Islands).

[77] Environment and Conservation Division (Solomon Islands Ministry of Environment, Climate Change, Disaster Management, and Meteorology) *Lake Tegano World Heritage Site, East Rennell, Rennell-Bellona Province: A Report on Community Consultation Visit on the Status of East Rennell World Heritage Site, 5–12 October 2011* (SIG, 2012); John Marnell,

community support for World Heritage is likely to continue to ebb and flow, so it is both inaccurate and unhelpful to assume that they possess a unified or constant opinion about the conservation or development of their land. Rather, the diversity and fluidity of the views of the East Rennellese people must be acknowledged in the design of any World Heritage initiatives involving them. Peoples' opinions will inevitably change, and ongoing discussions and negotiations will be required to maintain community support for conservation.

Efforts should be made to support and strengthen the decision-making processes of the East Rennellese people, to help them deal with diverse and changing community attitudes towards World Heritage conservation. This has been recognised elsewhere. For example, Denham has noted that the Kawelka (the customary owners of the Kuk Early Agricultural Site in Papua New Guinea) are not a homogenous unit with a single perspective on the site's significance, and are not represented by one leader. As such, strategies for the area's protection must try to accommodate their diverse opinions.[78] Trau, Ballard, and Wilson made a similar observation concerning the Chief Roi Mata's Domain site in Vanuatu, arguing that any meaningful understanding of local involvement in World Heritage protection must take into account the 'nuances, ambiguities and fluidities' of intra- and inter-community relations and interactions.[79] The same applies in Solomon Islands.

8.4.4.3 Supporting Pacific Islanders to Exercise Their Rights Under Relevant Legislation

Legislation for the protection of World Heritage is only effective if implemented and complied with. Supporting Pacific Islanders to implement and enforce legislation may therefore strengthen World Heritage protection. In Solomon Islands, for example, supporting the East Rennellese people to exercise their rights under logging and mining legislation and the *PA Act* could help them safeguard the site.

'Concerns Raised Over East Rennell Logging Application', *Sunday Isles*, 25 March 2012, 9; Hughes and Tuhanuku, above n 68, 12; Teddy Kafo, 'Proposed logging threatens World Heritage Lake Tegano', *The Solomon Star*, 24 February 2015; Paul Dingwall, *Report on the Reactive Monitoring Mission to East Rennell, Solomon Islands, 21–29 October 2012* (IUCN, 2013) 18.

[78] Tim Denham, 'Book review: Kuk Heritage: Issues and Debates in Papua New Guinea, Edited by Andrew Strathern and Pamela J Stewart' (1999) 34(2) *Archaeology in Oceania* 89, 90.

[79] Trau, Ballard and Wilson, above n 74, 98.

As discussed in Sect. 7.3.1, under the *Forest Resources and Timber Utilisation Act* and the *Mines and Minerals Act*, except in limited circumstances, logging and mining cannot occur on customary land without the consent of the landowners. While this suggests that Solomon Islanders have significant power to protect their heritage, in practice, this is rarely the case. Ambiguities in the drafting of the landowner consent provisions of relevant legislation, and their inconsistency with some customary laws, create uncertainty concerning whose consent is legally required. This situation is often manipulated by powerful people within landowning groups working in cohorts with resource companies to reap the benefits of land development. In the absence of significant government oversight and effective dispute resolution processes, logging and mining often occur without the consent of all people who have the customary right to make decisions with respect to the land. It is also very difficult for landowners to enforce their rights, given their limited access to legal services and the Honiara-centric nature of the State legal system.

There is a dire need for laws regulating these industries to be reformed. In addition to the issues referred to in Sect. 8.4.3.2, the legislation should be amended to incorporate new approaches to identifying the local people who are entitled to authorise developments. The legislation must be sufficiently flexible to accommodate the variety of land tenure systems that exist around Solomon Islands, including in Polynesian outlying islands like Rennell, where land ownership is more individualised than elsewhere in the country. There also needs to be greater government oversight over the agreement-making process, and dispute resolution processes must be strengthened. Additionally, the amendment of the *Forest Resources and Timber Utilisation Act* and the *Mines and Minerals Act* to give any person who may be affected by a logging or mining operation the right to object to the approval of that operation would give the East Rennellese greater power to protect the World Heritage site against these activities.

In lieu of such reforms, it is essential that the East Rennellese people are supported to reduce the chance of logging or mining occurring without full landowner approval. The SIG could assist by scrutinising agreements between landowners and resources companies more carefully to ensure that they meet the legislative requirements. Other groups could help by ensuring that the East Rennellese people are aware of any development proposals for their land, and improving their access to legal services. In addition, given that the East Rennellese people have little capacity to influence activities occurring in West Rennell, and the SIG's reluctance to refuse developments supported by the landowners, the West Rennellese people need to be involved with efforts to protect the World Heritage site. This might include encouraging and assisting them to oppose operations that will harm the site's OUV.

The East Rennellese people should also be supported to implement the *PA Act* at the World Heritage site. A protected area application must demonstrate compliance with the landowner consent process prescribed under the *Protected Area Regulations 2012*, and must include a management plan and details of the proposed management committee. The East Rennellese are likely to require assistance to navigate this application process, which raises several questions about the relationship between the *PA Act* regime and customary law (discussed in Sect. 7.2).

The East Rennellese will also need help to manage the protected area in accordance with the Act. It cannot be assumed that they will be willing to dedicate time and energy towards protected area conservation activities such as monitoring and enforcement, particularly in the absence of receipt of tangible benefits. The legislation will therefore not be successful unless the local communities are supported to implement the management plan measures and undertake the governance and enforcement roles available to them.

8.4.4.4 Supporting Local Development

While World Heritage is not the answer to all social and economic problems, efforts to implement the *Convention* must aim to assist local communities to obtain and maintain an adequate standard of living.[80] Furthermore, as many Pacific Islanders are not interested in participating in conservation programmes that are not accompanied by promises of development,[81] pursuing World Heritage protection through the framework of sustainable development is necessary for practical reasons. This is certainly the case in East Rennell, where food security and other livelihood issues are the dominant concern of much of the local population. Indeed, in the absence of local development, it is debateable whether the OUV of East Rennell can be protected in the long term.

[80] Gonzalo Oviedo and Tatjana Puschkarsky, 'World Heritage and Rights-Based Approaches to Nature Conservation' (2012) 18(3) *International Journal of Heritage Studies* 285, 291.

[81] Martha Macintyre and Simon Foale, 'Global Imperatives and Local Desires: Competing Economic and Environmental Interests in Melanesian Communities' in Victoria Lockwood (ed), *Globalisation and Culture Change in the Pacific Islands* (Pearson Prentice Hall, 2004) 149, 161.

A priority for Solomon Islands under both the *Pacific World Heritage Action Plan* and the *DSOCR* is the development of sustainable income-generating mechanisms for the East Rennellese communities.[82] While this should be supported, it must be preceded by a study of local development options. The establishment of income-generating projects in the area is very challenging. Indeed, in Rennell, almost all small-scale projects have failed, which is a common cause of community grievance.[83] The reasons behind the failure of past projects should therefore be analysed to ascertain whether any lessons can be learned. Opportunities for local development must also be assessed in light of detailed knowledge of land tenure and customary governance, both of which will influence the success of projects. One option that should be explored is the United Nations Programme on Reducing Emissions from Deforestation and Forest Degradation (UN-REDD).[84]

8.5   Conclusion

By exploring the implementation of the *World Heritage Convention* through a legal lens, this book provided new insights into World Heritage protection in Solomon Islands and the Pacific more broadly. It identified substantial opportunities for utilising the *Convention* to conserve the region's impressive cultural and natural places, stemming from the scope of the treaty, the Committee's broadening approach to heritage and its protection, and the legally plural nature of Pacific Island States. However, it recognised even more challenges, demonstrating that protecting Pacific Island heritage will rarely be easy.

East Rennell cannot be described as a success story, at least not yet. While its inscription on the World Heritage List was a milestone in the

---

[82] *Pacific World Heritage Action Plan 2016–2020* (2016) 16; *State of Conservation of Properties Inscribed on the World Heritage List*, WHC 41st sess, UN Doc WHC/17/41. COM/7A.Add (2 June 2017) 26 (East Rennell, Solomon Islands) 31–32.

[83] Allen et al, above n 73, 24.

[84] The potential for a REDD project to be implemented at East Rennell has been subject to some analysis: see Scott Alexander Stanley, *REDD Feasibility Study for East Rennell World Heritage Site, Solomon Islands* (Secretariat of the Pacific Community and Deutsche Gesellschaft für Internationale Zusammenarbeit, 2013). For a discussion of the implementation of REDD in Solomon Islands more generally, see Jennifer Corrin, *Background Analysis of REDD + and Forest Carbon Rights in Solomon Islands* (Secretariat of the Pacific Community and Deutsche Gesellschaft für Internationale Zusammenarbeit, 2012).

development of the *Convention* regime, it has not been protected to the level expected by the Committee. In addition, its listing has not generated the benefits anticipated by the SIG and the East Rennellese people, leaving them somewhat disenchanted with the World Heritage process and rendering conservation a low priority. In this context, it is unclear whether the island's incredible ecosystems and unique species can be conserved for future generations, in accordance with the goals of the *Convention*.

What is clear is that the East Rennellese people are the key to the island's future. It is their home, the basis of their livelihoods, and the foundation of their cultural identity. They are the main decision-makers concerning their land, so efforts to protect the site's OUV will always be intimately entwined with their needs and aspirations. Any resolutions or projects designed to strengthen the protection of the site that fail to recognise that are unlikely to succeed. Successful outcomes will only be achieved if the *Convention* bodies, the SIG, and the communities are able to agree upon and pursue common goals for the conservation of the area's heritage.

REFERENCES

ARTICLES, BOOKS AND REPORTS

Allen, Matthew, Sinclair Dinnen, Daniel Evans, Rebecca Monson, *Justice Delivered Locally: Systems, Challenges and Innovations in Solomon Islands* (World Bank, 2013)

Anderson, Inger, 'Today Defines Tomorrow: World Heritage as Litmus Test for Action on Agreements' (2016) 79 *World Heritage* 4

Aswani, Shankar, 'Customary Sea Tenure in Oceania as a Case of Rights-Based Fishery Management: Does it Work?' (2005) 15 *Reviews in Fish Biology and Fisheries* 285

Baines, Graham, *Solomon Islands is Unprepared to Manage a Minerals-Based Economy*, State, Society and Governance in Melanesia Discussion Paper 2015/6 (Australian National University, 2015)

Ballard, Chris and Meredith Wilson, 'Unseen Monuments: Managing Melanesian Cultural Landscapes' in Ken Taylor and Jane L Lennon (eds), *Managing Cultural Landscapes* (Routledge, 2012) 130

Bennett, Judith, 'Forestry, Public Land, and the Colonial Legacy in Solomon Islands' (1995) 7(2) *Contemporary Pacific* 243

Bennett, Judith, *Pacific Forest: A History of Resource Control and Contest in Solomon Islands, c 1800–1997* (Brill Academic Publishers Inc, 2000)

Bennett, Judith, *Roots of Conflict in Solomon Islands – Though Much is Taken, Much Abides: Legacies of Tradition and Colonialism*, State, Society and Governance in Melanesia Discussion Paper (Australian National University, 2002)

Boer, Ben, *Solomon Islands: Review of Environmental Law* (SPREP, 1993)

Breen, Colin, 'Advocacy, International Development and World Heritage Sites in Sub-Saharan Africa' (2007) 39(3) *World Archaeology* 355

Brown, Steve, 'Poetics and Politics: Bikini Atoll and World Heritage Listing' in Sue O'Connor, Denis Byrne and Sally Brockwell (eds), *Transcending the Culture-Nature Divide in Cultural Heritage: Views from the Asia-Pacific Region* (ANU E Press, 2012) 35

Clarke, Pepe and Charles Taylor Gillespie, *Legal Mechanisms for the Establishment and Management of Terrestrial Protected Areas in Fiji* (IUCN, 2009)

Corrin, Jennifer, *Background Analysis of REDD + and Forest Carbon Rights in Solomon Islands* (Secretariat of the Pacific Community and Deutsche Gesellschaft für Internationale Zusammenarbeit, 2012)

Deegan, Naomi, 'The Local-Global Nexus in the Politics of World Heritage: Space for Community Development?' in Marie-Theres Albert, Marielle Richon, Marie José Viñals and Andrea Witcomb (eds), *Community Development through World Heritage*, World Heritage Papers 31 (UNESCO, 2012) 77

Denham, Tim, 'Book review: Kuk Heritage: Issues and Debates in Papua New Guinea, Edited by Andrew Strathern and Pamela J Stewart' (1999) 34(2) *Archaeology in Oceania* 89

Dingwall, Paul, *Report on the Reactive Monitoring Mission to East Rennell, Solomon Islands, 21–29 October 2012* (IUCN, 2013)

Eboreime, Joseph, 'Nigeria's Customary Laws and Practices in the Protection of Cultural Heritage with Special Reference to the Benin Kingdom' in Webber Ndoro and Gilbert Pwiti (eds), *Legal Frameworks for the Protection of Immoveable Cultural Heritage in Africa*, ICCROM Conservation Studies 5 (ICCROM, 2005) 9

Environment and Conservation Division (Solomon Islands Ministry of Environment, Climate Change, Disaster Management, and Meteorology) *Lake Tegano World Heritage Site, East Rennell, Rennell-Bellona Province: A Report on Community Consultation Visit on the Status of East Rennell World Heritage Site, 5–12 October 2011* (SIG, 2012)

Foale, Simon, 'Where's Our Development? Landowner Aspirations and Environmentalist Agendas in Western Solomon Islands' (2001) 2(2) *Asia Pacific Journal of Anthropology* 44

Foale, Simon, Phillipa Cohen, Stephanie Januchowski-Hartley, Amelia Wenger and Martha Macintyre, 'Tenure and Taboos: Origins and Implications for Fisheries in the Pacific' (2011) 12 *Fish and Fisheries* 357

Forrest, Craig and Jennifer Corrin, 'A Model Law to Implement the Convention on the Protection of the Underwater Cultural Heritage and its Possible Application in Plural Legal Regimes in Pacific Small Island States: A Case Study of Solomon Islands' (Paper presented at Solomon Islands National University Workshop, Honiara, December 2014) http://www.themua.org/collections/files/original/602a7962da5dd01ceafc413b8ec2d8fe.pdf

Frazer, Ian, 'The Struggle for Control of Solomon Island Forests' (1997) 9(1) *Contemporary Pacific* 39

Gay, Daniel (ed), *Solomon Islands Diagnostic Trade Integration Study 2009 Report* (Solomon Islands Government, 2009)

Graham, Brian, Gregory J Ashworth and John E Tunbridge, *A Geography of Heritage: Power, Culture and Economy* (Arnold, 2000)

Hughes, Tony and Ali Tuhanuku, *Logging and Mining in Rennell: Lessons for Solomon Islands. Report to the World Bank and Solomon Islands Government* (2015)

Kabutaulaka, Tarcisius Tara, 'Rumble in the Jungle: Land, Culture and (Un)sustainable Logging in Solomon Islands' in Antony Hooper (ed), *Culture and Sustainable Development in the Pacific* (ANU E Press and Asia Pacific Press, 2005)

Kafo, Teddy, 'Proposed logging threatens World Heritage Lake Tegano', *The Solomon Star*, 24 February 2015

Lindsay, Jonathan M, *Creating Legal Space for Community-Based Management: Principles and Dilemmas* (Food and Agriculture Organisation of the United Nations, 1998)

Logan, William, 'Cultural Diversity, Cultural Heritage and Human Rights: Towards Heritage Management as Human Rights-Based Cultural Practice' (2012) 18(3) *International Journal of Heritage Studies* 231

Macintyre, Martha and Simon Foale, 'Global Imperatives and Local Desires: Competing Economic and Environmental Interests in Melanesian Communities' in Victoria Lockwood (ed), *Globalisation and Culture Change in the Pacific Islands* (Pearson Prentice Hall, 2004) 149

Marnell, John, 'Concerns Raised Over East Rennell Logging Application', *Sunday Isles*, 25 March 2012, 9

McDonald, Jan, *Marine Resource Management and Conservation in Solomon Islands: Roles, Responsibilities and Opportunities* (Griffith Law School, 2010)

McDonnell, Siobhan, Joseph Foukana and Alice Pollard, *Building a Pathway for Successful Land Reform in Solomon Islands* (2015)

Meskell, Lynn, Claudia Liuzza and Nicholas Brown, 'World Heritage Regionalism: UNESCO from Europe to Asia' (2015) 22 *International Journal of Cultural Property* 437

Mumma, Albert, 'Framework for Legislation on Immoveable Cultural Heritage in Africa' in Webber Ndoro, Albert Mumma and George Abungu (eds), *Cultural Heritage and the Law: Protecting Immoveable Heritage in English-Speaking Countries of Sub-Saharan Africa*, ICCROM Conservation Studies 8 (ICCROM, 2008) 97

Oviedo, Gonzalo and Tatjana Puschkarsky, 'World Heritage and Rights-Based Approaches to Nature Conservation' (2012) 18(3) *International Journal of Heritage Studies* 285

Pederson, Marianne, *Conservation Complexities: Conservationists' and Local Landowners' Different Perceptions of Development and Conservation in Dandaun Province, Papua New Guinea*, State, Society and Governance in Melanesia Discussion Paper 7 (The Australian National University, 2013)

Smith, Anita, 'The World Heritage Pacific 2009 Programme' in Anita Smith (ed), *World Heritage in a Sea of Islands: Pacific 2009 Programme*, World Heritage Papers 34 (UNESCO, 2012) 2

Smith, Anita, 'World Heritage and Outstanding Universal Value in the Pacific Islands' (2015) 21(2) *International Journal of Heritage Studies* 177

Solomon Islands Government, *State Party Report on the State of Conservation of the East Rennell World Heritage Area (Solomon Islands)* (SIG, 2012)

Solomon Islands Government, *State Party Report on the State of Conservation of the East Rennell World Heritage Area (Solomon Islands)* (SIG, 2013)

Solomon Islands Government, *State Party Report on the State of Conservation of the East Rennell World Heritage Area (Solomon Islands)* (SIG, 2014)

Solomon Islands Government, *State Party Report on the State of Conservation of the East Rennell World Heritage Site* (SIG, 2017)

Solomon Islands Office of the Auditor General, *An Auditor-General's Insights into Corruption in Solomon Islands Government*, National Parliament Paper 48 (2007)

Stanley, Scott Alexander, *REDD Feasibility Study for East Rennell World Heritage Site, Solomon Islands* (Secretariat of the Pacific Community and Deutsche Gesellschaft für Internationale Zusammenarbeit, 2013)

Trau, Adam M, 'The Glocalisation of World Heritage at Chief Roi Mata's Domain, Vanuatu' (2012) 24(3) *Historic Environment* 4

Trau, Adam M, Chris Ballard, Meredith Wilson, 'Bafa Zon: Localising World Heritage at Chief Roi Mata's Domain, Vanuatu' (2014) 20(1) *International Journal of Heritage Studies* 86

UNESCO/ICCROM/ICOMOS/IUCN, *Managing Natural World Heritage*, World Heritage Resource Manual (UNESCO, 2012)

van der Ploeg, Rick, 'Welcome Address by the Chair of the conference' in Eléonore de Merode, Rieks Smeets and Carol Westrik (eds), *Linking Universal and Local Values: Managing a Sustainable Future for World Heritage*, World Heritage Papers 13 (UNESCO, 2004) 24

Veitayaki, Joeli, Akosita D R Nakoro, Tareguci Sigarua and Nanise Bulai, 'On Cultural Factors and Marine Managed Areas in Fiji' in Jolie Liston, Geoffrey Clark and Dwight Alexander (eds), *Pacific Island Heritage: Archaeology, Identity and Community* (ANU E Press, 2011) 37

West, Paige and Dan Brockington, 'An Anthropological Perspective on Some Unexpected Consequences of Protected Areas' (2006) 20(3) *Conservation Biology* 609

Wilson, Meredith, Chris Ballard, Richard Matanik and Topie Warry, 'Community as the First C: Conservation and Development through Tourism at Chief Roi Mata's Domain, Vanuatu' in Anita Smith (ed), *World Heritage in a Sea of Islands: Pacific 2009 Programme*, World Heritage Papers 34 (UNESCO, 2012) 68

LEGISLATION AND BILLS: FIJI

Heritage Bill 2016 (Bill no. 10 of 2016)

LEGISLATION AND BILLS: KIRIBATI

Environment Act 1999
Phoenix Islands Protected Area Regulations 2008

LEGISLATION AND BILLS: SOLOMON ISLANDS

Biosecurity Act 2013
Environment Act 1998
Fisheries Management Act 2015
Forest Resources and Timber Utilisation Act (Cap. 40)
Mines and Minerals Act (Cap. 42)
Protected Areas Act 2010
Protected Areas Regulations 2012
Solomon Islands Independence Order 1978, sch (*Constitution of Solomon Islands*)

CONVENTIONS

Convention Concerning the Protection of the World Cultural and Natural Heritage, opened for signature 16 November 1972, 1037 UNTS 151 (entered into force 17 December 1975)

UNITED NATIONS DOCUMENTS

Policy for the Integration of a Sustainable Development Perspective into the Processes of the World Heritage Convention, WHC GA Res 20 GA 13, 20th sess, UN Doc WHC-15/20.GA/15 (20 November 2015)

Presentation of the World Heritage Programme for the Pacific, WHC 31st sess, UN Doc WHC-07/31.COM/11C (10 May 2007) annex I (Appeal to the World Heritage Committee from the Pacific Island State Parties)

Reports of the Advisory Bodies, WHC 39th sess, UN Doc WHC-15/39.COM/5B (15 May 2015)

State of Conservation of Properties Inscribed on the World Heritage List, WHC 41st sess, UN Doc WHC/17/41.COM/7A.Add (2 June 2017) 26 (East Rennell, Solomon Islands)

State of Conservation of the Properties Inscribed on the List of World Heritage in Danger, WHC 42nd sess, UN Doc WHC/18/42.COM/7A.Add.2 (15 June 2018) 17 (East Rennell, Solomon Islands)

UNESCO, *Operational Guidelines for the Implementation of the World Heritage Convention*, UN Doc WHC.16/01 (26 October 2016)

United Nations Declaration on the Rights of Indigenous Peoples, GA Res 61/295, UN GAOR, 61st sess, 107th plen mtg, Supp No 49, UN Doc A/RES/61/295 (13 September 2007)

WHC Res 29 COM 7B.10, WHC 29th sess, UN Doc WHC-05/29.COM/22 (9 September 2005) 45

WHC Res 31 COM 7B.21, WHC 31st sess, UN Doc WHC-07/31.COM/24 (31 July 2007) 58

WHC Res 33 COM 7B.19, WHC 33rd sess, UN Doc WHC-09/33.COM/20 (20 July 2009) 68

WHC Res 34 COM 7B.17, WHC 34th sess, UN Doc WHC-10/34.COM/20 (3 September 2010) 71

WHC Res 35 COM 12B, WHC 35th sess, UN Doc WHC-11/35.COM/20 (7 July 2011) 266

WHC Res 36 COM 7B.15, WHC 36th sess, UN Doc WHC-12/36.COM/19 (June–July 2012) 63

WHC Res 37 COM 7B.14, WHC 37th sess, UN Doc WHC-13/37.COM/20 (5 July 2013) 68

WHC Res 38 COM 7A.29, WHC 38th sess, UN Doc WHC-14/38.COM/16 (7 July 2014) 39

WHC Res 39 COM 5D, WHC 39th sess, UN Doc WHC-15/39.COM/19 (8 July 2015a) 7

WHC Res 39 COM 7A.16, WHC 39th sess, UN Doc WHC-15/39.COM/19 (8 July 2015b) 30

WHC Res 40 COM 7A.49, WHC 40th sess, UN Doc WHC-16/40.COM/19 (15 November 2016) 68

WHC Res 41 COM 7A.19, WHC 41st sess, UN Doc WHC-17/41.COM/18 (12 July 2017) 35

World Heritage and Sustainable Development, WHC 39th sess, UN Doc WHC-15/39.COM/5B (15 May 2015)

THESES

Tagini, Phillip Iro, *The Search for King Solomon's Gold: An Examination of the Policy and Regulatory Framework for Mining in Solomon Islands* (PhD Thesis, Australian National University, 2007)

Trau, Adam M, *World Heritage at Chief Roi Mata's Domain: The Global-Local Nexus of Community Heritage Conservation and Tourism Development in Vanuatu* (PhD Thesis, University of Western Sydney, 2013)

INTERNET MATERIALS

Alliance of Small Island States, *About AOSIS* http://aosis.org/about/

Model Law for the Protection of Traditional Ecological Knowledge, Innovations and Practices http://www.grain.org/system/old/brl_files/brl-model-law-pacific-en.pdf

Pacific Heritage Hub, *Who We Are* http://www.pacificheritagehub.org/about-us/who-we-are/

Pacific Islands Legal Information Institute http://www.paclii.org

UNESCO, *Model for a National Act on the Protection of Cultural Heritage* (2013) http://www.unesco.org/new/fileadmin/MULTIMEDIA/HQ/CLT/pdf/UNESCO_MODEL_UNDERWATER_ACT_2013.pdf

UNESCO, *UNESCO Database of National Cultural Heritage Laws* http://www.unesco.org/culture/natlaws/index.php

INTERVIEWS

Interview by the author with an officer in the Ministry of Education, who was formerly the focal point for World Heritage within the Solomon Islands National Commission for UNESCO (Honiara, 28 July 2013)

Interview by the author with Joe Horokou, Director of the Environment and Conservation Division of the Ministry of Environment (Honiara, 15 August 2013)

Interview by the author with Malchoir Mataki, Permanent Secretary of the Ministry of Environment (Honiara, 1 October 2013)

OTHER

Pacific World Heritage Action Plan 2016–2020 (2016)

Index[1]

[1] Note: Page numbers followed by 'n' refer to notes.

© The Author(s) 2018
S. C. Price, *World Heritage Conservation in the Pacific*,
Palgrave Series in Asia and Pacific Studies,
https://doi.org/10.1007/978-981-13-0602-0

Printed by Printforce, the Netherlands